The Joys of Haar Measure

The Joys of Haar Measure

Joe Diestel
Angela Spalsbury

Graduate Studies in Mathematics
Volume 150

American Mathematical Society
Providence, Rhode Island

2010 *Mathematics Subject Classification.* Primary 28-XX.

For additional information and updates on this book, visit
www.ams.org/bookpages/gsm-150

Library of Congress Cataloging-in-Publication Data

Diestel, Joe, 1943– author.
The joys of Haar measure / Joe Diestel, Angela Spalsbury.
pages cm. — (Graduate studies in mathematics ; volume 150)
Includes bibliographical references and index.
ISBN 978-1-4704-0935-7 (alk. paper)
1. Measure theory. 2. Integrals, Haar. 3. Numerical integration. I. Spalsbury, Angela, 1967– author. II. Title.

QA312.D429 2013
515′.42—dc23

2013027434

∞ The paper used in this book is acid-free and falls within the guidelines
established to ensure permanence and durability.
Visit the AMS home page at `http://www.ams.org/`

10 9 8 7 6 5 4 3 2 1   19 18 17 16 15 14

This book is dedicated to our little princes,

Sagen Diestel (the west coast prince)
and
Jackie Szigeti (the east coast prince)

Contents

Preface

From the earliest days of measure theory, invariant measures have held the interests of geometers and analysts alike. With Hausdorff's introduction of the measures that bear his name and the subsequent cementing of the relationships between measure theory and geometry, those interests attained a degree of permanency. Simultaneously, efforts at solving Hilbert's fifth problem (on recognizing Lie groups by their locally Euclidean structure) naturally found invariant measures an ally in analyzing the structure of topological groups, particularly compact and locally compact groups. Existence, uniqueness, and applications of invariant measures attracted the attention of many of the strongest mathematical minds. We hope in this volume to detail some of the highlights of those developments.

This book is aimed at an audience of people who have been exposed to a basic course in real variables, although we do review the development of Lebesgue's measure. As is usually the case, a certain amount of mathematical maturity is critical to the understanding of many of the topics discussed. We have included a few exercises; their occurrence is planned to coincide with sections where material somewhat divorced from typical experience is presented.

In the first chapter we develop Lebesgue measure in Euclidean spaces from a topological perspective. Roughly speaking, we start with knowledge of how big an open set is, pass from there to measuring the size of compact sets, and then, using regularity as a guide, determine which sets are measurable. Naturally, the details are more technical but the result—Lebesgue measure—is worth the effort.

We next discuss measures on metric spaces with special attention paid to Borel measures. We encounter the aforementioned Hausdorff measures and find that Hausdorff n-measure on Euclidean n-space is a multiple of Lebesgue measure on the same space. What's more, the constant of multiplicity is an apt rate of exchange between rectilinear measurements (Lebesgue measure) and spherical ones (Hausdorff's n-measure). Along the way we encounter and embrace Carathéodory's fundamental method of outer measures, a method we will return to throughout these deliberations.

We turn then to topological groups and give a brief introduction to this intriguing topic. A highlight of this chapter is the often surprising consequences one can draw about topological groups: the mixture of algebra and topology produces a sum in excess of what the summands hint. For instance, every (Hausdorff) topological group is completely regular, and if it satisfies the first axiom of countability (having a countable basis for the open sets about each point), then it is metrizable, with a left invariant metric moreover. These follow from the beautiful theorem of Birkhoff and Kakutani. We also show that if the group is locally compact then it is paracompact.

Next, in the chapter on Banach and measure theory, we present Banach's proof on the existence of an invariant measure on a compact metrizable topological group. Banach's proof, which is plainly of geometric flavor, is more general than showing "just" that compact metrizable groups have invariant measures; indeed, his proof asserts the existence of a Borel measure on any compact metric space that is invariant under the action of a transitive group of homeomorphisms. As is to be expected, Banach's proof relies on the methods of functional analysis, a subject he was deeply active in developing—most particularly in his use of "Banach limits", the existence of which relies on the Hahn–Banach theorem. To put Banach's result in context, it's important to know that we have a Borel probability in hand, one that allows every continuous function to be integrated, and for this we present Saks' proof that positive linear functionals of norm-1 $C(K)$'s, where K is a compact metric space, correspond to Borel probabilities. We follow the presentation of Saks' proof with Banach's approach to the Lebesgue integral. This appeared as an appendix in Saks' classical monograph *Theory of the Integral* [**111, 113**]. It contains Banach's proof of the Riesz Representation of $C(K)^*$ for a compact metric space K.

Having discussed the situation of compact metrizable topological groups, we next present von Neumann's proof of the existence and uniqueness of normalized Haar measure on any compact topological group. The importance to this proof of the uniform continuity of continuous real-valued functions defined an a compact group and the classical theorem of Arzelá and Ascoli should be plain and clear. Von Neumann's proof shows, in quite a natural way, that the normalized Haar measure is simultaneously left and right invariant. We include several other proofs of the existence of a Haar measure in the *Notes and Remarks* to this chapter.

An all-too-short chapter on applications of Haar measure on compact groups follows. Homogeneous spaces are shown to have unique invariant measures, invariance being with respect to a transitive group of homeomorphisms. This is followed by a presentation of the Peter–Weyl theorem on the existence of a complete system of irreducible finite-dimensional unitary representations of the group. We then broach the topic of absolutely p-summing operators on Banach spaces; after showing the existence of a "Pietsch measure" for any absolutely p-summing operator, we use the uniqueness of Haar measure to show that under appropriate mild invariance assumptions on a p-summing operator on a space that has an invariant norm that Haar measure serves as a Pietsch measure.

A chapter detailing the existence and uniqueness of Haar measure on a general locally compact topological group is next. There appears to be no clever trick to pass from the compact case to the locally compact situation; only hard work will suffice. The measure theory is more delicate and the proofs of existence and uniqueness of Haar measure follow suit. We present Weil's proof of existence, followed by H. Cartan's simultaneous proof of existence and uniqueness. Our *Notes and Remarks* in Chapter 6 complement this with the more commonly known proof via the Fubini theorem.

The special character of Haar measure is the topic of our next chapter with a gorgeous theorem of Bandt center stage. The theorem calls on an ingenious use of Hausdorff-like measures in tandem with the uniqueness aspects of Haar measure to show that if we encounter a locally compact metrizable topological group with a left invariant metric in place, then subsets that are isometric with this metric have the same Haar (outer) measure.

Just when we feel that we've done all that can be done with regard to Haar measure in a locally compact setting, we present Steinlage's remarkable description of necessary and sufficient conditions that a G-invariant Borel content exists on a locally compact Hausdorff space, where G is a suitable group of homeomorphisms of the space onto itself. The proof of existence is reminiscent of Banach's proof with a touch of Weil thrown in.

We finish with an all-too-brief description of Oxtoby's work on invariant Borel measures on nonlocally compact Polish groups.

We have two appendices. In one we discuss Haar's original proof of the existence of Haar measure in the case where the group is a compact metric group. The other appendix discusses the remarkable result of Kakutani and

Oxtoby in which they show that Haar measure on an infinite compact metric group can be extended to an amazingly large sigma field in a countably additive, translation invariant manner.

Our presentation of this material was greatly influenced by the experiences of talking about the material in a classroom setting, either in seminars or graduate classes. We found often that presenting material at a slightly less general level aided in conveying the essential ideas without any serious sacrifice. This also had the beneficial effect of inspiring questions, leading to deeper understanding, for both the students and us.

In any undertaking like this, many friends and colleagues have contributed through discussions, lectures, and reading attempts at exposition in varying states of preparation. We rush to thank all who have helped. We extend particular thanks to (the late) Diomedes Barcenas, Floyd Barger, Jonathan Borwein, Geraldo Botelho, Bruno Braga, John Buoni, Antonia Cardwell, Neal Carothers, Charlotte Crowther, John Dalbec, Geoff Diestel, Rocco Duvenhage, (the late) Doug Faires, Paul Fishback, Ralph Howard, Jozsi Jalics, Hans Jarchow, Livia Karetka, Jay Kerns, Darci Kracht, Charles Maepa, (the late) Roy Mimna, Daniel Pellegrino, Zbginiew Piotrowski, David Pollack, Zach Riel, Nathan Ritchey, Sarah Ritchey, Stephen Rodabaugh, Pilar Rueda, Dima Ryabogin, Juan Seoane, Brailey Sims, Anton Stroh, Johan Swart, Jamal Tartir, Padraic Taylor, Andrew Tonge, Thomas Wakefield, Matt Ward, Eric Wingler, George Yates, and Artem Zvavitch.

Through the years we have had the opportunity to talk about Haar measure at various universities, including the Department of Mathematics and Applied Mathematics at the University of Pretoria, South Africa, the Department of Mathematics, University of the Andes, Merida, Venezuela CARMA, University of Newcastle, Newcastle, NSW, Australia (where Jon Borwein and Brailey Sims provided the audience with added entertainment by having Joe Diestel misuse exceptional modern technology!), the Department of Mathematical Sciences at Kent State University, the Department of Mathematics and Statistics at Youngstown State University, and the Institute for Applied Topology and Topological Structures at Youngstown State University.

Finally, an endeavor such as this would not be possible without the loving support of our families, especially Linda Diestel, Kelly Spalsbury, Sue Spalsbury, Julius Szigeti, and Peter Szigeti.

Chapter 1

Lebesgue Measure in Euclidean Space

In the first section of this chapter we give an introduction to Lebesgue measure.

In the second section we give the elegant Hadwiger–Ohlmann proof of the *Brunn–Minkowski theorem*, a geometric version of the *arithmetic-geometric mean inequality*. As a simple consequence of this we derive the *isodiametric inequality* which says that among Borel sets in $\mathbb{R}^n$ of the same diameter, the ball has the greatest volume.

In the third section we derive the still-wonderful *covering theorem of Vitali* for Vitali families of balls.

We close the chapter with a section entitled *Notes and Remarks*. This is a catch-all which contains historical comments and some miscellaneous results that we found interesting and/or important—results that add to the message contained in the chapter.

1. An Introduction to Lebesgue Measure

By an interval in $\mathbb{R}^k$ we mean any set I of the form

$$I = I_1 \times \cdots \times I_k,$$

where $I_1, \ldots, I_k$ are finite intervals in $\mathbb{R}$. We do not ask that $I_1, \ldots, I_k$ all be open, closed, or half-open/half-closed; mixtures are just fine. Each $I_j \subseteq \mathbb{R}$

has a length $l(I_j)$ and with this in mind we define the "volume" of I by

$$\operatorname{vol}(I) = \prod_{j \leq k} l(I_j).$$

Definition 1.1. Let $A \subseteq \mathbb{R}^k$. We define the *outer measure* or *Lebesgue outer measure* of A, $m^*(A)$ by

$$m^*(A) = \inf \left\{ \sum_n \operatorname{vol}(I_n) : I_n \text{ is an interval in } \mathbb{R}^k, A \subseteq \bigcup_n I_n \right\}.$$

Regarding edges. There is a great deal of latitude with regard to the nature of the edges of the intervals in the coverings of a set $A \subseteq \mathbb{R}^k$ that are used to compute $m^*(A)$. For instance, if we wish we can assume each edge has length less than δ for some fixed $\delta > 0$. This is plain since any interval I in $\mathbb{R}^k$ is the union of nonoverlapping intervals, all of whose edges have length less than δ and the sum of whose volume totals I's volume.

Further, we can assume we're covering A by *open* intervals, that is, all the edges are open. In fact, if (I_j) is a covering of A by intervals and $\epsilon > 0$, then for each j we can enlarge I_j to an open interval J_j, $I_j \subseteq J_j$, and

$$\operatorname{vol}(J_j) < \operatorname{vol}(I_j) + \epsilon/2.$$

It follows that in computing $m^*(A)$, if (I_j) is a covering of A by intervals, then we can find a sum $\sum \operatorname{vol}(J_j)$ that is as close as we please to $\sum \operatorname{vol}(I_j)$, where (J_j) is a covering of A by open intervals. Hence, we can restrict our attention to finding the infimum of such sums $\sum \operatorname{vol}(J_j)$, where (J_j) is an open covering of A by intervals.

Theorem 1.2. *Basic results about Lebesgue outer measure:*

(i) *If $A \subseteq B$, then*

$$m^*(A) \leq m^*(B).$$

(ii) *If $A = \bigcup_n A_n$, then*

$$m^*(A) \leq \sum_n m^*(A_n).$$

(iii) *For any interval I,*

$$m^*(I) = \operatorname{vol}(I).$$

Proof. (ii) We can, and do, assume that $\sum_n m^*(A_n) < \infty$. With this in mind, let $\epsilon > 0$ be given, and choose for each n a sequence $(I_{n,j})$ of intervals that cover A_n and satisfy

$$\sum_j \operatorname{vol}(I_{n,j}) < m^*(A_n) + \frac{\epsilon}{2^n 2^j}.$$

Since $A = \bigcup_n A_n \subseteq \bigcup_{n,j} I_{n,j}$,

$$m^*(A) \leq \sum_{n,j} \operatorname{vol}(I_{n,j}) \leq \sum_n \left(m^*(A_n) + \frac{\epsilon}{2^n} \right) \leq \sum_n m^*(A_n) + \epsilon.$$

(iii) Let $\epsilon > 0$ be given, and let (I_j) be a covering of I by open intervals such that

$$\sum_j \operatorname{vol}(I_j) < m^*(I) + \epsilon.$$

Take any closed subinterval J of I. Since J is compact there is a j_0 such that $J \subseteq I_1 \cup \cdots \cup I_{j_0}$. Let's look closely to the intervals $I_1, \ldots, I_{j_0}, J$. Each $(k-1)$-dimensional face of these intervals lies in a $(k-1)$-dimensional hyperplane in $\mathbb{R}^k$; in turn, these hyperplanes divide $\bar{I}_1, \ldots, \bar{I}_{j_0}$ into closed intervals $K_1, \ldots, K_{n_{j_0}}$; similarly J is divided into closed intervals $J_1, \ldots, J_{m_{j_0}}$ by the same hyperplane. (Think of the case $n = 3$.) Since $J \subseteq \bigcup_{j \leq j_0} I_j$, each J_m is one of the K_n's so that

$$\begin{aligned} \operatorname{vol}(J) &= \sum_{m \leq m_{j_0}} \operatorname{vol}(J_m) \\ &\leq \sum_{n \leq n_{j_0}} \operatorname{vol}(K_n) \\ &\leq \sum_{j \leq j_0} \operatorname{vol}(I_j) \\ &\leq m^*(I) + \epsilon. \end{aligned}$$

This is so for every closed subinterval J of I so

$$\operatorname{vol}(I) \leq m^*(I) + \epsilon;$$

epsilonics soon tell us that

$$\operatorname{vol}(I) \leq m^*(I).$$

The reverse is plain. □

Some ground work is needed to prepare the way for measurable sets.

Theorem 1.3. *More basic results about Lebesgue outer measure:*

(i) *If F_1 and F_2 are disjoint closed bounded sets, then*

$$m^*(F_1 \cup F_2) = m^*(F_1) + m^*(F_2).$$

(ii) *If G is a bounded open set, then for each $\epsilon > 0$ there is a closed set $F \subseteq G$ such that*

$$m^*(F) > m^*(G) - \epsilon.$$

(iii) *If F is a closed subset of an open bounded set G, then*

$$m^*(G \backslash F) = m^*(G) - m^*(F).$$

Proof. (i) Let $\delta > 0$ be chosen so that no interval of diameter less than δ meets both F_1 and F_2 (e.g., $\delta < \frac{1}{2}d(F_1, F_2)$, where $d(F_1, F_2)$ is the distance between the disjoint closed sets F_1 and F_2). Let $\epsilon > 0$ be given. Pick a sequence (I_i) of intervals of diameter less than δ such that

$$F_1 \cup F_2 \subseteq \bigcup_i I_i, \quad \text{and} \quad \sum_i \operatorname{vol}(I_i) < m^*(F_1 \cup F_2) + \epsilon.$$

Denote by $(I_k^{(1)})$ those intervals among the I_i's that meet F_1 and by $(I_k^{(2)})$ those that meet F_2. Then $F_1 \subseteq \bigcup_j I_j^{(1)}$ and $F_2 \subseteq \bigcup I_j^{(2)}$ by our judicious concerns over δ. Alas,

$$\begin{aligned} m^*(F_1) + m^*(F_2) &\leq \sum_j \operatorname{vol}(I_j^{(1)}) + \sum_j \operatorname{vol}(I_j^{(2)}) \\ &\leq \sum_i \operatorname{vol}(I_i) \leq m^*(F_1 \cup F_2) + \epsilon. \end{aligned}$$

Epsilonics to the rescue: $m^*(F_1) + m^*(F_2) \leq m^*(F_1 \cup F_2)$.

(ii) Represent $G = \bigcup_i I_i$ where I_i's are nonoverlapping open intervals, and let $\epsilon > 0$ be given; of course $m^*(G) \leq \sum_i \operatorname{vol}(I_i)$, and so there is an n_0 such that

$$\sum_{i \leq n_0} \operatorname{vol}(I_i) > m^*(G) - \frac{\epsilon}{2}.$$

For each $i \leq n_0$, let J_i be a closed subinterval of the interior of I_i with

$$\operatorname{vol}(J_i) > \operatorname{vol}(I_i) - \frac{\epsilon}{2^i}.$$

Then $F = \bigcup_{i \leq n_0} J_i$ is a closed subset of G and

$$\begin{aligned} m^*(F) &= m^*\Big(\bigcup_{i \leq n_0} J_i\Big) \\ &= \sum_{i \leq n_0} m^*(J_i) \quad \text{(by A)} \\ &= \sum_{i \leq n_0} \operatorname{vol}(J_i) \\ &> \sum_{i \leq n_0} \Big(\operatorname{vol}(I_i) - \frac{\epsilon}{2^i}\Big) \\ &\geq \sum_{i \leq n_0} \operatorname{vol}(I_i) - \frac{\epsilon}{2} \\ &> m^*(G) - \epsilon. \end{aligned}$$

Notice that the openness of G was used to represent G in an appropriate way.

(iii) Let $\epsilon > 0$ be given. Using (ii), choose a closed set $F_1 \subseteq G \backslash F$ so that

$$m^*(F_1) > m^*(G \backslash F) - \epsilon.$$

Notice that

$$\begin{aligned} m^*(F) + m^*(G \backslash F) &\leq m^*(F) + (m^*(F_1) + \epsilon) \\ &= m^*(F \cup F_1) + \epsilon \\ &\leq m^*(G) + \epsilon. \end{aligned}$$

Epsilonics take over to say that

$$m^*(F) + m^*(G \backslash F) \leq m^*(G).$$

Of course, Theorem 1.2(ii) takes care of the reverse inequality, and with it (iii). □

Definition 1.4. A subset E of $\mathbb{R}^k$ is *Lebesgue measurable* if given an $\epsilon > 0$ there is a closed set F and open set G such that

$$F \subseteq A \subseteq G \text{ and } m^*(G \backslash F) < \epsilon.$$

By the complementary nature of open and closed sets, E *is Lebesgue measurable if and only if* E^c *is*

$$F \subseteq E \subseteq G \Leftrightarrow G^c \subseteq E^c \subseteq F^c \text{ and } F^c \backslash G^c = G \backslash F.$$

Theorem 1.5. *Some basic results on measurable sets:*

(i) *If A and B are measurable, then so is $A \cap B$.*

(ii) *A bounded set B is measurable if for each $\epsilon > 0$ there is a compact set $K \subseteq B$ such that*

$$m^*(K) > m^*(B) - \epsilon.$$

(iii) *Intervals are measurable.*

(iv) *Sets of outer measure zero are measurable.*

(v) *If (A_n) is a sequence of disjoint measurable subsets of the interval I, then $\bigcup_n A_n$ is measurable also and*

$$m^*\left(\bigcup_n A_n\right) = \sum_n m^*(A_n).$$

(vi) *If (A_n) is any sequence of disjoint measurable subsets of $\mathbb{R}^k$, then $\bigcup_n A_n$ is measurable and*

$$m^*\left(\bigcup_n A_n\right) = \sum_n m^*(A_n).$$

Proof. (i) Pick F_A, F_B closed and G_A, G_B open such that

$$F_A \subseteq A \subseteq G_A \text{ and } m^*(G_A \backslash F_A) \leq \tfrac{\epsilon}{2},$$
$$F_B \subseteq B \subseteq G_B \text{ and } m^*(G_B \backslash F_B) \leq \tfrac{\epsilon}{2}.$$

Then $F = F_A \cap F_B$ is closed, $G = G_A \cap G_B$ is open, $F \subseteq A \cap B \subseteq G$, and $G \backslash F \subseteq (G_A \backslash F_A) \cup (G_B \backslash F_B)$ so that

$$m^*(G \backslash F) \leq m^*(G_A \backslash F_A) + m^*(G_B \backslash F_B) \leq \frac{\epsilon}{2} + \frac{\epsilon}{2} = \epsilon.$$

(ii) Suppose the bounded set B satisfies the conditions set forth, and let $\epsilon > 0$ be given. We can find a compact set $K \subseteq B$ such that

$$m^*(K) \geq m^*(B) - \frac{\epsilon}{2}.$$

But $m^*(B) < \infty$. (Why is that?) So we can cover B by a sequence (I_j) of open intervals each of diameter less than $\frac{1}{37}$ such that

$$\sum_j \operatorname{vol}(I_j) < m^*(B) + \frac{\epsilon}{2}.$$

Let G be the union of all those I_j's that meet B. $K \subseteq B \subseteq G$ and G is bounded. So our preparatory work in Theorem 1.3(iii) tells us that

$$\begin{aligned} m^*(G \backslash K) &= m^*(G) - m^*(K) \\ &\leq \sum_j m^*(I_j) - m^*(K) \\ &= \sum_j \mathrm{vol}(I_j) - m^*(K) \\ &\leq m^*(B) + \frac{\epsilon}{2} - m^*(K) \\ &= m^*(B) - m^*(K) + \frac{\epsilon}{2} \\ &\leq \frac{\epsilon}{2} + \frac{\epsilon}{2} = \epsilon. \end{aligned}$$

(iii) After all, if I is an interval, then $m^*(I) = \mathrm{vol}(I)$, and so we can plainly approximate I from the inside by *compact* intervals.

(iv) If $m^*(N) = 0$ and $\epsilon > 0$ is given, there must be a sequence (I_j) of open intervals so $N \subseteq \bigcup_j I_j$ and

$$\sum_j \mathrm{vol}(I_j) \leq m^*(N) + \epsilon = \epsilon.$$

Then $G = \bigcup_j I_j$ and $F = \emptyset$ soon show the way to N's measurability.

(v) Let $\epsilon > 0$ be given. Choose compact sets $F_n \subseteq A_n$ so

$$m^*(F_n) > m^*(A_n) - \frac{\epsilon}{2^{n+1}}.$$

Since

$$m^*\left(\bigcup_n A_n\right) \leq \sum_n m^*(A_n),$$

there is an $n_0 \in \mathbb{N}$ so that

$$\sum_{n<n_0} m^*(A_n) > m^*\left(\bigcup_n A_n\right) - \frac{\epsilon}{2}.$$

If $F = \bigcup_{n \leq n_0} F_n$, then F is compact (it's closed, and being a subset of I, bounded). Hence by Theorem 1.3(i)

$$m^*(F) = \sum_{n \leq n_0} m^*(F_n) > \sum_{n \leq n_0} m^*(A_n) - \frac{\epsilon}{2} > m^*\left(\bigcup_n A_n\right) - \epsilon.$$

We've just taken the bounded set $\bigcup_n A_n$ and, for each $\epsilon > 0$, found a compact set F, contained in $\bigcup_n A_n$ so that

$$m^*(F) > m^*\left(\bigcup_n A_n\right) - \epsilon.$$

Therefore $\bigcup_n A_n$ is measurable thanks to (ii). Let's check the sums: for any $n_0 \in \mathbb{N}$

$$\begin{aligned}\sum_{n\le n_0} m^*(A_n) &< \sum_{n\le n_0}\left(m^*(F_n)+\frac{\epsilon}{2^{n+1}}\right)\\ &\le \sum_{n\le n_0} m^*(F_n)+\frac{\epsilon}{2}\\ &= m^*\left(\bigcup_{n\le n_0} F_n\right)+\frac{\epsilon}{2}\\ &\le m^*\left(\bigcup_n A_n\right)+\epsilon.\end{aligned}$$

Epsilonics assure us that

$$\sum_{n\le n_0} m^*(A_n) \le m^*\left(\bigcup_n A_n\right),$$

and this is so for each n_0. It follows that

$$\sum_n m^*(A_n) \le m^*\left(\bigcup_n A_n\right).$$

We've already seen the reverse, so we have (v).

(vi) We'll bootstrap our way from (v) to (vi). To start, let (I_m) be a sequence of disjoint intervals whose union is $\mathbb{R}^k$ and such that any bounded set in $\mathbb{R}^k$ is covered by finitely many I_m's.

For each $m, n \in \mathbb{N}$, let

$$A_{m,n} = I_m \cap A_n$$

be that part of A_n inside I_m. Each $A_{m,n}$ is measurable ((iii) and (i)) and the $A_{m,n}$'s are pairwise disjoint. Look to

$$\widetilde{A_m} = \bigcup_n A_{m,n},$$

the part of $\bigcup_n A_n$ in I_m. By (v), $\widetilde{A_m}$ is measurable. Further, the $\widetilde{A_m}$'s are pairwise disjoint and $\bigcup_n \widetilde{A_m} = \bigcup_n A_n$.

Let $\epsilon > 0$ be given. For each m, choose a closed set $F_m \subseteq \widetilde{A_m}$ and an open set G_m, which is a bounded open set, $\widetilde{A_m} \subseteq G_m$ such that

$$m^*(G_m \backslash F_m) < \frac{\epsilon}{2^m}.$$

Look at $F = \bigcup_m F_m$ and $G = \bigcup_m G_m$. F is closed (if (x_n) is a convergent sequence of points of F, then (x_n) is bounded, and so for some m_0, (x_n) is a sequence in $\bigcup_{m \le m_0} F_m$, hence visits one of $F_1, \ldots, F_{m_0}$ infinitely often—whichever F_j it visits so often contains its limit) and G is open.

$$F = \bigcup_m F_m \subseteq \bigcup_m \widetilde{A_m} = \bigcup_m A_m = \bigcup_m \widetilde{A_m} \subseteq \bigcup_m G_m = G.$$

Further,

$$G \backslash F = \bigcup_m (G_m \backslash F) \subseteq \bigcup_m (G_m \backslash F_m),$$

so

$$m^*(G \backslash F) \le \sum_m m^*(G_m \backslash F_m) \le \sum_m \frac{\epsilon}{2^m} = \epsilon,$$

and $\bigcup_n A_n$ is measurable.

Now

$$A_n = \bigcup A_{m,n},$$

so

$$m^*(A_n) \le \sum_m m^*(A_{m,n}).$$

It follows that

$$\begin{aligned} \sum_n m^*(A_n) &\le \sum_n \sum_m m^*(A_{m,n}) \\ &= \sum_m \sum_n m^*(A_{m,n}) \\ &= \sum_m m^*(\widetilde{A_m}), \end{aligned}$$

by (v). Take $m \in \mathbb{N}$. Then

$$\begin{aligned} \sum_{j \le m} m^*(\widetilde{A_j}) &\le \sum_{j \le m} (m^*(F_j) + m^*(G_j \backslash F_j)) \\ &= m^*\left(\bigcup_{j \le m} F_j \right) + \sum_{j \le m} m^*(G_j \backslash F_j) \text{ (by A)} \\ &\le m^*\left(\bigcup_{j \le m} F_j \right) + \sum_{j \le m} \frac{\epsilon}{2^j} \\ &\le m^*\left(\bigcup_n A_n \right) + \epsilon. \end{aligned}$$

The usual epsilonics leads us to conclude that

$$\sum_m m^*(\widetilde{A_m}) \le m^*\left(\bigcup_n A_n \right)$$

and, in tandem with what gone on before, we see

$$\sum_n m^*(A_n) \leq m^*\left(\bigcup_n A_n\right).$$

Again the reverse holds without assumption, so (vi) is proved. □

Theorem 1.6 (The Fundamental Theorem of Lebesgue Measure). *In summary*

(i) m^* *is a nonnegative, extended real-valued function defined for every subset of* $\mathbb{R}^k$ *which assigns*

- *to each interval, a value equal to its volume,*
- *to each set, a value common to all its translates,*
- *to bigger sets, bigger values,*
- *to compact sets, finite values,*
- *to nonempty sets, nonnegative values,*

and is countably subadditive in doing so.

For any $A \subseteq \mathbb{R}^k$,

$$m^*(A) = \inf\{m^*(G) : G \text{ is open } A \subseteq G\}.$$

(ii) *The Lebesgue measurable subsets of* $\mathbb{R}^k$ *form a* σ*-field* $\mathcal{M}$ *of sets containing every open set, closed set, interval, and set of outer measure zero;* $E \in \mathcal{M}$ *if and only if* E*'s translates are members of* $\mathcal{M}$.

(iii) m^* *is countably additive on* $\mathcal{M}$ *and for* $E \in \mathcal{M}$,

$$m^*(E) = \sup\{m^*(K) : K \text{ is compact } K \subseteq E\}.$$

2. The Brunn–Minkowski Theorem

Theorem 1.7 (The Brunn–Minkowski Theorem). *Let* $n \geq 1$, *and let* λ_n *denote Lebesgue measure on* $\mathbb{R}^n$. *If* A, B, *and* $A + B$ *are measurable subsets of* $\mathbb{R}^n$, *then*

$$\text{(BM)} \qquad (\lambda_n(A+B))^{1/n} \geq (\lambda_n(A))^{1/n} + (\lambda_n(B))^{1/n},$$

where

$$A + B = \{a + b : a \in A, b \in B\}.$$

Notice that (BM) is a geometric generalization of the *arithmetic-geometric mean inequality* since if A and B are rectangles with sides of length $(a_j)_{j=1}^n$ and $(b_j)_{j=1}^n$, respectively, then (BM) looks like

$$\text{(BM}'\text{)} \qquad \left[\prod_i^n (a_j + b_j)\right]^{1/n} \geq \left(\prod_1^n a_j\right)^{1/n} + \left(\prod_1^n b_j\right)^{1/n}.$$

Homogeneity lets us reduce this to the case where $a_j + b_j = 1$ for each j. But now the arithmetic-geometric mean inequality assures that

$$\frac{1}{n}\sum_{j=1}^n a_j \geq \left(\prod_1^n a_j\right)^{1/n} \qquad \text{and} \qquad \frac{1}{n}\sum_{j=1}^n b_j \geq \left(\prod_1^n b_j\right)^{1/n}.$$

So (noting that $\sum_{j=1}^n a_j + \sum_{j=1}^n b_j = \sum_{j=1}^n (a_j + b_j) = \sum_{j=1}^n 1 = n$)

$$1 = \frac{1}{n} \cdot n \geq \left(\prod_1^n a_j\right)^{1/n} + \left(\prod_1^n b_j\right)^{1/n},$$

which is (BM$'$). Thus we have proved (BM) for boxes, rectangular parallelepipeds whose sides are parallel to the coordinate hyperplanes.

Proof. To prove (BM), suppose that each of A and B is the union of finitely many rectangles whose interiors are disjoint. We proceed by induction on the total number of rectangles in A and B. It is important to realize that the inequality is unaffected if we translate A and B independently: in fact, replacing A by $A + h$ and B by $B + k$ replaces $A + B$ by $A + B + h + k$, and the corresponding measures are the same as what we started with. If R_1 and R_2 are essentially disjoint rectangles in the collection making up A, then they can be separated (a translation may be necessary) by a coordinate hyperplane, $\{x_j = 0\}$ say. Thus we may assume that R_1 lies in $A^- = A \cap \{x_j \leq 0\}$ and R_2 lies in $A^+ = A \cap \{x_j \geq 0\}$. Notice that A^+ and A^- each contain at least one rectangle less than does A and $A = A^+ \cup A^-$.

What to do with B? Well, slide B over so if $B^+ = B \cap \{x_j \geq 0\}$, then

$$\frac{\lambda_n(B^+)}{\lambda_n(B)} = \frac{\lambda_n(A^+)}{\lambda_n(A)};$$

of course this entails

$$\frac{\lambda_n(B^-)}{\lambda_n(B)} = \frac{\lambda_n(A^-)}{\lambda_n(A)}$$

as well. But

$$(A^+ + B^+) \cup (A^- + B^-) \subseteq A + B,$$

and the union on the left hand side is essentially disjoint. Moreover, the total number of rectangles in either A^+ and B^+ or in A^- and B^- is less than that in A and B. Our induction hypothesis applies.

The result?

$$\begin{aligned}
\lambda_n(A+B) &\geq \lambda_n(A^+ + B^+) + \lambda_n(A^- + B^-) \\
&\geq (\lambda_n(A^+)^{1/n} + \lambda_n(B^+)^{1/n})^n + ((\lambda_n(A^-)^{1/n} + \lambda_n(B^-)^{1/n})^n \\
&\qquad\qquad\qquad\qquad\qquad\qquad \text{(induction hypothesis)} \\
&= \lambda_n(A^+)\left(1+\left(\frac{\lambda_n(B)}{\lambda_n(A)}\right)^{1/n}\right)^n + \lambda_n(A^-)\left(1+\left(\frac{\lambda_n(B)}{\lambda_n(A)}\right)^{1/n}\right)^n \\
&= \lambda_n(A)\left(1+\left(\frac{\lambda_n(B)}{\lambda_n(A)}\right)^{1/n}\right)^n \\
&= (\lambda_n(A)^{1/n} + \lambda_n(B)^{1/n})^n,
\end{aligned}$$

and that's all she wrote. Thus we have (BM) for finite unions of boxes.

If A and B are open sets of finite measure, then once given a margin of error, $\epsilon > 0$, we can find unions A_ϵ, B_ϵ of essentially disjoint rectangles such that $A_\epsilon \subseteq A, B_\epsilon \subseteq B$, and

$$\lambda_n(A) \leq \lambda_n(A_\epsilon) + \epsilon, \ \lambda_n(B) \leq \lambda_n(B_\epsilon) + \epsilon.$$

Once done, $A_\epsilon + B_\epsilon \subseteq A + B$, and we can apply the work of the previous paragraphs:

$$\begin{aligned}
\lambda_n(A+B)^{1/n} &\geq \lambda_n(A_\epsilon + B_\epsilon)^{1/n} \\
&\geq (\lambda_n(A) - \epsilon)^{1/n} + (\lambda_n(B) - \epsilon)^{1/n}.
\end{aligned}$$

Let $\epsilon \to 0$. It follows that (BM) holds for open sets A and B of finite measure.

If A and B are compact sets, then $A + B$ is compact as well. Look at $A_\epsilon = [d(x, A) < \epsilon]$. Then A_ϵ is open, contains A and $A_\epsilon \searrow A$. Similarly B_ϵ is defined analogously. Plainly, A_ϵ and B_ϵ are open and bounded. What can we say about $(A+B)_\epsilon$? The same conclusions can be drawn. Further,

$$A + B \subseteq A_\epsilon + B_\epsilon \subseteq (A+B)_{2\epsilon}.$$

Applying (BM) to A_ϵ, B_ϵ, we get

$$\begin{aligned}
\lambda_n((A+B)_{2\epsilon})^{1/n} &\geq \lambda_n(A_\epsilon + B_\epsilon)^{1/n} \\
&\geq \lambda_n(A_\epsilon)^{1/n} + \lambda_n(B_\epsilon)^{1/n} \\
&\geq \lambda_n(A)^{1/n} + \lambda_n(B)^{1/n}.
\end{aligned}$$

Letting $\epsilon \to 0$ gives (BM) for $A, B, A+B$.

The general situation of A, B and $A+B$ measurable follows by approximating from within: if $\tilde{A}, \tilde{B}$ are compact sets with $\tilde{A} \subseteq A, \tilde{B} \subseteq B$, then $\tilde{A} + \tilde{B}$

is a compact set inside $A+B$, and

$$\lambda_n(A+B) \geq \lambda_n(\tilde{A}+\tilde{B}) \geq (\lambda_n(\tilde{A})^{1/n} + \lambda_n(\tilde{B})^{1/n})^n.$$

Let $\lambda_n(\tilde{A}) \nearrow \lambda_n(A), \lambda_n(\tilde{B}) \nearrow \lambda_n(B)$, and be done with it. □

With the Brunn–Minkowski inequality in hand we can easily find our way to the *isodiametric inequality.*

Theorem 1.8. *For any Borel set $B \subseteq \mathbb{R}^n$ we have*

$$\lambda_n(B) \leq \left(\frac{\operatorname{diam} B}{2}\right)^n \lambda_n(B_{\ell_n^2}),$$

where as usual $B_{\ell_n^2}$ denotes the closed unit ball of $\mathbb{R}^n$.

Proof. Without loss of sleep we can assume that

$$d := \operatorname{diam} B < \infty.$$

This in mind, realize that if $x, y, \in B$, then $||x-y|| \leq d$. Hence

$$B - B \subseteq d\, B_{\ell_n^2},$$

and so

$$\begin{aligned} 2\lambda_n(B)^{1/n} &= \lambda_n(B)^{1/n} + \lambda_n(B)^{1/n} \\ &= \lambda_n(B)^{1/n} + \lambda_n(-B)^{1/n} \\ &\leq \lambda_n(B-B)^{1/n} \quad \text{(Brunn–Minkowski to the rescue!)} \\ &\leq \lambda_n(d\, B_{\ell_n^2})^{1/n} \\ &= d\lambda_n(B_{\ell_n^2})^{1/n}. \end{aligned}$$

It's easy to deduce the conclusion of the theorem from this inequality:

$$2\lambda_n(B)^{1/n} \leq \operatorname{diam}(B)\lambda_n(B_{\ell_n^2})^{1/n}.$$

□

3. Covering Theorem of Vitali

Theorem 1.9 (Vitali). *Let $\mathcal{F}$ be a family of closed balls in $\mathbb{R}^n$ that covers a set E in the sense of Vitali, that is, given an $\epsilon > 0$ and $x \in E$, there is a $B \in \mathcal{F}$ such that the diameter of B is less than ϵ and $x \in B$. Then $\mathcal{F}$ contains a disjoint sequence that covers λ_n—almost all of E.*

Proof. First we suppose E is bounded and contained in the bounded open set G. We disregard any members of $\mathcal{F}$ that aren't contained in G as well as those that don't intersect E. The resulting family, which we will still refer to as $\mathcal{F}$ covers E in the sense of Vitali. Our proof will be by "exhaustion". To start, let R_1 be

$$R_1 := \sup\{\text{radius}(B) : B \in \mathcal{F}\}.$$

Choose $B_1 \in \mathcal{F}$, say centered at a_1, with radius r_1 so that

$$\frac{R_1}{2} \leq r_1.$$

Next let R_2 be

$$R_2 = \sup\{\text{radius } (B) : B \in \mathcal{F}, B_1 \cap B = \emptyset\}.$$

Choose $B_2 \in \mathcal{F}$, say, centered at a_2 with radius r_2 so that $B_2 \cap B_1 = \emptyset$ and

$$\frac{R_2}{2} \leq r_2.$$

Continue down this primrose lane, after k steps, let R_{k+1} be

$$R_{k+1} = \sup\{\text{radius}(B) : B \in \mathcal{F}, B \cap (B_1 \cup \cdots \cup B_k) = \emptyset\}.$$

Choose $B_{k+1} \in \mathcal{F}$ with center a_{k+1}, radius r_{k+1} so

$$B_{k+1} \cap (B_1 \cup \cdots \cup B_k) = \emptyset$$

and

$$\frac{R_{k+1}}{2} \leq r_{k+1}.$$

Naturally,

$$R_1 \geq R_2 \geq \cdots .$$

More is so. The B_k's are pairwise disjoint so

$$\sum_k \lambda_n(B_k) = \lambda_n\left(\bigcup_k B_k\right) \leq \lambda_n(G) < \infty;$$

hence

$$\lim_k \lambda_n(B_k) = 0.$$

But

$$\lambda_n(B_k) = c_n r_k^n$$

(while we can tell what c_n is, this precise bit of information is unnecessary at this juncture), so

$$\lim_k r_k = 0.$$

Since $0 \leq R_k \leq 2r_k$, we see

$$\lim_k R_k = 0.$$

The B_k's eat up E. How? What does it mean for $x \in E$ *not* to be in $\bigcup_{k \leq K} B_k$? Well $x \notin \bigcup_{k \leq K} B_k$ means

$$d\left(x, \bigcup_{k \leq K} B_k\right) = \delta > 0.$$

Choose $B \in \mathcal{F}$ so $x \in B$ and B (which is centered at say a with radius r) has radius less than $\delta/2$.

Each point of B is within $2r$ of x and $2r < \delta$. Hence B is disjoint from $B_1 \cup \cdots \cup B_K$. B is one of those balls belonging to $\mathcal{F}$ that take part in defining R_{K+1}; in particular, $r \leq R_{K+1}$. But

$$\lim_j R_j = 0,$$

and the R_j's are monotone and nonincreasing. Hence there is a j (which is necessarily bigger than K) so that

$$R_{j+1} < r \leq R_j.$$

Since $r > R_{j+1}$, B must meet one of the balls $B_1, \ldots, B_j$. Suppose $B \cap B_m \neq \emptyset$. Since B does *not* meet $B_1 \cup \cdots \cup B_K$, $m > K$. It follows that for any $c \in B \cap B_m$,

$$\begin{aligned} \|x - a_m\| &\leq \|x - a\| + \|a - c\| + \|c - a_m\| \\ &\qquad \text{(where } a_m \text{ is the center of } B_m \text{ and } a \text{ is the center of } B\text{)} \\ &\leq r + r + r_m \\ &= 2r + r_m. \end{aligned}$$

So

$$\|x - a_m\| \leq 2r + r_m,$$

where

$$r \leq R_j \leq R_m \leq 2r_m.$$

Hence

$$\|x - a_m\| \leq 2r + r_m \leq 2 \cdot (2r_m) + r_m \leq 5r_m.$$

To summarize: if $x \in E \backslash \bigcup_{k \leq K} B_k$, then x belongs to the union $\bigcup_{k \leq K} \tilde{B}_k$ of those closed balls $\tilde{B}_k$, where $\tilde{B}_k$ shares the center a_k with B_k but $\tilde{B}_k$ has five times the radius of B_k.

In other words,

$$\begin{aligned} \lambda_n\Big(E \backslash \bigcup_{k \leq K} B_k\Big) &\leq \sum_{k > K} \lambda_n(\tilde{B}_k) \\ &= \sum_{k > K} 5^n \lambda_n(B_k) \to 0, \text{ as } K \to \infty, \end{aligned}$$

since, as we have already noted,

$$\sum_k \lambda_n(B_k) = \lambda_n\Big(\bigcup_k B_k\Big) \leq \lambda_n(G) < \infty.$$

This finishes the proof for E's that are bounded.

For general E, look at the sets

$$E_m = \{x \in E : m - 1 \leq ||x|| < m\}, \ m \in \mathbb{N}.$$

Of course

$$\lambda_n(E \setminus \bigcup_m E_m) = 0,$$

and each E_m is as we've discussed above. What's more, the G we chose at the very start of our proof can be chosen to be the open set

$$\{x \in \mathbb{R}^n : m - 1 < ||x|| < m\},$$

so the disjointness is achieved by passing from one E_m to the next. □

4. Notes and Remarks

Our treatment of Lebesgue measure is an elaboration of Oxtoby's development as presented in his wonderful volume *Measure and category* [**96**].

4.1. Bernstein's Proof of the Existence of Nonmeasurable Sets. Bernstein's topological proof [**10**] of the existence of a non-Lebesgue measurable subset of the real line has many points of interest, not least of which the window it opens to the study of descriptive set theory.

A first step is to notice the following:

Theorem 1.10. *Any uncountable $\mathcal{G}_\delta$ subset of $\mathbb{R}$ contains a homeomorphic copy of the Cantor set Δ.*

Proof. Let E be such a set. Then there exists a descending sequence (G_n) of open sets in $\mathbb{R}$ so that $E = \bigcap_n G_n$. Let F be the set of all $x \in E$ for which given any open set U containing x, $U \cap E$ is uncountable. Notice that F is nonempty: after all, if $F = \emptyset$, then the family of open intervals with rational endpoints and having a countable number of points in common with E would cover F; it would follow that E is itself countable, which it is not. Similarly, F has no isolated points. In particular, F is infinite.

Now to start the construction of Δ inside E, let $I(0)$ and $I(1)$ be disjoint closed intervals of length at most $\frac{1}{3}$ whose interiors meet F and whose union is contained in G_1.

Now the interior of $I(0)$ has an infinite number of points in common with F, and so we can find disjoint closed intervals $I(0,0)$ and $I(0,1)$ of length no more than $\frac{1}{3^2}$ whose interiors meet F and whose union is contained in $G_2 \cap I(0)$.

Likewise the interior of $I(1)$ has an infinite number of points in common with F, and so we can find disjoint closed intervals $I(1,0)$ and $I(1,1)$ of length no more than $\frac{1}{3^2}$ whose interiors meet F and whose union is contained in $G_2 \cap I(1)$.

Etc., etc., etc.

The (compact) set

$$\bigcap_n \bigcup_{(i_1,\ldots,i_n)\in\{0,1\}^n} I(i_1,\ldots,i_n)$$

is a homeomorphic copy of the Cantor set and lies inside $\bigcap_n G_n = E$. □

A consequence of this is that *the collection of uncountable closed subsets of* $\mathbb{R}$ *has cardinality* c*, the cardinality of the continuum.* Since the collection of open intervals with rational endpoints is a countable collection and each open set is the union of some subfamily of this collection, there is *at most* c *open sets (and so at most* c *closed sets)* in $\mathbb{R}$. On the other hand, there are already c many distinct closed intervals.

Now we're ready to present the still-beautiful result of F. Bernstein.

Theorem 1.11 (F. Bernstein). *There exists a set* B *of real numbers such that both* B *and its complement* B^c *meet every uncountable closed set in* $\mathbb{R}$.

Proof. We call on the *well-ordering principle.* Well-order $\mathbb{R}$.

Using facts established above, we can also well-order the collection $\mathcal{F}$ of all uncountable closed subsets of $\mathbb{R}$ (each of which, is, by the way, of cardinality c); we can index this well-ordering by the ordinals $\alpha < \omega_c$, where ω_c is the first ordinal of cardinality c, also known as the first ordinal with c predecessors. So $\mathcal{F} = (F_\alpha : \alpha < \omega_c)$.

We put the points of $\mathbb{R}$ into two baskets B and B': look at F_1, the first of $\mathcal{F}$'s members, and take two distinct points from F_1 (there are lots of points to choose from) and put one in B and the other in B'.

Now go to the second member of $\mathcal{F}$, F_2. Take two points of F_2, neither of which is one of the points from the previous paragraph. Put one of these points in B, the other in B'.

Continue in this fashion. At each stage (for each ordinal $\alpha < \omega_c$), pick two points from F_α that as yet are not in either B or B' (there are still plenty of choices). Put one in B and the other in B'.

So B is in place, B^c contains B', and we have our "Bernstein set" B.

The point to be made here is that any Bernstein set B is non-Lebesgue measurable. After all, if F is a closed subset of $\mathbb{R}$ that is contained in B, then $F \bigcap B^c = \emptyset$, and so F must be countable, and so is of Lebesgue measure zero. The regularity of Lebesgue measure tells us that B must have measure zero if it is measurable. But then the same reasoning applies to B^c if B is measurable. □

Again we have followed Oxtoby's direction in the presentation of Bernstein's result from F. Bernstein [**10**].

4.2. Vitali's Proof of the Existence of Nonmeasurable Sets. Bernstein's proof is topological in nature. There is another proof of the existence of a nonmeasurable set, due to G. Vitali, that is more group theoretic. It proceeds as follows.

Let $U = (-1, 1)$, and $V = (-\frac{2}{3}, \frac{2}{3})$. Denote by $\mathbb{Q}$ the *subgroup* of rational numbers, and look at $\mathbb{R}/\mathbb{Q}$ which consists of cosets $\mathbb{Q} + x_\alpha$ of $\mathbb{R}$ modulo $\mathbb{Q}$. Notice that each coset $\mathbb{Q} + x_\alpha$ is dense in $\mathbb{R}$. From each coset $\mathbb{Q} + x_\alpha$, pick *exactly one* y_α that is also in V. Then the set $E = \{y_\alpha\}$ is nonmeasurable.

Indeed, suppose E is measurable. Look at the set

$$S = \left(\mathbb{Q} \cap \left(-\frac{2}{3}, \frac{2}{3}\right)\right) + E.$$

It is the countable union

$$S = \bigcup_{q \in \mathbb{Q} \cap \left(-\frac{2}{3}, \frac{2}{3}\right)} (q + E)$$

of pairwise disjoint sets. To see the disjointness, suppose q_1 and q_2 are distinct rationals from $\left(-\frac{2}{3}, \frac{2}{3}\right)$ and

$$(q_1 + E) \cap (q_2 + E) \neq \emptyset.$$

Then

$$q_1 + y_\alpha = q_2 + y_{\alpha'}$$

for some $y_\alpha, y_{\alpha'} \in E$ so

$$y_\alpha - y_{\alpha'} = q_2 - q_1 \in \mathbb{Q},$$

which means

$$y_\alpha = y_{\alpha'}$$

by construction of E, and that's a "no-no".

So supposing that E is measurable, S is also measurable and is the countable union of sets of the form $q + E$, each of which is measurable and of the same measure. So S has two choices in life: either it is of measure zero or it has infinite measure.

Each choice is closed to S! After all,

$$\left(-\frac{1}{3}, \frac{1}{3}\right) \subseteq S \subseteq (-1, 1).$$

Therefore E is nonmeasurable.

The construction is still more subtle than one might at first suspect. Indeed, if Σ is a any σ-field of subsets of $\mathbb{R}$ which contains all the Lebesgue measurable sets, and if μ is a countably additive nonnegative extended real-valued measure extending Lebesgue measure defined on Σ, and both Σ and μ are translation invariant, then $E \notin \Sigma$. After all, if $E \in \Sigma$ and Σ is translation invariant, then $S \in \Sigma$ as well; but $\mu(S)$ is faced with the same choices as the Lebesgue measure of S: $\mu(S)$ is either 0 or $+\infty$, yet it's still so that

$$\left(-\frac{1}{3}, \frac{1}{3}\right) \subseteq S \subseteq (-1, 1).$$

G. Vitali [**132**] gave the first example of a nonmeasurable set. His use of the *axiom of choice* led to considerable controversy in its day, and no less an enthusiast about Lebesgue measure than Henri Lebesgue himself was dismayed by the consequences.

The attentive student will notice that Vitali's construction relies on the existence of a Hamel basis for $\mathbb{R}$ as a vector space over the subspace $\mathbb{Q}$ of rationals. It is of more than passing interest to notice that *No Hamel basis over $\mathbb{Q}$ is a Borel set or even analytic.* As Nadkarni and Sunder [**90**] note, a Hamel basis $\mathbb{R}$ over $\mathbb{Q}$ can be Lebesgue measurable. Indeed the usual Cantor set Δ sitting inside $[0, 1]$ contains a Hamel basis since

$$\Delta + \Delta = \{x + y : x, y \in \Delta\}$$

coincides with the interval $[0, 2]$.

4.3. Isoperimetric Inequality. Among the many consequences of the Brunn–Minkowski theorem, we find the following classical result.

Theorem 1.12 (Isoperimetric Inequality). *Among convex bodies of a given volume, Euclidean balls have the least surface area.*

Proof. Let C be a convex, compact set in $\mathbb{R}^n$ whose n-dimensional volume is that of

$$B := B_{l_n^2}.$$

The surface area of C can be expressed by

$$\text{(1)} \qquad \mathrm{sa}(\partial C) = \lim_{\epsilon \searrow 0} \frac{\mathrm{vol}_n(C + \epsilon B) - \mathrm{vol}_n(C)}{\epsilon}.$$

The Brunn–Minkowski inequality tells us that

$$\mathrm{vol}_n(C + \epsilon B)^{\frac{1}{n}} \geq (\mathrm{vol}_n(C)^{\frac{1}{n}} + \epsilon \mathrm{vol}_n(B)^{\frac{1}{n}},$$

so

$$\begin{aligned} \mathrm{vol}_n(C + \epsilon B) &\geq \left((\mathrm{vol}_n(C))^{\frac{1}{n}} + \epsilon \mathrm{vol}_n(B)^{\frac{1}{n}} \right)^n \\ &\geq \mathrm{vol}_n(C) + n\epsilon \mathrm{vol}_n(C)^{\frac{n-1}{n}} \mathrm{vol}_n(B)^{\frac{1}{n}}. \end{aligned}$$

Hence

$$\mathrm{vol}_n(C + \epsilon B) - \mathrm{vol}_n(C) \geq n\epsilon \mathrm{vol}_n(C)^{\frac{n-1}{n}} \mathrm{vol}_n(B)^{\frac{1}{n}},$$

and so

$$\frac{\mathrm{vol}_n(C + \epsilon B) - \mathrm{vol}_n(C)}{\epsilon} \geq n \mathrm{vol}_n(C)^{\frac{n-1}{n}} \mathrm{vol}_n(B)^{\frac{1}{n}}.$$

Taking the expression (1) into account, we see that

$$\mathrm{sa}(\partial C) \geq n \mathrm{vol}_n(C)^{\frac{n-1}{n}} \mathrm{vol}_n(B)^{\frac{1}{n}}.$$

Further factoring into account that we're supposing C and B have the same volume, we have

$$\mathrm{sa}(\partial C) \geq n \mathrm{vol}_n(B) = \mathrm{sa}(\partial B)!$$

Isn't life beautiful? Isn't life gay? Isn't life the perfect way to pass the time of day? □

The above proof is taken from Keith Ball's beautiful essay "An Elementary Introduction to Modern Convex Geometry" [**3**].

Let X be a bounded compact metric space (the closure of any ball is compact), and let $\mathcal{B}$ be a family of closed balls in X with

$$\sup\{\mathrm{diam}(B) : B \in \mathcal{B}\} < \infty.$$

Then there is a countable subfamily $\mathcal{C}$ of $\mathcal{B}$ consisting of pairwise disjoint balls such that the union of all $B \in \mathcal{B}$ is contained in the union of balls with the same centers as those in $\mathcal{C}$ but five times the radii.

The proof of this and similar variations can be found in the book *Geometry of Sets and Measures in Euclidean Spaces* by Pertti Mattila [**80**].

More general versions can be studied in Federer's *Geometric Measure Theory* [**38**, section 2.8].

Chapter 2

Measures on Metric Spaces

1. Generalities on Outer Measures

Definition 2.1. A function μ defined on the sets of a space Ω is called an *outer measure* on Ω if it satisfies the following conditions:

(i) $0 \leq \mu(E) \leq +\infty$ for each $E \subseteq \Omega$;

(ii) $\mu(\emptyset) = 0$;

(iii) $\mu(E) \leq \mu(F)$ if $E \subseteq F$;

(iv) if (E_n) is a sequence of subsets of Ω, then

$$\mu\left(\bigcup_n E_n\right) \leq \sum_n \mu(E_n),$$

i.e., μ is countably subadditive.

If μ is an outer measure on the space Ω, then $E \subseteq \Omega$ is said to be μ-*measurable* if whenever $A \subseteq E$ and $B \subseteq E^c$, then

$$\mu(A \cup B) = \mu(A) + \mu(B).$$

Notice that $E \subseteq \Omega$ is μ-measurable precisely when given $A \subseteq E$ and $B \subseteq E^c$,

$$\mu(A \cup B) \geq \mu(A) + \mu(B); \tag{2}$$

after all, (iv) assures that the reverse inequality holds regardless of E, A, or B. In fact, we really need only show (2) for sets A, B of finite μ-measure, since if either $\mu(A)$ or $\mu(B)$ is infinite, then so is $\mu(A \cup B)$.

Theorem 2.2 (Carathéodory). *Let μ be an outer measure on the set Ω. Then*

(i) *if $\mu(N) = 0$, then N is μ-measurable;*

(ii) *E is μ-measurable if and only if E^c is;*

(iii) *if (E_n) is a sequence of μ-measurable sets, then $\bigcup_n E_n$ and $\bigcap_n E_n$ are too;*

(iv) *if (E_n) is a sequence of pairwise disjoint μ-measurable sets, then*

$$\mu\left(\bigcup_n E_n\right) = \sum_n \mu(E_n).$$

Proof. (i) Suppose $A \subseteq N$ and $B \subseteq N^c$. Of course $\mu(A) = 0$. Hence

$$\mu(B) \leq \mu(A \cup B) \leq \mu(A) + \mu(B) = \mu(B).$$

Squeezy says all are equal and so N is μ-measurable.

(ii) is plain.

To start the proof of (iii) and (iv), we'll show that the union of two μ-measurable sets E, F is μ-measurable. Take A and B to be sets of finite μ-measure with

$$A \subseteq E \cup F, \quad B \subseteq (E \cup F)^c.$$

Notice

$$A \cup B = (A \cap E) \cup ((A \cup B) \cap E^c).$$

Since

$$A \cap E \subseteq E \text{ and } (A \cup B) \cap E^c \subseteq E^c,$$

by E's measurability,

$$\mu(A \cup B) = \mu((A \cap E) \cup ((A \cup B) \cap E^c)) = \mu(A \cap E) + \mu((A \cup B) \cap E^c).$$

At the same time

$$(A \cup B) \cap E^c = (A \cap E^c) \cup B$$

with

$$A \cap E^c \subseteq F$$

and

$$B \subseteq (E \cup F)^c;$$

so F's μ-measurability assures that

$$\mu((A \cup B) \cap E^c) = \mu((A \cap E^c) \cup B) = \mu(A \cap E^c) + \mu(B).$$

But E's measurability is still in effect and says that

$$\mu(A \cap E) + \mu(A \cap E^c) = \mu(A),$$

so

$$\begin{aligned} \mu(A \cup B) &= \mu(A \cap E) + \mu((A \cup B) \cap E^c) \\ &= \mu(A \cap E) + \mu(A \cap E^c) + \mu(B) \\ &= \mu(A) + \mu(B). \end{aligned}$$

Therefore if E and F are μ-measurable, then so is $E \cup F$.

Next we consider a sequence (E_n) of pairwise disjoint μ-measurable sets and write $E = \bigcup_n E_n$. Let A and B be sets (of finite μ-measure) with

$$A \subseteq E, \quad B \subseteq E^c.$$

Each of $\bigcup_{m=1}^n E_m$ is μ-measurable and

$$\begin{aligned} \mu(A \cup B) &\geq \mu\left(\left[A \cap \bigcup_{m=1}^n E_m\right] \cup B\right) \\ &= \mu\left(A \cap \left(\bigcup_{m=1}^n E_m\right)\right) + \mu(B), \end{aligned}$$

because, after all,

$$A \cap \left(\bigcup_{m=1}^n E_m\right) \subseteq \bigcup_{m=1}^n E_m,$$

and

$$B \subseteq \left(\bigcup_n E_n\right)^c = \bigcap_n E_n^c \subseteq \bigcap_{m=1}^n (E_m^c) = \left(\bigcup_{m=1}^n E_m\right)^c.$$

Since the sets $E_n, E_{n-1}, \ldots, E_1$ are pairwise disjoint and μ-measurable,

$$\begin{aligned} \mu\left(A \cap \left(\left(\bigcup_{m=1}^n E_m\right)\right)\right) &= \mu\left(\left[A \cap \left(\bigcup_{m=1}^{n-1} E_m\right)\right]\right) \cup [A \cap E_n] \\ &= \mu\left(A \cap \left(\bigcup_{m=1}^{n-1} E_m\right)\right) + \mu(A \cap E_n) \\ &\;\;\vdots \\ &= \mu\left(A \cap \left(\bigcup_{m=1}^{n-2} E_m\right)\right) + \mu(A \cap E_{n-1}) + \mu(A \cap E_n) \\ &\;\;\vdots \\ &= \mu(A \cap E_1) + \mu(A \cap E_2) + \cdots + \mu(A \cap E_n). \end{aligned}$$

So we see that

$$\begin{aligned}\mu(A \cup B) &\geq \mu\left(A \cap \left(\bigcup_{m=1}^{n} E_n\right)\right) + \mu(B)\\ &= \sum_{m=1}^{n} \mu(A \cap E_m) + \mu(B),\end{aligned}$$

and this is so for each n. Thus

$$\begin{aligned}\mu(A \cup B) &\geq \sum_n \mu(A \cap E_n) + \mu(B)\\ &\geq \mu\left(A \cap \left(\bigcup_n E_n\right)\right) + \mu(B) \quad \text{(since } \mu \text{ is countably subadditive)}\\ &= \mu(A \cup E) + \mu(B)\\ &= \mu(A) + \mu(B)\\ &\geq \mu(A \cup B).\end{aligned}$$

All are equal; in particular, $\mu(A \cup B)$ and $\mu(A) + \mu(B)$ are equal and $\bigcup_n E_n$ is μ-measurable. Next if $B = \emptyset$, we get for any subset A of $\bigcup_n E_n$,

$$\mu(A) = \sum_n \mu(A \cap E_n).$$

Letting $A = \bigcup_n E_n$ gives

$$\mu\left(\bigcup_n E_n\right) = \sum_n \mu(E_n).$$

Alas from (ii) and the fact that any finite union of μ-measurable sets is μ-measurable, we see that any finite intersection of μ-measurable sets is also μ-measurable. This gives (iv) and (iii), at least in the case the sets are pairwise disjoint.

To establish (iv) is now clear sailing: let (E_n) be a sequence of (not necessarily pairwise disjoint) μ-measurable sets. Write $\bigcup_n E_n$ as follows:

$$\bigcup_n E_n = E_1 \cup (E_1 \backslash E_2) \cup (E_3 \backslash (E_1 \cup E_2)) \cup \cdots .$$

The result is what we wanted: $\bigcup_n E_n$ is the union of a sequence of pairwise disjoint μ-measurable sets, and so is itself μ-measurable. □

Definition 2.3. A function τ defined on a class $\mathcal{C}$ of subsets of a space Ω is a *premeasure* if

(i) $\emptyset \in \mathcal{C}$,

(ii) $0 \leq \tau(C) \leq \infty$ for all $C \in \mathcal{C}$,

(iii) $\tau(\emptyset) = 0$.

Theorem 2.4. *If τ is a premeasure defined on a class $\mathcal{C}$ of subsets of the space Ω, then the set function μ*

$$\mu(E) := \inf\left\{\sum_i \tau(C_i) : C_i \in \mathcal{C}, E \subseteq \bigcup_i C_i\right\}$$

is an outer measure.

Here, as usual, the infimum over an empty set is $+\infty$. The measure μ so constructed from τ is said to be constructed by *Method I*.

Proof. It is plain and easy to see that the set function μ as defined from τ satisfies properties (i)–(iii) of the definition of a measure. Only "countable subadditivity" requires a bit more proof. So let (E_n) be a sequence of subsets of the space Ω, and let's show that

$$\mu\left(\bigcup_n E_n\right) \leq \sum_n \mu(E_n).$$

For sure we can assume $\sum_n \mu(E_n) < \infty$, and so, in particular, each $\mu(E_n) < \infty$ as well. Now for each n, choose a covering of E_n, say $\{C_j(n)\}_j$ from $\mathcal{C}$, so that for the always present $\epsilon > 0$ we have

$$\mu(E_n) \leq \sum_j \tau(C_j(n)) \leq \mu(E_n) + \frac{\epsilon}{2^n}.$$

Now $\{C_j(n) : j, n \in \mathbb{N}\}$ covers $\bigcup_n E_n$, and so

$$\begin{aligned}\mu\left(\bigcup_n E_n\right) &\leq \sum_{j,n} \tau(C_j(n)) \\ &\leq \sum_n \sum_j \tau(C_j(n)) \\ &\leq \sum_n \left(\mu(E_n) + \frac{\epsilon}{2^n}\right) \\ &= \sum_n \mu(E_n) + \epsilon.\end{aligned}$$

As expected, $\epsilon > 0$ was arbitrary and its arbitrariness leads us to the conclusion that

$$\mu\left(\bigcup_n E_n\right) \leq \sum_n \mu(E_n). \qquad \square$$

Let τ be a premeasure on the family $\mathcal{C}$ of subsets of the **metric** space (Ω, ρ). Let $\delta > 0$ and denote by

$$\mathcal{C}_\delta = \{C \in \mathcal{C} : \text{ diameter } C \leq \delta\}.$$

Denote by τ_δ the restriction of τ to C_δ. The result is a premeasure that generates an outer measure μ_δ on Ω by Method I. It is worth taking a look at μ_δ:

$$\mu_\delta := \inf\left\{\sum_i \tau(C_i) : E \subseteq \bigcup_i C_i, C_i \in \mathcal{C},\ \text{diam}\ C_i \le \delta\right\}.$$

Notice that μ_δ is an outer measure by Method I. It is plain that as δ gets smaller, there are fewer sets with diameter $\le \delta$, so $\mu_\delta(E)$ gets bigger. Hence we can define $\mu(E)$ by

$$\mu(E) := \sup_{\delta>0} \mu_\delta(E),$$

and we say that τ generates μ by *Method II.*

Method I applies in a general setting, as opposed to Method II, which relies on the metric structure present in the underlying set Ω.

Theorem 2.5. *μ constructed by Method II is an outer measure.*

Proof. The only possible stumbling point to this is the countable subadditivity, so let's see why μ is countably subadditive. To this end, let (E_n) be a sequence of subsets of Ω and consider the quantities

$$\mu\left(\bigcup_n E_n\right) \quad \text{and} \quad \sum_n \mu(E_n).$$

Obviously the latter exceeds the former if it is ∞, so we may as well assume $\sum_n \mu(E_n) < \infty$.

Now for each $\delta > 0$, μ_δ is a known outer measure so

$$\mu_\delta\left(\bigcup_n E_n\right) \le \sum_n \mu_\delta(E_n),$$

which in turn is

$$\le \sum_n \mu(E_n).$$

The countable subadditivity follows from this. □

A key ingredient to our mix is provided in the following.

Theorem 2.6. *Let μ be an outer measure on a metric space Ω derived from the premeasure τ by Method II. If A and B are nonempty subsets of Ω that are "positively separated", then*

$$\mu(A \cup B) = \mu(A) + \mu(B).$$

Here A and B *positively separated* means that there is a $\delta > 0$ so that for any $a \in A$ and $b \in B$,

$$\rho(a, b) \geq \delta.$$

Outer measures that enjoy this additive property are called *metric outer measures* and their importance lies in the fact that these are precisely the measures on a metric space for which every Borel set is measurable. Here and throughout when we refer to the Borel σ-field $\mathcal{B}o(X)$ we mean the smallest σ-field that contains all the open sets and members of $\mathcal{B}o(X)$ as Borel sets.

Proof. We need to show that if A and B are positively separated, then

$$\mu(A \cup B) \geq \mu(A) + \mu(B),$$

where all the terms involved are finite. The idea of the proof is to cover A, B and $A \cup B$ with very fine covers from the domain $\mathcal{C}$ of τ, so fine that we can distinguish between which sets are needed to cover A and which to cover B.

More precisely, suppose

$$\rho(a, b) \geq \delta > 0$$

for any $a \in A$ and $b \in B$. Let $\epsilon > 0$ announce its presence. By the rules of engagement

$$\mu(A \cup B) = \sup_{d>0} \inf \left\{ \sum_i \tau(C_i) : \operatorname{diam} C_i \leq d, C_i \in \mathcal{C}, A \cup B \subseteq \bigcup_i C_i \right\}.$$

So we can choose a sequence (C_i) in $\mathcal{C}$ so that

- diam C_i are all *small*,
- $A \cup B \subseteq \bigcup_i C_i$,
- $\sum \tau(C_i) \leq \mu(A \cup B) + \epsilon$.

"Small"? Well to set things up for computation of the μ-measure of A and of B through the intermediaries $\mu_\delta(A), \mu_\delta(B)$, we let δ_1, δ_2 be positive numbers each less than diam Ω, and let η be the number

$$\eta = \min \left\{ \delta_1, \delta_2, \frac{\delta}{3} \right\}.$$

Choose η as our model of small.

Here's the first punchline: *if each C_i has diameter $\leq \eta$, then each has diameter $\leq \frac{\delta}{3}$, and so a given C_i can intersect A or B but not both.* It follows that

$$\sum_{C_i \cap A \neq \emptyset} \tau(C_i) + \sum_{C_i \cap B \neq \emptyset} \tau(B_i) \leq \sum_i \tau(C_i) \leq \mu(A \cup B) + \epsilon.$$

But each C_i that intersects A has diameter $\leq \eta \leq \delta_1$, so knowing, as we do, that A is a subset of

$$\bigcup_{C_i \cap A \neq \emptyset} C_i,$$

we get

$$\mu_{\delta_1}(A) \leq \sum_{C_i \cap A \neq \emptyset} \tau(C_i).$$

Similarly,

$$\mu_{\delta_2}(B) \leq \sum_{C_i \cap B \neq \emptyset} \tau(C_i).$$

The result:

$$\mu_{\delta_1}(A) + \mu_{\delta_2}(B) \leq \sum_{C_i \cap A \neq \emptyset} \tau(C_i) + \sum_{C_i \cap B \neq \emptyset} \tau(C_i) \leq \mu(A \cup B) + \epsilon,$$

and so Method II leads us to believe

$$\mu(A) + \mu(B) \leq \mu(A \cup B) + \epsilon.$$

Since $\epsilon > 0$ was arbitrary, this proof is done. □

Metric measures enjoy some very strong continuity properties. Here's one of them.

Proposition 2.7. *Let μ be a metric measure on a metric space Ω. Suppose (A_n) is an increasing sequence of subsets of Ω so that A_n and A_{n+1}^c are positively separated. Then*

$$\mu\left(\bigcup_n A_n\right) = \sup_n \mu(A_n).$$

We defer the proof until after giving the proposition a chance to "show off".

Theorem 2.8. *If μ is a metric outer measure on the metric space Ω, then every closed subset of Ω is μ-measurable. Consequently if μ is a metric outer measure on the metric space Ω, then every Borel subset of Ω is μ-measurable.*

Proof. Suppose F is a closed subset of the metric space Ω with metric ρ, and let $A \subseteq F$ and $B \subseteq F^c$ be nonempty sets. For each n let

$$B_n := \left\{ x \in B : \inf_{y \in F} \rho(x, y) > \frac{1}{n} \right\}.$$

Notice that $B_n \nearrow B$.

By their very definition, each B_n is positively separated from A. We'll show that each B_n is also positively separated from B^c_{n+1}. To this end, let $x \in B_n$ and $u \in B^c_{n+1}$. Now $u \notin B_{n+1}$, so

$$\frac{1}{n+1} \geq \inf_{y \in F} \rho(x, y).$$

There is a $y_0 \in F$ so that

$$\frac{1}{n + \frac{1}{2}} \geq \rho(u, y_0).$$

Now if

$$\rho(x, u) \leq \frac{\frac{1}{2}}{n\left(n + \frac{1}{2}\right)},$$

then

$$\begin{aligned} \inf_{y \in F} \rho(x, y) &\leq \rho(x, y_0) \\ &\leq \rho(x, u) + \rho(u, y_0) \\ &\leq \frac{\frac{1}{2}}{n\left(n + \frac{1}{2}\right)} + \frac{1}{\left(n + \frac{1}{2}\right)}, \end{aligned}$$

which, by the method of common denominators, is

$$= \frac{\frac{1}{2} + n}{n\left(n + \frac{1}{2}\right)} = \frac{1}{n}.$$

Hence $x \notin B_n$. Our conclusion? If $x \in B_n$ and $u \in B^c_{n+1}$, then

$$\rho(x, u) > \frac{\frac{1}{2}}{n\left(n + \frac{1}{2}\right)} = \frac{1}{n(2n+1)},$$

i.e., B_n and B^c_{n+1} are positively separated.

Let's compute $\mu(A \cup B)$:

$$\begin{aligned}
\mu(A \cup B) &\geq \sup_n \mu(A \cup B_n) \\
&= \sup_n (\mu(A) + \mu(B_n)) \\
&= \mu(A) + \sup_n \mu(B_n) \\
&= \mu(A) + \mu\left(\bigcup_n B_n\right) \quad \text{(by Proposition 2.7)} \\
&= \mu(A) + \mu(B)
\end{aligned}$$

and F is measurable. Carathéodory's Theorem 2.2 assures us that the family of μ-measurable sets is a σ-field and, since every closed set is μ-measurable, must contain the smallest σ-field containing all the closed sets, also known as the Borel σ-field. □

Now we look to the rather elegant proof of Proposition 2.7:

Proof. We have

$$A_1 \subseteq A_2 \subseteq \cdots \subseteq A_n \subseteq \cdots$$

with A_n and A_{n+1}^c positively separated for each n. We want to show that $\mu(\bigcup_n A_n) \leq \sup_n \mu(A_n)$ and in this effort we may plainly suppose $\sup_n \mu(A_n) < \infty$ since otherwise all is okay.

First we look at the difference sequence

$$D_1 = A_1, \quad D_2 = A_2 \backslash A_1, \quad D_3 = A_3 \backslash A_2, \ldots, \quad D_n = A_n \backslash A_{n-1}, \ldots .$$

Of course, $D_n \subseteq A_n$ but further

$$D_2 \subseteq A_1^c, \ D_3 \subseteq A_2^c \subseteq A_1^c, \ D_4 \subseteq A_3^c \subseteq A_2^c \subseteq A_1^c, \ldots .$$

So

$$\begin{aligned}
&D_1 \subseteq A_1 \text{ and } D_3, D_4, \ldots \subseteq A_2^c, \\
&D_2 \subseteq A_2 \text{ and } D_4, D_5, \ldots \subseteq A_3^c,
\end{aligned}$$

etc. In particular

$$\begin{array}{rlrl}
D_1 &\subseteq A_1, & D_3 &\subseteq A_2^c, \\
D_2 &\subseteq A_2, & D_4 &\subseteq A_3^c, \\
D_1 \cup D_3 &\subseteq A_3, & D_5 &\subseteq A_4^c, \\
D_2 \cup D_4 &\subseteq A_4, & D_6 &\subseteq A_5^c, \\
& & \vdots & \\
D_1 \cup D_3 \cup \cdots \cup D_{2n-1} &\subseteq A_{2n-1}, & D_{2n+1} &\subseteq A_{2n}^c, \\
D_2 \cup D_4 \cup \cdots \cup D_{2n} &\subseteq A_{2n}, & D_{2n} &\subseteq A_{2n+1}^c.
\end{array}$$

It follows (inductively if you must know) that

$$\mu(D_1) + \mu(D_3) + \cdots + \mu(D_{2n-1}) + \mu(D_{2n+1}) = \mu\left(\bigcup_{k=1}^{n+1} D_{2k-1}\right),$$

and

$$\mu(D_2) + \mu(D_4) + \cdots + \mu(D_{2n}) + \mu(D_{2n+2}) = \mu\left(\bigcup_{k=1}^{n+1} D_{2k}\right).$$

Each of $\bigcup_{k=1}^{n+1} D_{2k-1}$ and $\bigcup_{k=1}^{n+1} D_{2k}$ are subsets of A_{2n+2} and so all above find themselves

$$\leq \mu(A_{2n+2}) \leq \sup_n \mu(A_n) < \infty.$$

Conclusion: both series $\sum_n \mu(D_{2n-1})$ and $\sum_n \mu(D_{2n})$ converge. Now

$$\begin{aligned}
\mu\left(\bigcup_n A_n\right) &= \mu(A_n \cup D_{n+1} \cup D_{n+2} \cup \cdots) \\
&\leq \mu(A_n) + \mu(D_{n+1}) + \mu(D_{n+2}) + \cdots \\
&\leq \sup_n \mu(A_n) + \sum_{k=n+1}^{\infty} \mu(D_k);
\end{aligned}$$

if $\epsilon > 0$ is given, then there is an n so that the latter sum is less than ϵ so

$$\mu\left(\bigcup_n A_n\right) \leq \sup_n \mu(A_n) + \epsilon.$$

Enough said. □

We've seen that every Borel set in a metric space Ω is μ-measurable whenever μ is a metric outer measure on Ω. It's worthwhile to note that this is *not* accidental; it's part and parcel of being a metric outer measure. Indeed if we suppose μ is an outer measure on the metric space Ω for which every closed subset of Ω is μ-measurable and if $A, B \subseteq \Omega$ are positively separated (so $\overline{A} \cap \overline{B} = \emptyset$), then

$$\mu(A \cup B) = \mu((A \cup B) \cap \overline{A}) + \mu((A \cup B) \cap \overline{A}^c)$$

by the measurability of $\overline{A}$. But

$$(A \cup B) \cap \overline{A} = A \quad \text{and} \quad (A \cup B) \cap (\overline{A})^c = B$$

by the positive separation of A and B. So

$$\mu(A \cup B) = \mu((A \cup B) \cap \overline{A}) + \mu((A \cup B) \cap (\overline{A})^c) = \mu(A) + \mu(B),$$

and μ is a metric outer measure.

2. Regularity

We turn now to regularity properties of measures on a metric space, that is, how well we can approximate a typical value $\mu(E)$ of μ by values at "good" sets—closed, open, $\mathcal{F}_\sigma, \mathcal{G}_\delta$, or even compact. To be sure, we emphasize that if a measure is finite, then good things happen (at least in the world of Borel sets).

Here's (some of) what's so.

Theorem 2.9. *Let (X, ρ) be a metric space. Let μ be a countably additive map from the Borel σ-field $\mathcal{B}o(X)$ to $[0, \infty)$.*

(i) *For any Borel set B in X, we have*

$$\sup\{\mu(F) : F \subseteq B, F \textit{ closed}\} = \mu(B) = \inf\{\mu(U) : B \subseteq U, U \textit{ open}\}.$$

(ii) *If X is Polish (that is, there is a complete metric that generates X's topology and X is separable), then for any Borel set B in X*

$$\mu(B) = \sup\{\mu(K) : K \subseteq B, K \textit{ compact}\}.$$

This result finds so much use in what we do, we submit a proof in detail.

Proof. (i) Here's the trick to find an approach to general Borel sets; we consider the collection $\mathcal{S}$ of Borel sets in X that satisfy the conclusion of (i).

Open sets belong to $\mathcal{S}$. If U is open, then it's trivial to approximate $\mu(U)$ from above by μ's values on open super sets of U. How to approximate from within by μ's values at closed subsets? Well, notice that if U is an open set in X, then U^c is closed and so $x \in U^c$ precisely when $d(x, U^c) = 0$ where

$$d(x, U^c) = \inf\{\rho(x, y) : y \in U^c\}.$$

But $d(x, U^c)$ is a continuous function of $x \in X$ and so we see that

$$U = \bigcup_{n \in \mathbb{N}} \left[d(\cdot, U^c) \geq \frac{1}{n}\right],$$

and U is an $\mathcal{F}_\sigma$-set. If we let $F_n = \left[d(\cdot, U^c) \geq \frac{1}{n}\right]$, then each F_n is closed,

$$F_1 \subseteq F_2 \subseteq \cdots \subseteq F_n \subseteq \cdots \nearrow U,$$

and so

$$\mu(U) = \lim_n \mu(F_n).$$

Open sets are $\mathcal{F}_\sigma$'s and so each open set belongs to $\mathcal{S}$.

Now $\mathcal{S}$, by it's very definition, can be described in a manner suited to epsilonics. Indeed, a Borel set $B \in \mathcal{S}$ precisely when B satisfies

> given $\epsilon > 0$ there is an open set U containing B and a closed set F contained in B such that $\mu(U \backslash F) < \epsilon$.

It's plain from this that $\mathcal{S}$ is closed under complements.

What about countable unions? We'll suppose (B_k) is a sequence of Borel sets each one of which is in $\mathcal{S}$. For each n, pick U_n and F_n so that U_n is open, F_n is closed, $F_n \subseteq B_n \subseteq U_n$, and

$$\mu(U_n \backslash F_n) < \frac{\epsilon}{2^{n+1}}.$$

Let $U = \bigcup_n U_n$, $F = \bigcup_n F_n$. Then

$$F \subseteq \bigcup_n B_n \subseteq U.$$

Moreover

$$\begin{aligned} \mu(U \backslash F) &= \mu\left(\bigcup_n U_n \backslash \bigcup_n F_n\right) \leq \mu\left(\bigcup_n (U_n \backslash F_n)\right) \\ &\leq \sum_n \mu(U_n \backslash F_n) \leq \sum_n \frac{\epsilon}{2^{n+1}} = \frac{\epsilon}{2}. \end{aligned}$$

Now U is open and F is *almost* closed. In fact, $F = \bigcup_n F_n$ so there is an n_0 so that

$$\mu\left(F \backslash \bigcup_{n > n_0} F_n\right) < \frac{\epsilon}{2}.$$

The result is

$$\bigcup_{n=1}^{n_0} F_n \subseteq \bigcup_n B_n \subseteq U,$$

where $\bigcup_{n=1}^{n_0} F_n$ is closed, U is still open, and

$$\mu\left(U \backslash \bigcup_{n=1}^{n_0} F_n\right) < \epsilon.$$

Therefore $\mathcal{S}$ is a collection of Borel sets containing X's topology, closed under complementation and the taking of countable unions. Thus $\mathcal{S} = \mathcal{B}o(X)$.

(ii) Item (ii) hints at the plenitude of compact sets in Polish spaces. This comes from the (existence of a) complete metric—in complete metric spaces, subsets are relatively compact precisely when they're totally bounded. So to prove (ii), we'll use separability to construct good closed and totally bounded approximants.

We start with a complete separable metric space (X, ρ) and show that (ii) holds for the Borel set X itself.

Let $\{x_n : n \in \mathbb{N}\}$ be a countable dense subset of X. For each $n \in \mathbb{N}$

$$X = \bigcup_k B_{1/n}(x_k),$$

where

$$B_\epsilon(x) = \{y \in X : \rho(x,y) < \epsilon\}.$$

For each n we can find a k_n so that

$$\mu\left(\bigcup_{k=1}^{k_n} B_{1/n}(x_k)\right)$$

is within $\frac{\epsilon}{2^n}$ of $\mu(X)$.

Look at

$$K = \bigcap_n \left(\bigcup_{k=1}^{k_n} \overline{B_{1/n}(x_k)}\right).$$

Since K is closed, it's complete. K is totally bounded, too: After all, for any n,

$$K \subseteq \bigcup_{k=1}^{k_n} \overline{B_{1/n}(x_k)}.$$

So each point of K is within $2/n$ of an $x_k, k = 1, \ldots, k_n$. Therefore K is compact.

$$\begin{aligned}
\mu(K^c) &= \mu\left(\left(\bigcap_n \left[\bigcup_{k=1}^{k_n} \overline{B_{1/n}(x_k)}\right]\right)^c\right) \\
&= \mu\left(\bigcup_n \left[\bigcup_{k=1}^{k_n} \overline{B_{1/n}(x_k)}\right]^c\right) \\
&\leq \sum_n \mu\left(\left[\bigcup_{k=1}^{k_n} \overline{B_{1/n}(x_k)}\right]^c\right) \\
&\leq \sum_n \mu\left(\left[\bigcup_{k=1}^{k_n} B_{1/n}(x_k)\right]^c\right) \\
&\leq \sum_n \frac{\epsilon}{2^n} = \epsilon.
\end{aligned}$$

Generally if we apply (i) we can find inside any Borel subset B of X a closed set F with $\mu(F)$ within $\epsilon/2$ of $\mu(B)$; then define $\tilde{\mu}$ on the Borel sets of X by

$$\tilde{\mu}(E) = \mu(E \cap F).$$

Our opening salvo applied to $\tilde{\mu}$ gives a compact $K \subseteq X$ so $\tilde{\mu}(K)$ is within $\epsilon/2$ of $\tilde{\mu}(X) = \mu(F)$. Then $K \cap F$ does the dirty deed since $\mu(K \cap F) = \tilde{\mu}(K)$, which is within ϵ of $\mu(B)$. □

3. Invariant Measures on $\mathbb{R}^n$

3.1. Introduction. As we proceed in our study of invariant measures, we will encounter theorems that assert the uniqueness of such measures in varying degrees of generality and in differing senses.

On compact objects the uniqueness will be with regard to *Borel probabilities* (positive Borel measures of total mass 1).

If G is a locally compact group, then we'll see that there is a unique left invariant regular Borel measure (acting on the Borel subsets of G) with uniqueness taken to mean "up to multiplicative constants".

In concrete cases it occurs that those multiplicative constants can themselves be of considerable interest, representing the rate of exchange between various natural geometric viewpoints of the groups under consideration.

In this chapter we encounter an early example of seemingly multiple invariant measures on a concrete group and compute the constants that allow one to convert from one viewpoint to another.

Our setting will be $\mathbb{R}^n$. We will be dealing with Lebesgue n-measure λ_n on $\mathbb{R}^n$ and Hausdorff n-measure $\mu^{(n)}$ on $\mathbb{R}^n$. Each is translation invariant and, sure enough, for each n, there is a constant κ_n such that

$$\mu^{(n)} = \kappa_n \lambda_n.$$

The constant κ_n is in fact the precise rate of exchange that allows one to move from a rectilinear view of $\mathbb{R}^n$ (as expressed by λ_n) to a spherical view (as found in $\mu^{(n)}$).

Along the way to establishing the existence of κ_n and of computing it, we meet some of the real stalwarts of measure theory.

We follow this with a short introduction to Hausdorff measure on $\mathbb{R}^n$. This leads to the main point of this chapter—the proof that for each n there is a $\kappa_n > 0$ such that

$$\mu^{(n)} = \kappa_n \lambda_n.$$

3.2. Hausdorff n-measure on $\mathbb{R}^n$. A Hausdorff gauge function is a map

$$h : [0, \infty] \to [0, \infty]$$

such that $h(0) = 0, h(t) > 0$ whenever $t > 0$, $h(\infty) \leq \infty$, h is increasing and right-continuous from $[0, \infty)$ to $[0, \infty)$.

Let (X, ρ) be a metric space. For $E \subseteq X$, let

$$h(E) = h(\text{diam } E),\ h(\emptyset) = 0.$$

Notice that h so defined is a premeasure on any class $\mathcal{C}$ containing $\emptyset$. Let $\delta > 0$. Consider the following classes $\mathcal{C}$ of subsets of X:

$$\begin{aligned}\mathcal{C} &= \mathcal{G} = \{G \subseteq X : G \text{ is open}\},\\ \mathcal{C} &= \mathcal{F} = \{F \subseteq X : F \text{ is closed}\},\\ \mathcal{C} &= \mathcal{P} = \{S \subseteq X\}.\end{aligned}$$

On each of these classes, we have intermediate outer-measures

$$\mu_\delta^h, \nu_\delta^h, \sigma_\delta^h, \text{ and } \tau_\delta^h,$$

$$\mu_\delta^h(E) := \inf\left\{\sum_n h(G_n) : G_n \in \mathcal{G}, E \subseteq \bigcup_n G_n, \text{diam } G_n \leq \delta\right\},$$

$$\nu_\delta^h(E) := \inf\left\{\sum_n h(F_n) : F_n \in \mathcal{F}, E \subseteq \bigcup_n F_n, \text{diam } F_n \leq \delta\right\},$$

$$\sigma_\delta^h(E) := \inf\left\{\sum_n h(S_n) : E \subseteq \bigcup_n S_n, \text{diam } S_n \leq \delta\right\},$$

and

$$\tau_\delta^h(E) := \inf\left\{\sum_n h(S_n) : E = \bigcup_n S_n, \text{diam } S_n \leq \delta\right\}.$$

Proposition 2.10. *If* $0 < \delta < \epsilon$, *then*

$$\mu_\epsilon^h \leq \nu_\delta^h = \sigma_\delta^h = \tau_\delta^h \leq \mu_\delta^h.$$

Corollary 2.11. *If* μ^h, ν^h *and* σ^h *are the measures derived from* h *and the classes* $\mathcal{G}, \mathcal{F}$ *and* $\mathcal{P}$ *à la Method II, then*

$$\mu^h = \nu^h = \sigma^h.$$

Moreover

$$\mu^h(E) = \sup_{\delta > 0} \tau_\delta^h(E). \tag{3}$$

We rush to add that this measure is called Hausdorff h-measure *and is defined for any metric space* (X, ρ).

As C. A. Rogers noted [**105**], equation (3) shows that the μ^h-measure of E can be defined in terms of the diameters and covering properties of *subsets* of E. Thus the μ^h-measure of E is an intrinsic property of E as a set with the metric ρ—so $\mu^h(E)$ doesn't change when E is put (isometrically) inside another metric space (X', ρ') than (X, ρ).

Proof. Several assertions of the proposition are clear or, at the very least, easy to see. Since ν_δ^h requires covering by *closed* sets of diameter less than or equal to δ,

$$\sigma_\delta^h \leq \nu_\delta^h.$$

On the other hand, diam $S =$ diam $\bar{S}$ so in fact, $\sigma_\delta^h = \nu_\delta^h$.

Again it's plain that $\sigma_\delta^h \leq \tau_\delta^h$; yet if (S_n) is a sequence of sets that covers E with each S_n having diameter less than or equal to δ and $T_n = S_n \cap E$, then $\{T_n\}$ also covers E, diam $T_n \leq \delta$ and $\bigcup_n T_n = E$. It soon follows that

$$\tau_\delta^h(E) \leq \sum_n h(T_n) \leq \sum_n h(S_n),$$

and so $\tau_\delta^h \leq \sigma_\delta^h$. In sum,

$$\tau_\delta^h = \sigma_\delta^h.$$

Naturally, $\sigma_\delta^h \leq \mu_\delta^h$, and so the only missing ingredient to a complete proof is that $\mu_\epsilon^h \leq \nu_\delta^h$ when $\delta < \epsilon$. Here some epsilonics are called for. Let $\eta > 0$. Cover E by a sequence (F_n) of closed sets with diameter less than or equal to δ so that

$$\nu_\delta^h(E) \leq \sum_n h(F_n) \leq \nu_\delta^h(E) + \eta$$

(we clearly need only consider what happens when $\nu_\delta^h(E) < \infty$).

Choose $\eta_n > 0$ so that

$$h(\text{diam } F_n + 2\eta_n) < h(F_n) + \frac{\eta}{2^n}$$

and

$$\delta + 2\eta_n < \epsilon.$$

After all, $\delta < \epsilon$ and we're choosing $2\eta_n$ to be very small.

Let

$$U_n = [d(x, F_n) < \eta_n].$$

Each U_n is open, contains F_n, and

$$\text{diam } U_n \leq \text{diam } F_n + \eta_n + \eta_n \leq \delta + 2\eta_n < \epsilon.$$

Further

$$h(U_n) \leq h(\text{diam } F_n + 2\eta_n) \leq h(F_n) + \frac{\eta}{2^n}.$$

Hence $E \subseteq \bigcup_n U_n$, each U_n is open, has diameter less than ϵ, and

$$\begin{aligned} \sum_n h(U_n) &\leq h(\text{diam } F_n + 2\eta_n) \\ &\leq \sum_n h(F_n) + \frac{\eta}{2^n} \leq \nu_\delta^h(E) + \eta + \eta. \end{aligned}$$

It follows that

$$\mu_\epsilon^h(E) \leq \nu_\delta^h + 2\eta.$$

Since $\eta > 0$ was arbitrary, the proof is complete. □

Let $h(t) = t^n$. *The Hausdorff measure generated by h is denoted by $\mu^{(n)}$.* Keep in mind that "the Hausdorff measure" is a misnomer; after all $\mu^{(n)}$ makes sense in any metric space (X, ρ).

A comparison of $\mu^{(n)}$ and n-dimensional Lebesgue measure λ_n on $\mathbb{R}^n$ is worth our while. The comparison proceeds in two steps. In the first, we'll argue to their proportionality; in the second step we'll compute the constant of proportionality.

Theorem 2.12. *Let*

$$C_0 = \{x = (x_1, \ldots, x_n) \subseteq \mathbb{R}^n : 0 \leq x_i < 1, i = 1, 2, \ldots, n\}$$

and

$$\kappa_n = \mu^{(n)}(C_0).$$

Then

$$\mu^{(n)}(E) = \kappa_n \lambda_n(E).$$

Proof. By definition,

$$\mu^{(n)}(C_0) = \kappa_n \lambda_n(C_0).$$

Since both $\mu^{(n)}$ and λ_n are invariant under translations and are homogeneous of order n (i.e., for any $t \geq 0$, $\mu^{(n)}(tE) = t^n \mu^{(n)}(E)$, and $\lambda_n(tE) = t^n \lambda_n(E)$), it follows that

$$\mu^{(n)}(C) = \kappa_n \lambda_n(C)$$

for any cube of the form

$$C = \{x = (x_1, \ldots, x_n) : a_i \leq x_i < a_i + s\},$$

where $(a_1, \ldots, a_n) \in \mathbb{R}$ and $s > 0$.

Consider the collection $\mathcal{C}$ of cubes of the form

$$\left\{x = (x_1, \ldots, x_n) : \frac{r_{i-1}}{2^k} \leq x_i < \frac{r_i}{2^k}\right\},$$

where $k \geq 0$ is an integer and $r_1, \ldots, r_n \in \mathbb{Z}$. $\mathcal{C}$ enjoys the following property: if two members of $\mathcal{C}$ have a point in common, then one is a subset of the other.

So what? Well if G is any open set, then each point $g \in G$ lies in a cube $C \in \mathcal{C}$ that is contained entirely within G and is maximized in the sense that C is not a subset of any larger member of $\mathcal{C}$ that is entirely inside G. Hence

G is the union of the maximal cubes from $\mathcal{C}$, and these cubes are disjoint. Let's write

$$G = \bigcup_m C_m,$$

where (C_m) is the sequence of maximal cubes just described. Everything we're talking about is measurable and so

$$\mu^{(n)}(G) = \sum_m \mu^{(n)}(C_m) = \sum_m \kappa_n \lambda_n(C_m) = \kappa_n \lambda_n(G).$$

So

$$\mu^{(n)}(G) = \kappa_n \lambda_n(G)$$

for open sets $G \subseteq \mathbb{R}^n$.

Next we bootstrap this equality's verity to $\mathcal{G}_\delta$-sets. Let H be a $\mathcal{G}_\delta$-subset of $\mathbb{R}^n$. If $\mu^{(n)}(H)$ and $\lambda^{(n)}(H)$ are both equal to $+\infty$, then there's little that need be said. Naturally, they're also finite at the same time. In fact, if $\mu^{(n)}(H) < \infty$, then we can cover H by a sequence (G_m) of open sets with $\sum_m (\text{diam } G_m)^n < \infty$; if we enclose each G_m in an open cube U_m where each edge of U_m is twice the diameter of G_m, then $H \subseteq \cup_m U_m$ and

$$\begin{aligned} \lambda_n\left(\bigcup_m C_m\right) &\leq \sum_m \lambda_n(C_m) \leq \sum_m (2 \cdot \text{diam } G_m)^n \\ &\leq 2^n \sum_m (\text{diam } G_m)^n < \infty. \end{aligned}$$

On the other hand, if $\lambda_n(H) < \infty$, then H can be covered by an open set of finite λ_n-measure and so (since for open sets $\mu^{(n)}$ and $\kappa_n \lambda_n$ agree)

$$\mu^{(n)}(H) \leq \mu^{(n)}(G) = \kappa_n \lambda_n(G)$$

for any appropriate open set $H \subseteq G$, and $\mu^{(n)}(H) < \infty$ follows.

Okay, suppose $\mu^{(n)}(H)$ and $\kappa_n \lambda_n(H)$ are both finite. Since H is a $\mathcal{G}_\delta$-set, we can write H in the form

$$H = \bigcap_m G_m,$$

where (G_m) is a descending sequence of open sets each of finite measure (be that measure $\mu^{(n)}$ or λ_n!). It follows that

$$\mu^{(n)}(H) = \inf_m \mu^{(n)}(G_m) = \inf_m \kappa_n \lambda_n(G_m) = \kappa_n \lambda_n(H),$$

thanks to the measurability of all sets involved. So

$$\mu^{(n)}(H) = \kappa_n \lambda_n(H)$$

for any $\mathcal{G}_\delta$-set $H \subseteq \mathbb{R}^n$.

Now look at any $E \subseteq \mathbb{R}^n$. Choose $\mathcal{G}_\delta$-sets $H_1, H_2 \in \mathbb{R}^n$ with $E \subseteq H_1 \cap H_2$ so

$$\mu^{(n)}(E) = \mu^{(n)}(H_1) \text{ and } \lambda_n(E) = \lambda_n(H_2).$$

Of course

$$\mu^{(n)}(E) \leq \mu^{(n)}(H_1 \cap H_2) \leq \mu^{(n)}(H_1) = \mu^{(n)}(E)$$

and

$$\lambda_n(E) \leq \lambda_n(H_1 \cap H_2) \leq \lambda_n(H_2) = \lambda_n(E),$$

so

$$\mu^{(n)}(E) = \mu^{(n)}(H_1 \cap H_2) \text{ and } \lambda_n(E) = \lambda_n(H_1 \cap H_2).$$

It soon follows that

$$\mu^{(n)}(E) = \mu^{(n)}(H_1 \cap H_2) = \kappa_n \lambda_n(H_1 \cap H_2) = \kappa_n \lambda_n(E)$$

since $H_1 \cap H_2$ is a $\mathcal{G}_\delta$ set. □

Theorem 2.13.

$$\kappa_n = \left(\frac{4}{\pi}\right)^{n/2} \Gamma\left(\frac{n+2}{2}\right).$$

Proof. To compute κ_n we need to know the volume, $\lambda_n(B_{\ell_2^n})$, of the closed unit ball in $\mathbb{R}^n$. Cutting to the quick, this is given by

$$\lambda_n(B_{\ell_2^n}) = \frac{\pi^{n/2}}{\Gamma\left(\frac{n+2}{2}\right)}.$$

Let's see what is involved. Let (S_k) be a sequence of subsets of $\mathbb{R}^n$ that cover the cube C_0. Then

$$\begin{aligned} 1 &= \lambda_n(C_0) \leq \lambda_n\left(\bigcup_k S_k\right) \\ &\leq \sum_k \lambda_n(S_k) \\ &\leq \sum_k \left(\frac{\pi}{4}\right)^{n/2} \frac{(\mathrm{diam}(S_k))^n}{\Gamma\left(\frac{n+2}{2}\right)} \end{aligned}$$

by the isodiametric inequality. It follows that

$$1 \leq \left(\frac{\pi}{4}\right)^{n/2} \frac{\mu^{(n)}(C_0)}{\Gamma\left(\frac{n+2}{2}\right)} = \left(\frac{\pi}{4}\right)^{n/2} \frac{\kappa_n}{\Gamma\left(\frac{n+2}{2}\right)}.$$

On the other hand, if $\delta > 0$ is given to us, then Vitali's covering theorem provides us with a sequence (B_k) of disjoint balls each of diameter less than δ and each contained in C_0 with

$$\lambda_n(C_0) = \sum_k \lambda_n(B_k) \text{ and } \lambda_n\left(C_0 \setminus \bigcup_k B_k\right) = 0.$$

It follows that

$$\mu^{(n)}\left(C_0 \setminus \bigcup_k B_k\right) = 0.$$

Hence

$$\begin{aligned}
\mu_\delta^{(n)}(C_0) &\leq \mu_\delta^{(n)}\left(\bigcup_k B_k\right) + \mu_\delta^{(n)}\left(C_0 \setminus \bigcup_k B_k\right) \\
&= \mu_\delta^{(n)}\left(\bigcup_k B_k\right) \\
&\leq \sum_k (\operatorname{diam} B_k)^n \\
&= \sum_k \left(\frac{4}{\pi}\right)^{n/2} \Gamma\left(\frac{n+2}{2}\right) \lambda_n(B_k) \\
&= \left(\frac{4}{\pi}\right)^{n/2} \Gamma\left(\frac{n+2}{2}\right) \sum_k \lambda_n(B_k) \\
&= \left(\frac{4}{\pi}\right)^{n/2} \Gamma\left(\frac{n+2}{2}\right) \lambda_n(C_0) \\
&= \left(\frac{4}{\pi}\right)^{n/2} \Gamma\left(\frac{n+2}{2}\right).
\end{aligned}$$

So

$$\kappa_n = \mu^{(n)}(C_0) \leq \left(\frac{4}{\pi}\right)^{n/2} \Gamma\left(\frac{n+2}{2}\right),$$

as well. Therefore

$$\kappa_n = \left(\frac{4}{\pi}\right)^{n/2} \Gamma\left(\frac{n+2}{2}\right).$$

□

The proportionality of $\mu^{(n)}$ and λ_n on $\mathbb{R}^n$ opens the way to observations about Lebesgue measure that are well worth recording.

For instance, suppose E and F are isometric measurable subsets of $\mathbb{R}^n$. Then E's measurability is tantamount to its $\mu^{(n)}$-measurability. Further if E is Lebesgue measurable, then E can be written as $E = B \cup N$, where B is the countable union of an increasing sequence of compact sets and N is a set of Lebesgue (or $\mu^{(n)}$-) measure zero; it follows that the same can be said of F. Furthermore E and F have the same $\mu^{(n)}$-measure thanks to the feature of Hausdorff measure being intrinsically defined. By the proportionality of $\mu^{(n)}$ and λ_n, E and F have the same Lebesgue measure as well.

Fact 2.14. If $m > n$, then

$$\mu^{(m)}(\mathbb{R}^n) = 0.$$

Proof. Suppose $a = (a_1, \ldots, a_n), b = (b_1, \ldots, b_n) \in \mathbb{R}^n$, and let $C_{a,b}$ be the interval

$$\prod_{j=1}^{n} [a_j, b_j],$$

where $0 \leq a_j \leq b_j \leq m$, $b_j - a_j = \frac{1}{m}$. Then the diameter of C is $\frac{\sqrt{n}}{m}$. If the unit circle $C = [0,1]^n$ is covered by nonoverlapping cubes of the sort $C_{a,b}$, then the number of such congruent cubes is m^n; so for any $s > n$,

$$\mu^s_{\frac{\sqrt{n}}{m}}(C) \leq m^n \left(\frac{\sqrt{n}}{m}\right)^s = n^{\frac{s}{2}} \frac{m^n}{m^s}.$$

Hence if $s > n$, then $\mu^s(C) = 0$. In particular, for any $k > n$, the k-dimensional Hausdorff measure of $\mathbb{R}^n$ is zero. □

The situation of how $\mu^{(m)}$ behaves on $\mathbb{R}^n$ when $m < n$ is quite different and both interesting and important. In fact, $\mu^{(m)}$ picks out those subsets of $\mathbb{R}^n$ that are m-dimensional in character so, in particular, recognizes subsets E that lie in (isometric) copies of $\mathbb{R}^m$ that lie inside of $\mathbb{R}^n$ and assigns to them the Hausdorff m-measure $\kappa_m \lambda_m(E)$.

4. Notes and Remarks

The results of this chapter are well known and classical. We have borrowed heavily from C. A. Rogers' beautiful [**105**] book, *Hausdorff Measures*, wherever possible.

All presentations of outer measures follow a path cleared by C. Carathéodory when he lectured members of the Göttingen mathematics club in 1914. His lectures formed the basis of his paper [**14**].

Part (ii) of Theorem 2.9 is usually accredited to S. Ulam [**128**] and is sometimes stated as follows: "Each probability Borel measure on a Polish space is tight."

Both hypotheses (separability and completeness) of Theorem 2.9 are needed to some extent. Completeness is essential. If X is a subset of the unit interval $[0,1]$, then the *inner measure* of X, $\lambda_*(X)$, is given by

$$\lambda_*(X) = 1 - \lambda^*(X),$$

where $\lambda^*(X)$ is X's outer measure. A subset A of $[0,1]$ is Lebesgue measurable if and only if

$$\lambda_*(A) = \lambda^*(A).$$

There is a subset X of $[0,1]$ for which $\lambda_*(X) = 0$ and $\lambda^*(X) = 1$. Naturally, this set is hopelessly incomplete (though separable). If we look at the Borel structure of X, a subset of X is a Borel set in X precisely when it is of the form $B \cap X$ for some Borel set B in $[0,1]$. We can define a probability measure P on X by

$$P(B \cap X) = \lambda^*(B).$$

Note that P cannot be regular as in Theorem 2.9(ii). Indeed if K is a compact subset of X, then K is compact in $[0,1]$ as well; hence

$$P(K) = \lambda^*(K) = \lambda_*(K) = 0.$$

Notice that if P is a Borel probability measure on the metric space (X, ρ) that satisfies the conclusions of Theorem 2.9(ii), then there is an ascending sequence (K_n) of compact subsets of X so that

$$P\left(X \setminus \left(\bigcup_n K_n\right)\right) = 0.$$

The set

$$\overline{\left(\bigcup_n K_n\right)}$$

is separable. With this in mind we say that a probability Borel measure has *separable support* if there exists a separable Borel set B such that $P(B) = 1$. Plainly, for Theorem 2.9(ii) to hold, one must be dealing with measures having separable support. This is a very delicate issue as the following result of S. Ulam attests.

Theorem 2.15 (S. Ulam [**128**]). *Let (X, ρ) be a metric space. Then in order that every Borel probability measure on X have separable support, it is both necessary and sufficient that each discrete subspace of X have* nonmeasurable cardinal.

A set has *measurable cardinal* if one can find a probability measure defined on the collection of all subsets of S which vanishes at singletons. The set S has *nonmeasurable cardinal* if it doesn't have measurable cardinal. It is unknown if sets with measurable cardinal exist; it is known that if they exist they must be humongous.

Bernstein's argument for the existence of a non-Lebesgue measurable set has a natural generalization to the situation of Polish spaces.

The cornerstones of this generalization are two theorems due to P. Alexandrov and F. Hausdorff [**54**]:

- Every uncountable Borel subset of a Polish space contains a homeomorphic copy of the Cantor set.

- Every nonvoid compact metric space is the continuous image of the Cantor set.

To bring these two classical gems into play, we take note of an important aspect of the usual continuous surjection of the Cantor set onto the closed unit interval $[0,1]$. This map can be given by the formula: for $\epsilon = (\epsilon_n) \in \{0,1\}^{\mathbb{N}}$,

$$\phi(\epsilon) = \sum_n \frac{\epsilon_n}{2^n}.$$

The only points $y \in [0,1]$ for which $\phi^{\leftarrow}(\{y\})$ is *not* a singleton are dyadic rationals, a countable set. Consequently the collection $\mathcal{F}$ of uncountable closed subsets of the Cantor set has cardinality c, that of the continuum. As a consequence of Bernstein's theorem, we have the following generalization:

Theorem 2.16. *If X is an uncountable Polish space, then X contains a subset B such that both B and B^c intersect each uncountable closed subset of X.*

Using Theorem 2.16 and Ulam's theorem (Theorem 2.9(ii)), we can mimic Bernstein's construction of a nonmeasurable set to get the following:

Theorem 2.17. *If X is an uncountable Polish space and μ is a positive Borel measure on X that vanishes on all singletons, then X contains a subset that is not μ-measurable.*

4.1. Dimension and Measure. In the theory of topological dimension, if X is a separable metric space we say:

- $\dim X = 0$, if there is a basis $\mathcal{U}$ for X's topology consisting of open sets U with empty boundary;
- if $n \in \mathbb{N}$, then $\dim X \leq n$ if X has a basis $\mathcal{U}$ for its topology so each $U \in \mathcal{U}$ has a boundary of dimension less than or equal to $n-1$.

Hausdoff measures can be used to bound the topological dimension. That this is so follows from

Theorem 2.18 (Szpilrajn [**126**] (Marczewski)). *If X is a separable metric space and $\mu^{(p+1)}X = 0$, then for any $x_0 \in X$*

$$\mu^{(p)}(\{x \in X : d(x, x_0) = s\}) = 0$$

for Lebesgue measure almost all $s > 0$.

Proof. Let $E \subseteq X$, and define r_E, R_E as follows:

$$r_E = \inf\{d(x,x_0) : x \in E\} \text{ and } R_E = \sup\{d(x,x_0) : x \in E\}.$$

If $\operatorname{diam}(E) = \infty$, then $R_E - r_E \leq \operatorname{diam}(E)$ is plain to see. Now suppose that $\operatorname{diam}(E) < \infty$. Then simple epsilonics are on call: if $\epsilon > 0$, pick $x, y \in E$ so that

$$R_E \leq d(x,x_0) + \frac{\epsilon}{3} \text{ and } d(y,x_0) \leq r_E + \frac{\epsilon}{3}.$$

Then

$$R_E - r_e \leq d(x,x_0) + \frac{\epsilon}{3} - d(y,x_0) + \frac{\epsilon}{3} \leq d(x,y) + \frac{2\epsilon}{3} < \operatorname{diam}(E) + \epsilon.$$

Now realize that for any $s \geq 0$

$$\operatorname{diam}\{x \in E : d(x,x_0) = s\} \leq \operatorname{diam}(E) \cdot \chi_{[r_E,R_E]}(s),$$

so

$$\operatorname{diam}\{x \in E : d(x,x_0) = s\}^p \leq (\operatorname{diam} E)^p \chi_{[r_E,R_E]}(s).$$

Suppose $\mu^{(p+1)}(X) = 0$. Then for any $k \in \mathbb{N}$, we can cover X by a sequence $(A_n(k))_n$ such that

$$\sum_n [\operatorname{diam}(A_n(k))]^{p+1} \leq \frac{1}{k}.$$

But for each n, k

$$\operatorname{diam}\{x \in A_n(k) : d(x,x_0) = s\}^p \leq (\operatorname{diam}(A_n(k)))^p \chi_{[r_{A_n(k)},R_{A_n(k)}]}(s).$$

Since

$$\begin{aligned}\int_0^\infty \operatorname{diam}(A_n(k))^p \chi_{[r_{A_n(k)},R_{A_n(k)}]}(s)\,ds &\leq (\operatorname{diam} A_n(k))^p (R_{A_n(k)} - r_{A_n(k)})\\ &\leq (\operatorname{diam} A_n(k))^{p+1},\end{aligned}$$

we have

$$\sum_n \operatorname{diam}\{x \in A_n(k) : d(x,x_0) = s\}^p \leq \sum_n (\operatorname{diam} A_n(k))^p \chi_{[r_{A_n(k)},R_{A_n(k)}]}(s)$$

and

$$\int_0^\infty \sum_n \operatorname{diam}(A_n(k))^p \chi_{[r_{A_n(k)},R_{A_n(k)}]}(s)\,ds \leq \sum_n (\operatorname{diam}) A_n(k))^{p+1} \leq \frac{1}{k}.$$

If we now let

$$f_k(s) = \sum_n (\operatorname{diam} A_n(k))^p \chi_{[r_{A_n(k)},R_{A_n(k)}]}(s),$$

then $f_k \in L^1[0,\infty)$ with $\|f_k\|_1 \leq \frac{1}{k}$. F. Riesz tells us that (f_k) has a subsequence (g_j) that tends to zero for Lebesgue–almost all s. It follows from this that

$$\mu^{(p)}(\{x \in X : d(x,x_0) = s\}) = 0. \qquad \square$$

A dimension-theoretic conclusion is

Theorem 2.19. *If X is a separable metric space and* $\dim(X) = n$, *then* $\mu^{(n)}(X) > 0$.

We rush to add that whenever X is a separable metric space with $\dim(X) \leq n$, then X is homeomorphic to a subset of $[-1,1]^{2n+1}$ of $\mu^{(n+1)}$-measure zero. In fact, there is a homeomorphism $\phi : X \to [-1,1]^{2n+1}$ so that for all real $p > n$,

$$\mu^{(p)}(\overline{\phi(X)}) = 0.$$

As a matter of fact the collection of all such homeomorphisms is a dense $\mathcal{G}_\delta$-set inside the space of continuous $[-1,1]^{2n+1}$-valued functions defined on X, equipped with the topology of uniform convergence!

We mention here that if X is any separable metric space, then the Hausdorff dimension of X is defined by

$$\text{Hausdorff dimension}(X) = \sup\{p \geq 0 : \mu^{(p)}(X) > 0\}.$$

Using the discussion above, one can show that if X is a separable metric space, then the Hausdorff dimension of X is at least as large as the topological dimension of X. A word of warning: the Hausdoff dimension uses all p's greater than or equal to zero, and so it need not be an integer. Indeed, F. Hausdorff [**53**] showed that the Hausdorff dimension of the Cantor set is $\frac{\log 2}{\log 3}$, a known noninteger.

These comments about dimension and measure are an elaboration of Chapter VII: Dimension and Measure of the classic monograph *Dimension Theory* of W. Hurewicz and H. Wallman [**58**].

Hausdorff measures are not the only measures one can call upon to study the dimension of a separable metric space—H. Federer, [**37**] has proved similar results (much more delicate in detail) using integral geometric measures.

Chapter 3

Introduction to Topological Groups

1. Introduction

For us, a *topological group* is a group G that is equipped with a topology that makes the functions $(x, y) \mapsto xy$ from $G \times G$ to G and $x \mapsto x^{-1}$ from G to G continuous. Here are some basic observations regarding topological groups; they follow simply and directly from the definition

- Given $a, b \in G$, should ab find itself in an open set U, then there are open sets V and W such that $a \in V, b \in W$ and $V \cdot W = \{xy : x \in V, y \in W\} \subseteq U$.
- Given $a \in G$, should U be an open set containing a^{-1}, then there is an open set V containing a so that $V^{-1} = \{v^{-1} : v \in V\} \subseteq U$.
- Given $a, b \in G$, should U be an open set containing ab^{-1}, then there are open sets V and W such that $a \in V, b \in W$ and $V \cdot W^{-1} \subseteq U$.
- Given $a, b \in G$, should U be an open set containing $a^{-1}b$, then there are open sets V and W such that $a \in V, b \in W$ and $V^{-1} \cdot W \subseteq U$.
- Each of the mappings from G to G

$$\begin{aligned} l_a &: G \to G, & l_a(x) &= ax, \\ r_a &: G \to G, & r_a(x) &= xa, \\ \mathrm{inv} &: G \to G, & \mathrm{inv}(x) &= x^{-1}, \end{aligned}$$

  is a homeomorphism of G onto G.

- If F is a closed subset of G, then so are aF, Fa, and F^{-1} for any $a \in G$.
- If U is an open subset of G and S is a nonvoid subset of G, then the sets $S \cdot U, U \cdot S$, and U^{-1} are open subsets of G.
- G *is homogeneous*: if $p, q \in G$, then there is a homeomorphism ϕ of G onto G such that $\phi(p) = q$. Indeed $\phi = l_{qp^{-1}}$ will do the trick.

The homogeneity of topological groups has consequences regarding its topological structure. Here's one such

Theorem 3.1. *Let G be a topological group. If U is an open set containing the identity e, then there is an open set V containing e such that $e \in V \subseteq \overline{V} \subseteq U$. Consequently, a T_0 topological group is regular and so Hausdorff.*

Proof. First things first: Let U be an open set that contains the identity e. By continuity of multiplication, there is an open set W containing e such that $W \cdot W \subseteq U$. If we set $V = W \cap W^{-1}$, then we have an open set that contains e, is symmetric ($V = V^{-1}$), and satisfies $V \cdot V \subseteq U$. We claim $\overline{V} \subseteq U$. Take $x \in \overline{V}$. Then xV is an open set that contains x and so $xV \cap V \neq \emptyset$, so there must be $v_1, v_2 \in V$ so that $xv_1 = v_2$. But then

$$x = v_2 v_1^{-1} \in V \cdot V^{-1} = V \cdot V \subseteq U.$$

The homogeneous structure of G now tells us that whenever $x \in G$ and U is an open set containing x, then there is an open set V such that $x \in V \subseteq \overline{V} \subseteq U$. Regularity follows from this and G's homogeneity.

Again, homogeneity of G tells us that if $x, y \in G$ are distinct and there's a neighborhood of x that doesn't contain y, then we have a neighborhood of y that doesn't contain x. In other words, T_0 topological groups satisfy the T_1 axiom. □

Henceforth, we assume that all topological groups are Hausdorff.

The surprising conclusion reached in Theorem 3.1 is a typical product of the mix of the algebra and topology in topological groups. Here's another:

Proposition 3.2. *Every open subgroup of a topological group is closed.*

Proof. Let H be an open subgroup of the topological group G. Take $g \in \overline{H}$. Every open set that contains g intersects H; gH is such an open set.

Therefore $gH \cap H \neq \emptyset$. Since cosets are either the same or disjoint, $gH = H$. Thus

$$g = ge \in gH = H$$

and $\overline{H} \subseteq H$. □

Exercise 3.3. Let G be a topological group, let F be a closed subset of G, and let K be a compact subset of G. Then FK and KF are closed in G.

Exercise 3.4. If G is a connected topological group, then any neighborhood of the identity is a system of generators for G.

Exercise 3.5. Let G be a topological group, and let N be the connected component of the identity in G.

(i) N is a closed subgroup of G.

(ii) N is an invariant (normal) subgroup of G.

(*Hint*: Consider N', the set of inverses of N, which is connected, and so if x is a member of N, then xN' is connected and contains the identity, so xN' is contained in N for any x in N.)

Exercise 3.6. Let G be a locally compact group. Then G contains an open σ-compact subgroup. (*Hint*: Choose an open symmetric subset U containing G's identity with compact closure, and look at the union of powers of U.)

2. The Classical (Locally Compact) Groups

$\mathbb{R}^n$ and $\mathbb{C}^n$ will denote, as usual, real and complex n-spaces, respectively. M_n will denote the linear algebra of all $n \times n$ matrices with complex entries. We can associate with any $(a_{ij}) \in M_n$ the point $(b_1, \ldots, b_{n^2}) \in \mathbb{C}^{n^2}$, where

$$b_{i+(j-1)n} = a_{ij}.$$

This establishes a bijective correspondence between M_n and $\mathbb{C}^{n^2}$, a correspondence we use to equip M_n with the Euclidean topology of $\mathbb{C}^{n^2}$.

Of course if $\alpha, \beta \in M_n$, then $\alpha \cdot \beta \in M_n$, too, with

$$c_{ij} = \sum_{k=1}^{n} a_{ik} b_{kj},$$

whenever $\alpha = (a_{ij}), \beta = (b_{ij})$, and $\alpha \cdot \beta = (c_{ij})$. It is easy to see that the operation $(\alpha, \beta) \mapsto \alpha \cdot \beta$ is continuous from $M_n \times M_n$ to M_n.

In addition to addition and multiplication, M_n is endowed by nature with a couple of other natural operations—transposition and conjugation: if

$\alpha = (a_{ij}) \in M_n$, then $\alpha^t = (a_{ji})$ and $\bar{\alpha} = (\bar{a}_{ij})$, where $\bar{a}$ is the complex conjugate of the complex number a. Both of these operations are homeomorphisms of M_n onto itself and each is of "order two", that is, $\alpha^{tt} = \alpha$ and $\bar{\bar{\alpha}} = \alpha$.

Some members of M_n have a multiplicative inverse; the collection of all such matrices will be denoted by $GL(n, \mathbb{C})$ and called the *general linear group*. Of course, $\alpha \in M_n$ belongs to $GL(n, \mathbb{C})$ precisely when $\det \alpha \neq 0$. Since $\det \alpha$ is a polynomial in the coordinates of α, we know that $\alpha \to \det \alpha$ is a continuous function, and so $GL(n, \mathbb{C})$ is an open subset of M_n. If $\alpha, \beta \in GL(n, \mathbb{C})$, then $\alpha \cdot \beta \in GL(n, \mathbb{C})$ and $(\alpha \cdot \beta)^{-1} = \beta^{-1} \cdot \alpha^{-1}$. This is elementary linear algebra; further if $\alpha \in GL(n, \mathbb{C})$, then α^{-1} has coordinates b_{ij} where

$$b_{ij} = \frac{p_{ij}(\alpha)}{\det(\alpha)},$$

where $p_{ij}(\alpha)$ is a polynomial of the coordinates of α. It follows that the operation $\alpha \to \alpha^{-1}$ of $GL(n, \mathbb{C})$ onto itself is also a homeomorphism (of order two).

Corollary 3.7. *$GL(n, \mathbb{C})$ is a locally compact metrizable topological group.*

Proof. After all,

$$GL(n, \mathbb{C}) = \det^{\leftarrow}(\{z \in \mathbb{C} : z \neq 0\}),$$

and so $GL(n, \mathbb{C})$ is homeomorphic to an open subset of a locally compact metric space, $\mathbb{C}^{n^2}$. Our comments about continuity of the operations $(\alpha, \beta) \to \alpha \cdot \beta$ and $\alpha \to \alpha^{-1}$ finish the proof. □

Inside $GL(n, \mathbb{C})$ we can find other *classical* topological groups. Here are a few of them.

O(n) : $\alpha \in GL(n, \mathbb{C})$ is *orthogonal* if $\alpha = \bar{\alpha} = (\alpha^{-1})^t$.

O(n, ℂ) : $\alpha \in GL(n, \mathbb{C})$ is *complex orthogonal* if $\alpha = (\alpha^{-1})^t$.

U(n) : $\alpha \in GL(n, \mathbb{C})$ is *unitary* if $\bar{\alpha} = (\alpha^{-1})^t$. Since the mappings $\alpha \to \alpha^{-1}$ and $\alpha \to (\alpha^{-1})^t$ are continuous in $GL(n, \mathbb{C})$, *each of the groups $O(n), O(n, \mathbb{C})$ and $U(n)$ is a closed subgroup of $GL(n, \mathbb{C})$.*

GL(n, ℝ) : $\alpha \in M_n$ is real if $\alpha = \bar{\alpha}$; denote the set of real members of M_n by $M_n(\mathbb{R})$ so

$$GL(n, \mathbb{R}) = GL(n, \mathbb{C}) \cap M_n(\mathbb{R}).$$

SL(n, ℂ) : the members of $GL(n, \mathbb{C})$ with determinant 1, called the *special linear group.*

SL(n, ℝ) : $SL(N, \mathbb{C}) \cap M_n(\mathbb{R})$.

SO(n) : $SL(n, \mathbb{C}) \cap O(n)$.

SU(n) : $SL(n, \mathbb{C}) \cap U(n)$.

Again, $SL(n,\mathbb{C}), SL(n,\mathbb{R}), SO(n), SU(n)$ *are closed subgroups of* $GL(n,\mathbb{C})$). Actually more can be said.

Theorem 3.8. *The groups* $U(n), O(n), SU(n)$, *and* $SO(n)$ *are compact metric topological groups.*

Proof. Each of $O(n), SU(n), SO(n)$ are closed subgroups of $U(n)$, so it's enough to establish that $U(n)$ is compact. Now $\alpha \in U(n)$ precisely when $\alpha^t\bar{\alpha} = \mathrm{id}_{\mathbb{C}^n}$. This in turn is the same as saying of $\alpha = (a_{ij})$ that for each $1 \le i, k \le n$,

$$\sum_j a_{ji}\bar{a}_{jk} = \delta_{ik}.$$

Now the left side is a continuous function of α so $U(n)$ is closed not just in $GL(n,\mathbb{C})$ but even in M_n. Moreover

$$\sum_j a_{ji}\bar{a}_{ji} = 1$$

ensures that $|a_{ij}| \le 1$ for $1 \le i, j \le n$. So the entries of any $\alpha \in U(n)$ are bounded. But this ensures that $U(n)$ is homeomorphic to a closed bounded subset of $\mathbb{C}^{n^2}$. □

3. The Birkhoff–Kakutani Theorem

Our next result is technical, but it's an investment well worth the price.

Theorem 3.9 (Garrett Birkhoff and Shizuo Kakutani). *Let* (U_n) *be a sequence of symmetric open sets each containing the identity* e *of the topological group* G. *Suppose that for each* $k \in \mathbb{N}$,

$$U_{k+1} \cdot U_{k+1} \subseteq U_k.$$

Let $H = \bigcap_k U_k$. *Then there is a left invariant pseudo-metric* σ *on* G *such that*

(i) σ *is* left uniformly continuous *on* $G \times G$;

(ii) $\sigma(x, y) = 0$ *if and only if* $x \in yH$;

(iii) $\sigma(x, y) \le 4 \cdot 2^{-k}$, *if* $x \in yU_k$;

(iv) $2^{-k} \le \sigma(x, y)$, *if* $x \notin yU_k$.

Proof. There is no harm in assuming that the sequence $\{U_n\}$ is decreasing. Otherwise we can replace U_n with $\bigcap_{k=1}^n U_k$, which will leave H unchanged and allow $U_{n+1} \cdot U_{n+1} \subseteq U_n$ to hold for each n.

We start by reinterpreting the sequence's descending character with an eye toward defining σ. For each k, set

$$V_{2^{-k}} = U_k.$$

Next define V_r for r a dyadic rational number with $0 < r < 1$ as follows. If

$$r = 2^{-l_1} + 2^{-l_2} + \cdots + 2^{-l_n},$$

with $0 < l_1 < \cdots < l_n$, all positive integers, let

$$V_r = V_{2^{-l_1}} \cdot V_{2^{-l_2}} \cdots V_{2^{-l_n}}.$$

For dyadic r's greater than or equal to 1, set $V_r = G$. Thus,

$$r < s \Rightarrow V_r \subseteq V_s; \tag{4}$$

further for any $l \in \mathbb{N}$, we have

$$V_r \cdot V_{2^{-l}} \subseteq V_{r+2^{-l+2}}. \tag{5}$$

We put off the somewhat tedious yet clever proofs of the indicated relationships between the V's (equations (4) and (5)) until the end of our general discussion.

With these in hand we go forth to define σ. First, for $x \in G$, let

$$\phi(x) = \inf\{r : x \in V_r\}.$$

Plainly, $\phi(x) = 0$ if and only if $x \in H$. Now for $x, y \in G$ define

$$\sigma(x, y) = \sup\{|\phi(zx) - \phi(zy)| : z \in G\}.$$

Plainly, $\sigma(x, y) = \sigma(y, x)$ and $\sigma(x, x) = 0$. It's easy to see that $\sigma(x, u) \leq \sigma(x, y) + \sigma(y, u)$ for any $x, y, u \in G$, and the fact that $\sigma(ax, ay) = \sigma(x, y)$ for all $a \in G$ is obvious. So σ is a left-invariant pseudo-metric.

Now we join the hunt.

Let $l \in \mathbb{N}$, and suppose $u \in V_{2^{-l}}$ and $z \in G$. If $z \in V_r$, then $z \cdot u \in V_{r+2^{-l+2}}$ thanks to (5). Hence

$$\phi(z \cdot u) \leq r + 2^{-l+2};$$

this is true whenever $z \in V_r$, so using the definition of ϕ,

$$\phi(z \cdot u) \leq \phi(z) + 2^{-l+2}.$$

Similarly, if $z \cdot u \in V_r$, then

$$z \in V_r \cdot u^{-1} \subseteq V_r V_{2^{-l}}^{-1} = V_r V_{2^{-l}} \subseteq V_{r+2^{-l+2}},$$

again by (5). It follows that

$$\phi(z) \leq r + 2^{-l+2},$$

so we see that

$$\phi(z) \leq \phi(z \cdot u) + 2^{-l+2}.$$

So

$$\phi(z) \le \phi(z \cdot u) + 2^{-l+2} \quad \text{and} \quad \phi(z \cdot u) \le \phi(z) + 2^{-l+2}.$$

The only conclusion that we can make is that for $u \in V_{2^{-l}}$ and $z \in G$,

$$|\phi(z) - \phi(z \cdot u)| \le 2^{-l+2}.$$

From this we see that

$$\sigma(u, e) \le 2^{-l+2} \text{ for } u \in V_{2^{-l}}.$$

The third statement of the theorem follows from this and σ's left invariance: if $x \in yU_k$, then $y^{-1}x \in U_k = V_{2^{-k}}$ so

$$\sigma(x, y) = \sigma(y^{-1}x, e) \le 2^{-k+2} = 4 \cdot 2^{-k}.$$

Next we deal with σ's *uniform continuity*. Suppose $x, y, \tilde{x}$, and $\tilde{y}$ satisfy

$$y^{-1}x \in V_{2^{-l-1}} \quad \text{and} \quad \tilde{y}^{-1}\tilde{x} \in V_{2^{-l-1}}.$$

Then

$$\tilde{x}^{-1}\tilde{y}y^{-1}x \in V_{2^{-l-1}}^{-1} \cdot V_{2^{-l-1}} = V_{2^{-l-1}} \cdot V_{2^{-l-1}} \subseteq V_{2^{-l}},$$

and so

$$\begin{aligned}
|\sigma(x, y) - \sigma(\tilde{x}, \tilde{y})| &= |\sigma(y^{-1}x, e) - \sigma(\tilde{y}^{-1}\tilde{x}, e)| \\
&\le |\sigma(y^{-1}x, \tilde{y}^{-1}\tilde{x})| \\
&= |\sigma(\tilde{x}^{-1}\tilde{y}y^{-1}x, e)| \\
&\le 2^{-l+2} \text{ (by (iii))},
\end{aligned}$$

and this is what we mean by σ is *left uniformly continuous*.

For (iv), suppose $y^{-1}x \notin U_l = V_{2^{-l}}$. Then $\phi(y^{-1}x) \ge 2^{-l}$ and so

$$\sigma(x, y) = \sigma(y^{-1}x, e) \ge \phi(y^{-1}x) \ge 2^{-l},$$

where the last inequality follows since for any $a \in G$,

$$\begin{aligned}
\sigma(a, e) &= \sup\{|\phi(za) - \phi(z)| : z \in G\} \\
&\ge |\phi(a) - \phi(e)| - |\phi(a)| \text{ (since } e \in V_r \text{ for every } r)
\end{aligned}$$

Finally, (ii) is an easy consequence of (iii) and (iv). □

The hard work of Birkhoff and Kakutani pays off in a couple of fundamental consequences—consequences which underscore the special character of topological groups.

Corollary 3.10. *Let G be a topological group. If G has a countable neighborhood base at $\{e\}$, then G is metrizable. In this case the metric can be taken to be left invariant.*

Proof. Suppose $\{V_n : n \in \mathbb{N}\}$ is a countable open base at e. Let $U_1 = V_1 \cap V_1^{-1}$, and let U_2 be a symmetric open neighborhood of e such that $U_2 \subseteq U_1 \cap V_2$ and $U_2 \cdot U_2 \subseteq U_1$. Continuing, let U_n be a symmetric open neighborhood of e such that

$$U_n \subseteq U_1 \cap U_2 \cap \cdots \cap U_{n-1} \cap V_n,$$

and $U_n \cdot U_n \subseteq U_{n-1}$.

The family $\{U_k : k \in \mathbb{N}\}$ satisfies the conditions set forth in the Birkhoff–Kakutani theorem. Further, $H = \bigcap_n U_n = \{e\}$. Let σ be the left pseudo-metric introduced in the theorem. In fact, σ is a true metric: $\sigma(x, y) = 0$ if and only if $x = y$, since after all, $H = \{e\}$!

Since

$$\{x \in G : \sigma(x, e) < 2^{-k}\} \subseteq U_k \subseteq \{x \in G : \sigma(x, e) \leq 2^{-k+2}\},$$

the topology defined by σ coincides with the given topology of G. □

Corollary 3.11. *Let G be a topological group, let $a \in G$, and let F be a closed subset of G such that $a \notin F$. Then there is a continuous real function χ on G such that $\chi(a) = 0$ and $\chi(x) = 1$ for all $x \in F$. Consequently, every topological group is completely regular.*

Proof. Let U_1 be a symmetric neighborhood of e such that $(aU_1) \cap F = \emptyset$. Choose a sequence $(U_n : n \geq 2)$ of open neighborhoods of e such that each U_n is symmetric, $U_{n+1} \cdot U_{n+1} \subseteq U_n$, and let $H = \bigcap_n U_n$. Apply the Birkhoff–Kakutani theorem to the U_n's. For $x \in G$, define $\chi(x)$ by

$$\chi(x) = \min\{1, 2\sigma(a, x)\},$$

where σ is the left-invariant, uniformly continuous pseudo-metric produced in the Birkhoff–Kakutani theorem. Now χ is continuous by (i) of our theorem, and $\chi(a) = 0$. If $x \in F$, then $a^{-1}x \in a^{-1}F$, a set disjoint from U_1; consequently, $a^{-1}x \notin U_1$ and $\sigma(a, x) \geq 2^{-1}$ by (iv) of our theorem. It follows that $\chi(x) = 1$ for all $x \in F$. □

We'll be spending much of our time in a locally compact setting, and again there's more than meets the eye because of the group's structure.

Theorem 3.12. *Any locally compact topological group is paracompact, hence normal.*

Proof. Let G be a locally compact topological group, and let V be an open set in G containing the identity of G and having a compact closure $\overline{V}$. By Theorem 3.1 there exists a symmetric open set U such that

$$e \in U \subseteq \overline{U} \subseteq U \cdot U \subseteq V.$$

Look at $H = \bigcup_n U^n$, where $U^n = \underbrace{U \cdots U}_{n}$. Then H is an open subgroup (hence closed) of G. Also H is σ-compact since

$$H = \bigcup_n \underbrace{U \cdots U}_{2n} \supseteq \bigcup_n \underbrace{\overline{U} \cdots \overline{U}}_{n} \supseteq H,$$

a countable union of compact subsets. Hence H is Lindelöf and regular, so H is paracompact [**85**].

Let $\mathcal{U}$ be an open cover of G. Each coset xH of H is also σ-compact and so there is a countable subfamily $\{V_{xH}^{(n)} : n \in \mathbb{N}\}$ of $\mathcal{U}$ that covers xH.

Naturally,

$$\{V_{xH}^{(n)} \cap xH : n \in \mathbb{N}\}$$

is an open cover of xH and since xH is paracompact, there is a locally finite open cover of xH that refines $\{V_{xH}^{(n)} \cap xH : n \in \mathbb{N}\}$, call it $\{\mathcal{W}_{xH}^{(n)}\}_{n=1}^{\infty}$. Note that for each $n = 1, 2, \ldots,$

$$\mathcal{W}_{xH}^{(n)} \subseteq xH.$$

For each $n = 1, 2, \ldots,$ let

$$\mathcal{W}^n = \bigcup_{xH \in G/H} \mathcal{W}_{xH}^{(n)},$$

so $\{\mathcal{W}^n\}_{n=1}^{\infty}$ is a locally finite open cover. Therefore $\mathcal{W} = \bigcup_n \mathcal{W}^n$ is an open cover of G that's plainly σ–locally finite and refines $\mathcal{U}$.

Thus each open cover $\mathcal{U}$ of G admits a σ–locally finite open cover $\mathcal{W}$ that refines $\mathcal{U}$ so G is paracompact and therefore normal ([**66**], Theorem 5.28 and Corollary 5.32). □

Warning. Not every topological group is normal as exhibited by the following classical example.

Example 3.13. Let m be an uncountable cardinal number. Then $\mathbb{Z}^m$ is a T_0 group, hence completely regular. But $\mathbb{Z}^m$ is not normal.

Proof. We'll write $\mathbb{Z}^m$ as $\prod_{i\in I}\mathbb{Z}_i$ where each $\mathbb{Z}_i$ is $\mathbb{Z}$ and $|I| = m$. For the sake of the present efforts, let $A, B \subseteq \mathbb{Z}^m$ be given as follows

$$A = \{(x_i) \in \mathbb{Z}^m : \text{for any } n \neq 0, \quad \text{there is at most one index } i \text{ for which } x_i = n\}$$

and

$$B = \{(x_i) \in \mathbb{Z}^m : \text{for any } n \neq 1, \quad \text{there is at most one index } i \text{ for which } x_i = n\}.$$

Then A and B are disjoint. If $(x_i)_{i\in I} \notin A$, then there are $i_0, i_1 \in I, i_0 \neq i_1$ such that for some $n \in \mathbb{Z}, n \neq 0, x_{i_0} = x_{i_1} = n$. The set

$$\{(y_i)_{i\in I} \in \mathbb{Z}^m : y_{i_0} = y_{i_1} = n\}$$

is an open set containing $(x_i)_{i\in I}$ but no point of A, so A is a closed set. Similarly B is a closed set.

Let U, V be open subsets of $\mathbb{Z}^m$ such that $A \subseteq U$ and $B \subseteq V$. We claim that $U \cap V \neq \emptyset$, which will show that $\mathbb{Z}^m$ is not normal.

Let $(x_i^{(1)})_{i\in I} \in \mathbb{Z}^m$ be defined by $x_i^{(1)} = 0$ for each $i \in I$. Since $(x_i^{(1)}) \in A \subseteq U$, there are distinct indices $i_1, \dots, i_{m_1} \in I$ such that

$$(x_i^{(1)})_{i\in I} \in \{(x_i)_{i\in I} \in \mathbb{Z}^m : x_{i_1} = x_{i_2} = \cdots = x_{i_{m_1}} = 0\} \subseteq U.$$

Let $(x_i^{(2)})_{i\in I}$ be defined by

$$x_i^{(2)} = \begin{cases} k & \text{for } i = i_k, k = 1, \dots, m_1, \\ 0 & \text{otherwise.} \end{cases}$$

Since $(x_i^{(2)}) \in A \subseteq U$ there exists $i_{m_1+1}, \dots, i_{m_2} \in I$, distinct indices from each other and from $i_1, \dots, i_{m_1}$ such that

$$(x_i^{(2)})_{i\in I} \in \{(x_i)_{i\in I} \in \mathbb{Z}^m : x_{i_1} = 1, x_{i_2} = 2, \dots, x_{i_{m_1}} = m_1, \quad x_{i_{m_1}+1} = 0, \dots, x_{i_{m_2}} = 0\} \subseteq U,$$

Continue in this manner.

Define $(y_i)_{i\in I} \in \mathbb{Z}^m$ as follows: $y_{i_k} = k$ for any k and $y_i = 1$ if $i \neq i_k$. Plainly, $(y_i)_{i\in I} \in B$. Hence for some finite subset J of I,

$$\{(x_i)_{i\in I} \in \mathbb{Z}^m : x_i = y_i, \quad \text{for } i \in J\} \subseteq V.$$

But J is finite so there is an n_0 such that $i_k \notin J$ whenever $k > m_{n_0}$. Look at $(z_i)_{i\in I} \in \mathbb{Z}^m$, where

$$z_i = \begin{cases} k & \text{if } i = i_k, k \leq m_{n_0}, \\ 0 & \text{if } i = i_k, m_{n_0} + 1 \leq k \leq m_{n_0+1}, \\ 1 & \text{otherwise.} \end{cases}$$

So

$$(z_i)_{i\in I} \in \{(x_i)_{i\in I} \in \mathbb{Z}^m : x_i = y_i, \quad \text{for} \quad i \in J\} \subseteq V.$$

So

$$(z_i)_{i\in I} \in \{(x_i)_{i\in I} \in \mathbb{Z}^m : x_{i_1} = 1, x_{i_2} = 2, \ldots, x_{i_{m_0}} = m_{n_0},$$
$$x_{i_{m_{n_0}}+1} = x_{i_{m_{n_0}}+2} = \cdots = x_{i_{m_{n_0+1}}} = 0\} \subseteq U.$$

So $(z_i)_{i\in I} \in V$, $(z_i)_{i\in I} \in U$, and $U \cap V \neq \emptyset$. Therefore $\mathbb{Z}^m$ is not normal. $\square$

After Thoughts. The missing steps of the Birkhoff–Kakutani theorem. To prove (4)

$$r < s \Rightarrow V_r \subseteq V_s.$$

Assume that $r < s < 1$, where

$$r = 2^{-l_1} + 2^{-l_2} + \cdots + 2^{-l_n},$$
$$s = 2^{-m_1} + 2^{-m_2} + \cdots + 2^{-m_p},$$

and where $0 < l_1 < l_2 < \cdots < l_n$ and $0 < m_1 < m_2 < \cdots < m_p$ are integers. Since $r < s$, there is a positive integer k such that $l_1 = m_1$, $l_2 = m_2$, $\ldots$, $l_{k-1} = m_{k-1}$, but $l_k > m_k$. Let $W = V_{2^{-l_1}} \cdot V_{2^{-l_2}} \cdots V_{2^{-l_{k-1}}}$. Then

$$\begin{aligned} V_r &= W \cdot V_{2^{-l_k}} \cdot V_{2^{-l_{k+1}}} \cdots V_{2^{-l_{n-2}}} \cdot \underbrace{V_{2^{-l_{n-1}}} \cdot V_{2^{-l_n}}} \\ &\subseteq W \cdot V_{2^{-l_k}} \cdot V_{2^{-l_{k+1}}} \cdots V_{2^{-l_{n-2}}} \cdot \underbrace{V_{2^{-l_{n-1}}} \cdot V_{2^{-l_{n-1}}}} \\ &\subseteq W \cdot V_{2^{-l_k}} \cdot V_{2^{-l_{k+1}}} \cdots V_{2^{-l_{n-2}}} \cdot \underbrace{V_{2^{-l_{n-1}+1}}} \\ &\subseteq W \cdot V_{2^{-l_k}} \cdot V_{2^{-l_{k+1}}} \cdots V_{2^{-l_{n-2}}} \cdot \underbrace{V_{2^{-l_{n-2}}}}, \end{aligned}$$

where the first and third inclusions are due to the decreasing nature of the U_n's, and the second inclusion uses the fact that $U_{l_{n-1}} \cdot U_{l_{n-1}} \subset U_{l_{n-1}-1}$. In this way, in these successive inclusions, we can "clip off the ends" one at a time. Continuing in this way, we have that

$$\begin{aligned} V_r &\subseteq W \cdot V_{2^{-l_k}} \cdot V_{2^{-l_{k+1}}} \cdots V_{2^{-l_{n-2}}} \cdot V_{2^{-l_{n-2}}} \\ &\vdots \\ &\subseteq W \cdot V_{2^{-l_k}} \cdot V_{2^{-l_k}} \\ &\subseteq W \cdot V_{2^{-l_k+1}} \\ &\subseteq W \cdot V_{2^{-m_k}} \\ &\subseteq W \cdot V_{2^{-m_k}} \cdot V_{2^{-m_{k+1}}} \cdots V_{2^{-m_p}} \\ &= V_s, \end{aligned}$$

where the last inclusion uses the fact that the U_n's each contain the identity e.

With the same representation of r in mind, we set off to prove formula (5)

$$V_r \cdot V_{2^{-l}} \subseteq V_{r+2^{-l+2}}.$$

We suppose that $r + 2^{-l+2} < 1$. If $l > l_n$, then

$$V_r \cdot V_{2^{-l}} = V_{r+2^{-l}},$$

and all is well. So we look to the case that $l \leq l_n$.

Let k be the positive integer such that

$$l_{k-1} < l \leq l_k, \quad (l_0 = 0).$$

Let r_1 be given by

$$r_1 = 2^{-l+1} - 2^{-l_k} - 2^{-l_{k+1}} - \cdots - 2^{-l_n}$$

and

$$r_2 = r + r_1.$$

It's plain that

$$r < r_2 < r + 2^{-l+1},$$

so

$$V_r \cdot V_{2^{-l}} \subseteq V_{r_2} \cdot V_{2^{-l}} = V_{r_2+2^{-l}} \subseteq V_{r+2^{-l+1}+2^{-l}} \subseteq V_{r+2^{-l+2}},$$

and that's that. □

Exercise 3.14. (i) Let G be a locally compact group, let K be a compact subset of G, and let U be an open set containing the identity. Then there is a open set V containing the identity so that for each $x \in K$, $Vx \subseteq xU$.

(ii) Deduce from (i) that if G is a compact metrizable group, then G has a bi-invariant metric generating its topology.

Exercise 3.15. (i) Let G be a locally compact metrizable group with a left invariant metric d generating G's topology. Then (G, d) is a complete metric space.

(ii) Deduce from (i) and the Birkhoff–Kakutani theorem that every locally compact separable metrizable group is homeomorphic to a complete separable metric space.

(iii) Suppose H is a continuous homomorphism of the locally compact separable metrizable group G onto the locally compact separable metrizable group G'. Then H takes open sets to open sets.

4. Products of Topological Spaces

Of considerable importance in many topological affairs is the notion of a product space. Suppose $(S_\alpha, \tau_\alpha)_{\alpha \in \mathcal{A}}$ is a nonempty family of nonvoid topological spaces, and let

$$S = \prod_{\alpha \in \mathcal{A}} S_\alpha$$

be the Cartesian product of the sets $\{S_\alpha : \alpha \in \mathcal{A}\}$. We equip S with the weakest (i.e., smallest) topology τ such that for each $\alpha \in \mathcal{A}$, the projection $P_\alpha : S \to S_\alpha$ is continuous. So we ask of τ that

$$P_\alpha^{\leftarrow}(U_\alpha) \in \tau$$

for each $U_\alpha \in \tau_\alpha$ and each $\alpha \in \mathcal{A}$. The collection

$$\{P_\alpha^{\leftarrow}(U_\alpha) : U_\alpha \in \tau_\alpha, \alpha \in \mathcal{A}\}$$

is a subbase for the *product topology* τ. A base τ is the family of all finite intersections of the subbase members, i.e., all sets of the form

$$\bigcap_{\alpha \in F} P_\alpha^{\leftarrow}(U_\alpha),$$

where F is a *finite* subset of $\mathcal{A}$ and U_α is an open set in (S_α, τ_α). Here are some easily established facts and features (also known as "exercises" if you please) of the product topology.

- Each $P_\alpha : S \to S_\alpha$ is an open map, i.e., $P_\alpha(U)$ is open in S_α for each open set $U \subseteq (S, \tau)$.
- For any topological space T, a function $f : T \to S$ is continuous precisely when $P_\alpha \circ f : T \to S$ is continuous for each $\alpha \in \mathcal{A}$.
- If $(x_d)_{d \in \mathcal{D}}$ is a generalized sequence (i.e., a net) in S, then $(x_d)_{d \in \mathcal{D}}$ is convergent to $x_0 \in S$ if and only if for each $\alpha \in \mathcal{A}$, we have

$$\lim_{\mathcal{D}} x_{d,\alpha} = \lim_{\mathcal{D}} P_\alpha x_d = x_{d,0} = P_\alpha x_0.$$

Of course we are naturally interested in the stability of various topological invariants. Here is a basic result.

Theorem 3.16. *The product of Hausdorff spaces is a Hausdorff space.*

Proof. If x, y are distinct members of $S = \prod_{\mathcal{A}} S_\alpha$, then there is an $\alpha \in \mathcal{A}$ so that $x_\alpha \neq y_\alpha$; since S_α is a Hausdorff space there are open sets U and V in S_α so $x_\alpha \in U, y_\alpha \in V$ and $U \cap V \neq \emptyset$. The sets $P_\alpha^{\leftarrow}(U)$ and $P_\alpha^{\leftarrow}(V)$ are disjoint open sets in S with $x \in U$ and $y \in V$. $\square$

A result of great importance throughout mathematics is the following famous result of Tychonoff.

Theorem 3.17 (Tychonoff's Product Theorem). *The product of compact spaces is compact.*

Proof. (Chevalley and Frink [**17**]). Let $\mathcal{F}$ be a family of subsets of the product S of the compact spaces $S_\alpha, \alpha \in \mathcal{A}$, and suppose that any finite subfamily of $\mathcal{F}$ has a nonempty intersection. Our aim is to show

$$\bigcap_{\mathcal{F}} F \neq \emptyset.$$

Let $\mathcal{M}$ be a maximal system of subsets of S with the finite intersection property (FIP) containing $\mathcal{F}$. Consider

$$M_\alpha = \{F_\alpha = P_\alpha F : F \in \mathcal{M}\}.$$

Then each $\mathcal{M}_\alpha$ has the FIP. Each S_α is compact so for each $\alpha \in \mathcal{A}$ there is a point p_α common to

$$\bigcap_{F \in \mathcal{M}} \overline{F}_\alpha.$$

Let p be the point of S with $P_\alpha p = p_\alpha$. We want to show that

$$p \in \bigcap_{F \in \mathcal{F}} \overline{F}.$$

Now the very way we chose p_α ensures that if U_α is any open set containing p_α in S_α, then U_α has a point in common with any member of $\mathcal{M}_\alpha$. Hence, in S, the open set

$$W_\alpha = P_\alpha^{\leftarrow}(U_\alpha)$$

containing p has a point in common with each member of $\mathcal{M}$. But $\mathcal{M}$ is maximal so W_α must belong to $\mathcal{M}$. Similarly each and every finite intersection of open sets such as $P_\alpha^{\leftarrow}(U_\alpha)$ must belong to $\mathcal{M}$. But $\mathcal{M}$ is *still* maximal so $\mathcal{M}$ contains all such finite intersections. But sets of this sort form a basis for the open sets containing p: each open set in S that contains p is a member of $\mathcal{M}$. Therefore every open set containing p has a point in common with each member of $\mathcal{M}$. Hence p is in $\overline{M}$ for each $M \in \mathcal{M}$, and so

$$p \in \bigcap_{F \in \mathcal{F}} \overline{F}.$$ □

Exercise 3.18. Let S be a topological product of the spaces (S_α, τ_α).

- If each S_α is connected, then so is S.
- If each S_α is T_1, then so is S.

- The product of countably many spaces, each of which has a countable basis for its topology, has a countable basis for its topology.
- The product of locally compact spaces is locally compact if (and only if) each coordinate is locally compact and all but finitely many are compact.

What about topological groups? Well, it's easy to see that if (G_α, τ_α) is a topological group for each $\alpha \in \mathcal{A}$ and

$$G = \prod_{\mathcal{A}} G_\alpha$$

is the Cartesian product, then G is a group with coordinate-wise operations and a topological space. It is, in fact, a topological group as well. To see this, let $x = (x_\alpha), y = (y_\alpha)$ be points in G, and let U be any open set in G that contains xy. Then, by definition of the product topology, there are open sets U_α in G_α containing $x_\alpha y_\alpha$ for all α such that, except for finitely many α's, $U_\alpha = G_\alpha$. But each G_α is a topological group, and so for each α we can find open sets V_α and W_α such that $x_\alpha \in V_\alpha, y_\alpha \in W_\alpha$ and

$$V_\alpha \cdot W_\alpha \subseteq U_\alpha.$$

Moreover if we choose $V_\alpha = W_\alpha = U_\alpha$ whenever $U_\alpha = G_\alpha$, then

$$V = \prod_{\mathcal{A}} V_\alpha$$

is open in G, contains x, and

$$W = \prod_{\mathcal{A}} W_\alpha$$

is open in G, contains y and $V \cdot W \subseteq U$.

Exercise 3.19. Modify this argument to show that $x \mapsto x^{-1}$ is also continuous on G.

Of course, in tandem with Tychonoff's theorem, this provides us with a plethora of compact groups. If we take into account the last statement of Exercise 3.18 we also have a boat load of locally compact, noncompact groups.

5. Notes and Remarks

The definition of topological group, as we use it, is due to F. Leja [**69**]; soon after a large number of mathematicians were uncovering interesting and important features of the subject. The early development can be traced through the papers of D. van Dantzig [**129**], [**130**], H. Freudenthal [**40**] and J. von Neumann [**133**]. The monograph of L. Pontrjagin [**104**] organized much of what was known and presented it in a palatable form. Often the hypothesis that the topology satisfy the second axiom of countability (thereby ensuring that the group's topology is metrizable) was added.

A. Weil's monograph [**136**] rid the added hypothesis of second countability wherever possible.

A. Weil also gave us the theory of uniform spaces [**135**] isolated with the express purpose of providing a common ground for studying metric spaces and topological groups. In particular, the realization that every topological group is completely regular is due to Weil.

Our presentation of the classical locally compact groups was greatly influenced by H. Weyl's [**137**] monograph, *The Classical Groups.*

The Birkhoff–Kakutani theorem is due to G. Birkhoff [**11**] and S. Kakutani [**60**]. Kakutani noted that his metric could be chosen to be left invariant.

That locally compact topological groups are normal was pointed out (to E. Hewitt and K. Ross [**55**]), by E. A. Michael. It seems like the paracompactness was first noted by Hewitt and Ross in their first volume on *Abstract Harmonic Analysis.*

That $\mathbb{Z}^{\boldsymbol{m}}$ is not normal if $\boldsymbol{m}$ is an uncountable cardinal is due to A. H. Stone [**122**] who also noted that a product $\prod_{i\in\mathbb{Z}} G_i$ of topological groups being normal implies that all but a countable number of the groups are countably compact.

The subject of topological groups holds many fascinating avenues of investigation and has been given correspondingly fine expositions. In addition to the monographs already mentioned, the books of P. J. Higgins [**56**], K. H. Hoffman and S. A. Morris [**57**], D. Montgomery and L. Zippin [**84**] and S. A. Morris [**86**] should keep the reader busy and entertained.

Chapter 4

Banach and Measure

Like most abstract analysts of his day, Banach took a keen interest in developments related to invariant measures. In this chapter, we present *some* of Banach's work in this direction. We open with his discovery of Banach limits, a remarkable consequence of the Hahn–Banach theorem (which we also prove). We follow this with Banach's proof of the existence of a Borel measure on a compact metric space invariant under "congruence"; in particular, it is to be noted that there exists a left-invariant probability Borel measure on any compact metrizable topological group. After this we take a slight detour to present S. Saks' proof that the positive linear functionals in $C(K)^*$, the dual of the space $C(K)$ of continuous real-valued functions defined on a compact metric space K, consist of regular Borel measures on K.

The result on the dual of $C(K)$, when K is a compact metric space, was first derived by Banach, and we present his proof in Section 4. We have included Banach's Daniell-like description of the positive linear functionals in $C(K)^*$ for its elegance. To see that his abstract Lebesgue integral does indeed apply to the case of positive linear functionals on $C(K)$, where K is a compact metric space, requires us to present some of the most overlooked results appearing in his *Theorie des Operations Lineaires* [**7**], results that demonstrate his complete understanding of life with bounded, finitely additive measures.

Particular attention ought to be paid to Banach's approach to Lebesgue integrals, an approach taken to describe $C(K)^*$ in terms of (regular) Borel measures on K. Like his work on invariant measures, this appeared in the appendices of Saks' beautiful book *Theory of the Integral* [**111, 113**].

We follow Banach's view of the abstract Lebesgue integral with his solution of the so-called easy problem of measure for the real line and the plane. To our knowledge, Banach's proof has slipped through the cracks of mathematical knowledge. We follow with a very brief discussion of the Hausdorff paradox, and its role in solving the easy problem of measure *negatively* for Euclidean spaces of dimension greater than two. An all-too-brief discussion of amenability follows.

1. Banach Limits

Theorem 4.1 (Hahn–Banach Theorem). *Let X be a* real *linear space, and let S be a linear subspace of X. Suppose that $p : X \to \mathbb{R}$ is a subadditive, positively homogeneous functional and $f : S \to \mathbb{R}$ is a linear functional with $f(s) \leq p(s)$ for all $s \in S$. Then there is a linear functional F defined on all of X such that $F(x) \leq p(x)$ for all $x \in X$, and $F(s) = f(s)$ for all $s \in S$.*

Proof. Our first task is to see how to extend a functional like f one dimension at a time while preserving the domination by p. With this in mind, let $x \in X \backslash S$, and notice that for any linear combination $s + \alpha x$ of a vector in S and x, whatever the linear extension F's value at x (say that value is "c"), we must have

$$F(s + \alpha x) = F(s) + \alpha F(x) = f(s) + \alpha c.$$

So it must be (if we are to have F dominated by p) that

$$f(s) + \alpha c \leq p(s + \alpha x)$$

holds for all $s \in S$ and all real α's. We'll follow where this leads us. For all $\alpha \in \mathbb{R}$ and $s \in S$ we have to have

$$\alpha c \leq p(s + \alpha x) - f(s).$$

For $\alpha > 0$ this tells us that

$$\begin{aligned} c &\leq \frac{1}{\alpha}\Big(p(s + \alpha x) - f(s)\Big) \\ &= p\Big(\frac{s}{\alpha} + x\Big) - f\Big(\frac{s}{\alpha}\Big). \end{aligned}$$

While if $\alpha < 0$, then ($-\alpha > 0$ and)

$$-\alpha c \geq f(s) - p(s + \alpha x),$$

or since $-\alpha > 0$,

$$\begin{aligned} c &\geq -\frac{1}{\alpha}f(s) - \frac{1}{-\alpha}p(s+\alpha x) \\ &= f\left(\frac{s}{-\alpha}\right) - p\left(\frac{s}{-\alpha} - x\right). \end{aligned}$$

Taking into account the linearity of S, we see that what we seek is a $c \in \mathbb{R}$ (which will be $F(x)$) so that regardless of $s, s' \in S$ satisfies

$$f(s') - p(s' - x) \leq c \leq p(s+x) - f(s).$$

Does such a c exist? You bet! After all if $s, s' \in S$, then

$$f(s) + f(s') = f(s+s') \leq p(s+s') \leq p(s-x) + p(s'+x).$$

So for all $s, s' \in S$

$$f(s) - p(s-x) \leq p(s'+x) - f(s'),$$

and we *can* chose c in an appropriate manner.

Now we know that we can extend linear functionals one dimension at a time while preserving p's domination. It's time for some transfinite hijinks. We consider the collection of all linear functionals g defined on a linear subspace Y of X such that $S \subseteq Y$, $g|_S = f$, and, on Y, $g \leq p$. We partially order this collection by saying that "$g_1 \leq g_2$" if g_2 is an extension of g_1 (so g_1 is defined on a linear subspace Y_1 that is contained in g_2's domain).

Hausdorff's maximal principle ensures us that there is a maximal linearly ordered subfamily $\{g_\alpha\}$ of linear extensions of f so that on g_α's domain Y_α (which contains S), $g_\alpha \leq p$. We define F on the linear space that's the union of the domains of the g_α's as one might expect! So $F(x) = g_\alpha(x)$, if x is in g_α's domain. Because of the ordering described above, the domain of F is a linear subspace of X and on that domain F is linear and dominated by p. Of course, F is a linear extension of f and the domain of F *must be all of* X—this is assured us by the opening salvo. □

We put the Hahn–Banach theorem to immediate use by establishing the existence of "generalized limits" or Banach limits, as we'll refer to them henceforth. We will call on two spaces: l^∞, the space of bounded, real-valued sequences; and the linear subspace c of l^∞, consisting of all the convergent sequences. Typically if $x \in l^\infty$, then

$$||x||_\infty = \sup\{(x_n) : x \in \mathbb{N}\}.$$

Theorem 4.2 (Banach). *There exists a linear functional LIM on* l^∞ *such that*

(i) *LIM*$(x) \geq 0$ *if* $x = (x_n) \in l^\infty$ *and* $x_n \geq 0$ *for all* n;

(ii) $|LIM(x)| \leq ||x||_\infty$, *for all* $x \in l^\infty$;

(iii) *If* $x \in l^\infty$ *and* $Tx = (x_2, x_3, \dots)$ *for* $x = (x_1, x_2, \dots)$, *then*

$$LIM(Tx) = LIM(x);$$

(iv) *For any* $x \in l^\infty$

$$\liminf_{n\to\infty} x_n \leq LIM(x) \leq \limsup_{n\to\infty} x_n.$$

Proof. Let $p : l^\infty \to \mathbb{R}$ be given by

$$p(x) = \limsup_{n\to\infty} \frac{x_1 + \cdots + x_n}{n}.$$

It is easy to see that p is subadditive and positively homogeneous. Next, let $f : c \to \mathbb{R}$ be the linear functional

$$f(x) = \lim_n x_n.$$

Since f and p agree on c, $f(x) \leq p(x)$ is trivially satisfied. LIM is any Hahn–Banach extension of f to all of l^∞. LIM is a linear functional on l^∞ such that for any $x \in l^\infty$

$$\text{LIM}(x) \leq p(x).$$

To see item (i), take $x \in l^\infty$ and suppose $x_n \geq 0$ for all n. Then

$$-\text{LIM}(x) = \text{LIM}(-x) \leq p(-x),$$

so

$$\begin{aligned} \text{LIM}(x) &\geq -p(x) = -\limsup_{n\to\infty} \frac{-x_1 - x_2 - \cdots - x_n}{n} \\ &= \liminf_{n\to\infty} \frac{x_1 + x_2 + \cdots + x_n}{n} \geq 0, \end{aligned}$$

because $x_n \geq 0$ for all n.

Item (ii) is plain since $p(x) \leq ||x||_\infty$ for all $x \in l^\infty$. To see item (iii), take $x \in l^\infty$, then

$$p(x - Tx) = \limsup_{n\to\infty} \frac{x_1 - x_{n+1}}{n} = 0,$$

since x *is* bounded. It follows that

$$\text{LIM}(x - Tx) \leq p(x - Tx) = 0.$$

So for any $x \in l^\infty$

$$\text{LIM}(x) \leq \text{LIM}(Tx). \tag{6}$$

This applies as well to $-x$ so

$$-\text{LIM}(x) = \text{LIM}(-x) \leq \text{LIM}(-Tx) = -\text{LIM}(Tx),$$

and so

$$\text{(7)} \qquad \mathrm{LIM}(x) \geq \mathrm{LIM}(Tx)$$

as well. $\mathrm{LIM}(x) = \mathrm{LIM}(Tx)$ is the only conclusion that can be drawn from (6) and (7).

For item (iv), let $\epsilon > 0$. Find $N \in \mathbb{N}$ so

$$\inf_n x_n \leq x_N \leq \inf_n x_n + \epsilon.$$

Then

$$x_n + \epsilon - x_N \geq \inf_n x_n + \epsilon - x_N \geq 0.$$

Hence

$$0 \leq \mathrm{LIM}(x + \epsilon - x_N) = \mathrm{LIM}(x) + \epsilon - x_N$$

by (i). Hence

$$\inf_n x_n \leq x_N \leq \mathrm{LIM}(x) + \epsilon.$$

Since $\epsilon > 0$ was arbitrary,

$$\inf_n x_n \leq \mathrm{LIM}(x).$$

For any k

$$\inf_{n \geq k} x_n = \inf_{n \geq k} T^k x_n \leq \mathrm{LIM}(T^k x) = \mathrm{LIM}(x);$$

but

$$\liminf_n x_n = \sup_n \inf_{k \geq n} x_k \leq \sup_n \mathrm{LIM}(x) = \mathrm{LIM}(x).$$

Again using LIM's linearity, we see that

$$\limsup_n x_n = -\liminf_{n \to \infty} (-x_n) \geq -\mathrm{LIM}(-x) = \mathrm{LIM}(x).$$

The proof of the existence of Banach limits appears. □

2. Banach and Haar Measure

Let Q be a fixed compact metric space.

Definition 4.3. We suppose that for subsets of Q the notion of *congruence* is defined to satisfy the following conditions (here $A \cong B$ means A is congruent to B):

(i) $A \cong A$.

(ii) $A \cong B \Leftrightarrow B \cong A$.

(iii) $A \cong B,\ B \cong C \Rightarrow A \cong C$.

(iv) If A is an open set, then so is any set congruent to A.

(v) If A is congruent to B and A can be covered by a sequence (A_n) of open sets, then B can be covered by a sequence (B_n) so that $B_n \cong A_n$ for each n.

(vi) For any open set A the collection of sets congruent to A cover Q.

(vii) If (S_n) is a sequence of open concentric balls with radii tending to zero, and if $G_n \cong S_n$ and $a_n, b_n \in G_n$ with $\lim_n a_n$ and $\lim_n b_n$ existing, then these limits coincide.

Proposition 4.4. *If G is a compact metrizable group, then $A \cong B$ whenever $A = gB$ for some $g \in G$ is a congruence.*

Proof. Let (S_n) be a sequence of concentric open balls (with common center c) with radii tending to zero. Suppose (A_n) satisfies $A_n \cong S_n$ for each n and suppose $a_n, b_n \in A_n$ have $\lim_n a_n = a, \lim_n b_n = b$.

We claim that $a = b$. Since $A_n \cong S_n$, there exists $g_n \in G$ such that $A_n = g_n S_n$. Hence

$$a_n = g_n s_n \text{ and } b_n = g_n s_n^{'}$$

for some $s_n, s_n^{'} \in S_n$. Plainly $s_n \to c, s_n^{'} \to c, a_n \to a$, and $b_n \to b$ so

$$g_n = a_n s_n^{-1} \to ac^{-1} \text{ and } g_n = b_n (s_n^{'})^{-1} \to bc^{-1}.$$

Limit one to a customer, so $ac^{-1} = bc^{-1}$ and $a = b$ as we claimed. □

Proposition 4.5. *Let Q be a compact metric space, and let G be a transitive equicontinuous group of homeomorphisms of Q onto itself. If $A, B \in Q$, then $A \cong B$ whenever $A = g(B)$ for some $g \in G$ is a congruence.*

Proof. Transitivity of G is simply the feature that if $q_1, q_2 \in Q$, then there is a $g \in G$ so that $g(q_1) = q_2$. Because members of G are homeomorphisms of Q onto Q, (iv) and (v) of Definition 4.3 are so. Transitivity of G assures us of (vi).

Equicontinuity just says that given an $\epsilon > 0$, there is a $\delta > 0$ so if $x, y \in Q$ and $\delta(x, y) \leq \delta$, then $d(f(x), f(y)) \leq \epsilon$ for all $f \in G$. Because each $g \in G$ is a homeomorphism of Q onto itself, each such g is a uniformly continuous map on the compact set Q as is g^{-1}. So G is an equiuniformly continuous group of uniform isomorphisms of G onto G. Thus (vii) follows from this. □

Given two nonempty open sets A, B, by Definition 4.3(vi), the collection of sets congruent to A covers $\overline{B}$, a compact set. Hence there is a finite collection of sets congruent to A that still cover $\overline{B}$. This motivates Haar's covering function $h(B, A)$:

$$h(B, A) = \text{the least number of sets congruent to } A \text{ needed to cover } \overline{B}.$$

Proposition 4.6. *Suppose A, B, and C are nonempty open subsets of Q. Then*

(i) $C \subseteq B \Rightarrow h(C, A) \leq h(B, A)$.

(ii) $h(B \cup C, A) \leq h(B, A) + h(C, A)$.

(iii) $B \cong C \Rightarrow h(B, A) = h(C, A)$.

(iv) $h(B, A) \leq h(B, C)h(C, A)$.

(v) *If $d(A, B)$ = distance from A to B is positive (so $\bar{A} \cap \bar{B} = \emptyset$) and (S_n) is sequence of open concentric balls with radii tending to zero, then there is a number N so that for $n \geq N$*

$$h(A \cup B, S_n) = h(A, S_n) + h(B, S_n).$$

Proof. Item (v) requires some serious and careful attention. Suppose (v) fails. Then there is (n_k) so that

$$h(A \cup B, S_{n_k}) < h(A, S_{n_k}) + h(B, S_{n_k})$$

for each k. We can plainly suppose that the n_k's are chosen so large that $S_{n_k} \cap A \neq \emptyset$ and $S_{n_k} \cap B \neq \emptyset$ cannot both occur. It follows that there is a sequence (G_k) such that $G_k \cong S_{n_k}$ and $G_k \cap A \neq \emptyset$ and $G_k \cap B \neq \emptyset$. Why? We'll fix n_k momentarily and imagine that any G that is congruent to S_{n_k} could meet at most one of A and B. If we cover $A \cup B$ by $h(A \cup B, S_{n_k})$ many sets congruent to S_{n_k}, then this cover (call it $\mathcal{C}$) would be the disjoint union of the collection $\mathcal{A}$ (respectively $\mathcal{B}$) where $\mathcal{A}$ (respectively $\mathcal{B}$) consists of the members of $\mathcal{C}$ that meet only A (respectively B). Consequently,

$$h(A \cup B, S_{n_k}) = |\mathcal{C}| = |\mathcal{A}| + |\mathcal{B}| \geq h(A, S_{n_k}) + h(B, S_{n_k}),$$

which is not an option. So we get a sequence (G_k) of sets with $G_k \cong S_{n_k}$ and $G_k \cap A \neq \emptyset$, $G_k \cap B \neq \emptyset$. From each of the sets $G_k \cap A$, pick a point a_k and from each $G_k \cap B$, pick a point b_k. The sequences $(a_k), (b_k)$ lie inside Q so there is a $\mathbb{J} \in P_\infty(\mathbb{N})$ so that

$$a = \lim_{j \in \mathbb{J}} a_j, \quad b = \lim_{j \in \mathbb{J}} b_j$$

both exist. Of course, $a \in \bar{A}, b \in \bar{B}$. But now we're in precisely the position to which Definition 4.3(vii) is applicable: $a_k, b_k \in G_k, G_k \cong S_{n_k}$. Hence $a = b$. But $\overline{A} \cap \overline{B} = \emptyset$. OOPS! The denial of item (v) leads to unnecessary chaos. □

Let (S_n) be a sequence of open concentric balls with radii tending to zero. For any open set $A \subseteq Q$, define

$$l_n(A) = \frac{h(A, S_n)}{h(Q, S_n)}.$$

Then

$$h(A, S_n) \leq h(A, Q) \cdot h(Q, S_n)$$

and

$$h(Q, S_n) \leq h(Q, A) \cdot h(A, S_n)$$

tell us that

$$\frac{1}{h(Q, A)} \leq l_n(A) \leq h(A, Q).$$

Therefore $(l_n(A))$ is a bounded sequence of real numbers, each of whose terms exceeds the fixed positive number $1/h(Q, A)$. Let LIM be a Banach limit, that is,

$$\text{LIM} \in B_{l_\infty^*},$$

and LIM satisfies

$$\liminf x \leq \text{LIM}(x) \leq \limsup x \tag{8}$$

for each and every $x \in l_\infty$. Let

$$l(A) = \text{LIM}((l_n(A)))$$

for any open set $A \subseteq Q$. □

Note. We did not use the shift-invariance of LIM; rather we just used the fact that it satisfied (8).

Proposition 4.7. *If A and B are open sets, then*

(i) $0 < l(A) < \infty$, *as long as* $A \neq \emptyset$.

(ii) $A \subseteq B \Rightarrow l(A) \leq l(B)$.

(iii) $l(A \cup B) \leq l(A) + l(B)$.

(iv) $A \cong B \Rightarrow l(A) = l(B)$.

(v) *If* $d(A, B) > 0$, *then* $l(A \cup B) = l(A) + l(B)$.

Proof. To see Proposition 4.7(v), note that by Proposition 4.6(v), there exists an N such that for all $n \geq N$,

$$h(A \cup B, S_n) = h(A, S_n) + h(B, S_n).$$

From this we easily see that for all $n \geq N$,

$$l_n(A \cup B) = l_n(A) + l_n(B),$$

and (v) follows. □

Let $X \subseteq Q$. Define $\lambda(X)$ as follows:

$$\lambda(X) = \inf\left\{ \sum_n l(A_n) : X \subseteq \bigcup_n A_n,\ A_n \text{ open} \right\}.$$

Here are the fundamental properties of λ.

Theorem 4.8. *Let Q be a compact metric space. Then*

(i) $0 \leq \lambda(X)$.

(ii) *If X is a nonempty open subset of Q, then* $0 < \lambda(X) < \infty$.

(iii) $X \subseteq Y \subseteq Q \Rightarrow \lambda(X) \leq \lambda(Y)$.

(iv) $X \subseteq \bigcup_n X_n \Rightarrow \lambda(X) \leq \sum_n \lambda(X_n)$.

(v) $X \cong Y \Rightarrow \lambda(X) = \lambda(Y)$.

(vi) $d(X, Y) > 0 \Rightarrow \lambda(X \cup Y) = \lambda(X) + \lambda(Y)$.

Proof. Items (i)–(iv) tell us that λ is an outer measure, item (v) assures us that λ respects congruence, and item (vi) says that λ is a "metric outer measure". It is a known consequence of λ's metric outer measure character that every Borel set $B \subseteq Q$ is λ-measurable.

Item (i) deserves comment. If X is open, then

$$\lambda(X) \leq l(X) < \infty.$$

If, in addition, $X \neq \emptyset$, then for any $\epsilon > 0$, we can find a sequence (A_n) of open sets so that $X \subseteq \bigcup_n A_n$ and

$$\lambda(X) + \epsilon \geq \sum_n l(A_n).$$

Now if S is an open ball centered at a point in X and a subset of X, then only finitely many of the A_n's, say $A_1, A_2, \ldots A_N$, are needed to cover S. Then

$$0 < l(S) \leq l(\overline{S}) \leq l(A_1 \cup \cdots \cup A_N) \leq \sum_n l(A_n) < \lambda(X) + \epsilon,$$

and $0 < \lambda(X)$ follows.

Item (vi), too, deserves proof—after all, it's item (v) that ensures that λ is a metric outer measure. By Proposition 4.7(iv), it suffices to show that $\lambda(X) + \lambda(Y) \leq \lambda(X \cup Y)$. If $d(X, Y) > 0$, then there are disjoint open sets U, V such that $d(U, V) > 0$ with $X \subseteq U$ and $Y \subseteq V$; this is so thanks to normality of metric spaces, if you please. Let $\epsilon > 0$. Pick a sequence (A_n) of open sets such that

$$X \cup Y \subseteq \bigcup_n A_n$$

and

$$\sum_n l(A_n) \leq \lambda(X \cup Y) + \epsilon.$$

Now each of the sets $A_n \cap U$, $A_n \cap V$ are open and

$$d(A_n \cap U, A_n \cap V) \geq d(U, V) > 0.$$

Hence by Proposition 4.7(v)

$$l((A_n \cap U) \cup (A_n \cap V)) = l(A_n \cap U) + l(A_n \cap V) \leq l(A_n),$$

where the last inequality follows since $A_n \cap U$ and $A_n \cap V$ are disjoint open sets whose union is a subset of A_n. Further

$$X \subseteq \bigcup_n (A_n \cap U), \quad Y \subseteq \bigcup_n (A_n \cap V).$$

So

$$\lambda(X) \leq \sum_n l(A_n \cap U), \quad \lambda(Y) \leq \sum_n l(A_n \cap V).$$

This in turns ensures that

$$\begin{aligned} \lambda(X) + \lambda(Y) &\leq \sum_n l(A_n \cap U) + \sum_n l(A_n \cap Y) \\ &\leq \sum_n \Big(l(A_n \cap U) + l(A_n \cap V) \Big) \\ &\leq \sum_n l(A_n). \end{aligned}$$

It follows that $\lambda(X) + \lambda(Y) \leq \lambda(X \cup Y)$. □

From this we know that λ is a metric outer measure on Q which assigns the same value to any pair of congruent subsets of Q. Hence the collection of λ-measurable sets is a σ-field of subsets of Q which contains the Borel σ-field, and on this σ-field, λ is countably additive and assigns congruent measurable sets the same measure.

Exercise 4.9. Notice that the full strength of Banach limits was not used in the above proof. Indeed, what was used was that LIM satisfies that it is a linear functional on the space of all bounded real-valued sequences such that for any bounded real sequence x

$$\liminf x \leq \text{LIM}(x) \leq \limsup x.$$

(i) If F is a positive linear functional on the Banach space of bounded real-valued sequences, then F satisfies

(9) $$\liminf x \leq F(x) \leq \limsup x.$$

(ii) Denote by $\beta\mathbb{N}$ the Čech–Stone compactification of the (discrete) space $\mathbb{N}$ of natural numbers. The defining characteristic of the compact space $\beta\mathbb{N}$ is that it contains $\mathbb{N}$ as a dense subspace and every bounded real-valued function defined on $\mathbb{N}$ has a unique continuous real-valued extension to all of $\beta\mathbb{N}$. So $C(\beta\mathbb{N})$ is isometrically isomorphic to ℓ_∞.

If we take a point $z \in \beta\mathbb{N}$ that's not in $\mathbb{N}$, then it defines a continuous linear functional F_z on $C(\beta\mathbb{N})$ (or what's the same ℓ_∞) via the formula
$$F_z(f) = f(z).$$
The functional F_z satisfies condition (9).

Remark 4.10. The two parts of this exercise combine to establish the beautiful result of Nakamura and Kakutani [**93**] to the effect that there are 2^c many generalized Banach limits—after all, there are 2^c many points in $\beta\mathbb{N}$.

The opening steps of Banach's proof of the existence of a Haar measure on a compact metrizable group follow Haar's argument. One should be aware that both Haar and Banach were interested in "bigger fish", and each dealt with the case of locally compact, separable metrizable underlying space. We opted to follow Banach's approach because it broaches the subject of homogeneous spaces (and the groups that act on them), and because of our functional analytic prejudices.

3. Saks' Proof of $C(Q)^*, Q$ a Compact Metric Space

Soon after the appearance of Saks' monograph [**111**], Saks published an alternative proof of the theorem of Banach [**112**] regarding positive linear functionals on $C(Q)$, Q a compact metric space. This proof relies on the theory of metric outer measures to ensure that the resulting measure is a Borel measure.

We set our notation. Let Q be a compact metric space (with metric d), $C(Q)$ is the Banach space of continuous real-valued functions defined on Q, and (to be consistent with Saks) Φ is a positive linear functional on $C(Q)$. Recall that
$$|\Phi(x)| \leq \Phi(1)$$

whenever $x \in B_{C(Q)}$, so Φ is a member of $C(Q)^*$ with norm $\Phi(1)$.

For any $q \in Q, r > 0$, denote by $U_r(q)$ and $B_r(q)$ the sets

$$U_r(q) = \{y \in Q : d(q,y) < r\}, \quad B_r(q) = \{y \in Q : d(q,y) \leq r\}.$$

Stage I. For $E \subseteq Q$, define $\lambda(E)$ by

$$\lambda(E) := \inf\{\Phi(x) : x \in C(Q), x(q) \geq \chi_E(q), \text{ for all } q \in Q\}.$$

Here's what's so about λ:

(i) if $A \subseteq B$, then $\lambda(A) \leq \lambda(B)$;

(ii) if $A, B \subseteq Q$, then $\lambda(A \cup B) \leq \lambda(A) + \lambda(B)$;

(iii) if $d(A,B) > 0$ (which is the same as $\overline{A} \cap \overline{B} = \emptyset$), then $\lambda(A \cup B) = \lambda(A) + \lambda(B)$.

Item (iii) demands comment and proof, even. Let $\epsilon > 0$. Pick $x \in C(Q)$ so

$$x(q) \geq \chi_{A \cup B}(q)$$

for all $q \in Q$ (where $A, B \subseteq Q$ satisfy $d(A,B) > 0$) and so that

$$\Phi(x) \leq \lambda(A \cup B) + \epsilon.$$

Next chose $h \in C(Q)$ so that $0 \leq h(q) \leq 1$ for all q, with $h(q) = 1$ for $q \in B, h(q) = 0$ for $q \in A$. Let $x_A = (1-h)x$ and $x_B = hx$. Both $x_A, x_B \in C(Q)$; also

$$\begin{aligned} x_A(q) &= \begin{cases} x(q) & \text{if } q \in A \\ 0 & \text{if } q \in B \end{cases} \geq \begin{cases} \chi_{A \cup B}(x) & \text{if } q \in A \\ 0 & \text{if } q \in B \end{cases} \\ &= \begin{cases} 1 & \text{if } q \in A \\ 0 & \text{if } q \in B \end{cases} = \chi_A(q) \end{aligned}$$

and

$$x_B(q) = \begin{cases} 0 & \text{if } q \in A \\ x(q) & \text{if } q \in B \end{cases} \geq \chi_B(q)$$

for all $x \in Q$. It follows that

$$\begin{aligned} \lambda(A) + \lambda(B) &\leq \Phi(x_A) + \Phi(x_B) \\ &= \Phi(x_A + x_B) \\ &= \Phi(x) \\ &\leq \lambda(A \cup B) + \epsilon. \end{aligned}$$

Stage II. Let $E \subseteq Q$, and define $\mu(E)$ by

$$\mu(E) := \inf\left\{\sum_n \lambda(G_n) : G_n \text{ is open}, E \subseteq \bigcup_n G_n\right\}.$$

Then μ is an outer measure on Q, a metric outer measure, with the added property that for any closed subset F of Q, $\mu(F) = \lambda(F)$. We'll establish this last claim. Suppose G_n is a sequence of open sets so that

$$F \subseteq \bigcup_n G_n.$$

Since F is closed, Q's compactness is inherited by F, and so we can find N such that

$$F \subseteq G_1 \cup \cdots \cup G_N.$$

It follows that

$$\lambda(F) \leq \sum_{i=1}^{N} \lambda(G_i) \leq \sum_n \lambda(G_n).$$

It follows that

$$\lambda(F) \leq \mu(F).$$

Now let $\epsilon > 0$ be given. Pick $x \in C(Q)$ so that $x(q) \geq \chi_F(q)$ for all q and $\Phi(x) \leq \lambda(F) + \epsilon$. Look at the open set

$$G = \{q \in Q : x(q) > (1+\epsilon)\}.$$

So $F \subseteq G$ since if $q \in F$, then $x(q) \geq \chi_F(x) = 1$. So

$$\begin{aligned}
\mu(F) &\leq \lambda(G) \text{ (since } G \text{ is open and } F \subseteq G) \\
&\leq \Phi(x) \text{ (by definition of } \lambda(G)) \\
&= \Phi(x) \\
&\leq (\lambda(F) + \epsilon).
\end{aligned}$$

Let $\epsilon \searrow 0$. Then $\mu(F) \leq \lambda(F)$ follows.

Stage III. Like all *metric* outer measures, μ has among its μ-measurable sets each and every Borel subset of Q. In particular, each $x \in C(Q)$ is μ-measurable, and of course, bounded. Hence each $x \in C(Q)$ is μ-integrable. Let's check $\Phi(x)$ *vis-à-vis* $\int x d\mu$. We'll show that

$$\Phi(x) \leq \int x \, d\mu$$

for every $x \in C(Q)$.

Take $x \in C(Q)$. Let $\epsilon > 0$. Realizing that $\mu(Q) = \Phi(1)$, we assume that $x(q) > 0$ for all $q \in Q$—just add an appropriate constant to x and note that this has the exact same effect on $\Phi(x)$ and $\int x d\mu$, leaving their relationship unchanged. Choose $\eta > 0$ so that if $d(q, q') \leq \eta$, then $|x(q) - x(q'| \leq \epsilon$. Cover Q by open balls $U_{r_1}(q_1), \ldots, U_{r_n}(q_n)$, of radii $r_1, \ldots, r_n$, with each $< \eta/2$ centered at $q_1, \ldots q_n$, respectively, with the added feature that

$$\mu(\{y \in Q : d(y, q_i) = r_i\}) = 0.$$

This last feature can be assumed since $\mu(Q) < \infty$, and so for a fixed $q_0 \in Q$, only countably many of the sets $\{y \in Q : d(y, q_i) = r\}$ can have positive μ-measure. Now that the U's are in place, set

$$\begin{aligned}
E_1 &= \overline{U_{r_1}(q_1)} \\
E_2 &= \overline{U_{r_2}(q_2)\backslash E_1} \\
E_3 &= \overline{U_{r_3}(q_3)\backslash(E_1 \cup E_2)} \\
&\vdots \\
E_n &= \overline{U_{r_n}(q_n)\backslash(E_1 \cup \cdots \cup E_{n-1})}.
\end{aligned}$$

Each set $E_1, \ldots, E_n$ is closed with diameter $\leq \eta$ and

$$Q = E_1 \cup \cdots \cup E_n.$$

What's more—and this is crucial—the E_i's overlap only in a set of μ-measure 0! Let $m_i = \min\{x(q) : q \in E_i\}$, $i = 1, 2, \ldots, n$. Consider

$$a(q) = \sum_{i=1}^{n} m_i \chi_{E_i}(q);$$

notice that

$$x(q) \geq a(q)$$

with μ-almost everywhere. Hence

$$\begin{aligned}
\int x\, d\mu &\geq \int \sum_{i=1}^{n} m_i \chi_{E_i}(q)\, d\mu \\
&= \sum_{i=1}^{n} m_i \mu(E_i) = \sum_{i-1}^{n} m_i \lambda(E_i).
\end{aligned}$$

For each $k = 1, \ldots, n$, pick $x_k \in C(Q)$ so that for all $q \in Q$, $x_k(q) \geq \chi_{E_k}(q)$ and

$$\mu(E_k) = \lambda(E_k) \geq \Phi(x_k) - \frac{\epsilon}{n m_k}.$$

Put

$$u = m_1 x_1 + \cdots + m_k x_k + \epsilon.$$

Since the oscillation of x on E_k is no more than ϵ, then

$$u(q) = \sum_{i=1}^{n} m_i x_i(q) + \epsilon \geq x(q).$$

It follows that

$$\begin{aligned}
\int x\,d\mu &\geq \sum_{i=1}^{n} m_i \mu(E_i) \\
&\geq \sum_{i=1}^{n} m_i \left(\Phi(x_i) - \frac{\epsilon}{n m_i} \right) \\
&= \Phi\left(\sum_{i=1}^{n} m_i x_i \right) - \epsilon \\
&= \Phi(u - \epsilon) - \epsilon \\
&= \Phi(u) - \Phi(\epsilon) - \epsilon \\
&\geq \Phi(x) - \epsilon(\Phi(1) + 1).
\end{aligned}$$

Let $\epsilon \searrow 0$ and, be happy, don't worry—after all,

$$\int x d\mu \geq \Phi(x)$$

for all $x \in C(Q)$. □

4. The Lebesgue Integral on Abstract Spaces

In this section we present Banach's approach to the Lebesgue integral in abstract spaces. Banach's approach is a Daniell-like construction built using his clear and deep understanding of lim sup's and lim inf's. His starting point is a positive linear functional f acting on a vector lattice $\mathcal{C}$ of real-valued functions defined on some set K; we suppose (with Banach) that the functional satisfies a kind of *Bounded Convergence theorem* (BCθ) on $\mathcal{C}$. It is important to note that in the previous section we presented Banach's famous result characterizing weakly convergent sequences in spaces $C(Q)$, Q a compact metric space; it follows from this that should the initial vector lattice $\mathcal{C}$ be such a $C(Q)$, then every positive linear functional satisfies the (BCθ) hypothesis.

After an initial discussion of technical consequences of the (BCθ) hypothesis involving lim sup's and lim inf's of functions in $\mathcal{C}$, Banach introduces an upper and a lower integral. Were we doing measure theory, this piece of the puzzle would be concerned with properties of outer and inner measures generated from an initial set function.

Next the class of integrable functions is isolated; it is identified as those real-valued functions for which the upper and lower integrals coincide and are simultaneously finite. The classical *Monotone and Dominated Convergence Theorems* are derived, and all is well with the world.

We follow with a discussion of what the construction does in the all-important case that the initial vector lattice $\mathcal{C} = C(Q)$, with Q a compact metric space.

It is noteworthy that this construction of Banach led him to a description of $C(Q)^*$. It is impossible to tell for certain (but easy to imagine) what Saks thought of Banach's "construction" of $C(Q)^*$. It appeared after all, as an appendix to Saks' classic monograph *Theory of the Integral* [**111**], yet uses *practically* none of the material from the monograph! Whatever Saks thought, he soon presented an elegant proof of Banach's result about $C(Q)^*$ in the Duke Mathematics Journal [**112**]. We presented Saks' proof in Section 3.

4.1. A Start. (The numbering and notation throughout this section is consistent with Banach's [**111**].) Let $\mathcal{C}$ denote a vector lattice of real-valued functions defined on a set Q (i.e., if $x, y \in \mathcal{C}$, then so is $x \vee y = \inf\{x, y\}$ and $x \wedge y = \sup\{x, y\}$). A linear functional f on $\mathcal{C}$ is a *positive linear functional* if $f(x) \geq 0$ for any $x \in \mathcal{C}, x \geq 0$.

Throughout this section, we will suppose f is a positive linear functional on $\mathcal{C}$ satisfying

$$(\mathrm{BC}\theta) \quad \begin{cases} \text{if } (x_n) \subseteq \mathcal{C}, M \in \mathcal{C}, \text{ with } |x_n| \leq M \text{ and } \lim_n x_n(t) = 0 \\ \text{for all } t \in Q, \text{ then} \lim_n f(x_n) = 0. \end{cases}$$

If $\mathcal{C} = C(Q)$, Q a compact metric space, then any positive linear functional f on $\mathcal{C}$ satisfies (BCθ) thanks to Theorem 4.24. This is an example well worth keeping in mind.

1° If $(x_n) \subseteq \mathcal{C}, m \in \mathcal{C}, z \geq 0, x_n \geq m$, and $\liminf_n x_n \geq z$, then

$$\lim_n x_n - |x_n| = 0.$$

Now $\liminf_n x_n \geq z$ ensures $\liminf_n x_n(t) \geq 0$ for each $t \in Q$; that is, for each $t \in Q$,

$$\lim_{n \to \infty} \inf\{x_n(t), x_{n+1}(t), \ldots\} \geq 0.$$

So, given $\eta > 0$ there's $n = n(\eta)$ so that for all $k \geq n$,

$$x_k(t) \geq -\eta. \tag{10}$$

Look at $x_n - |x_n|$:

$$(x_n - |x_n|)(t) = \begin{cases} 0, & \text{if } x_n(t) \geq 0 \\ 2x_n(t), & \text{if } x_n(t) < 0. \end{cases}$$

Naturally, $(x_n - |x_n|)(t) \leq 0$ for all t. So, if $\lim_n(x_n - |x_n|)$ exists, it must be less than or equal to 0.

Let's check on contrary possibilities. Can it be that

$$\liminf_n (x_n - |x_n|)(t_0) < 0$$

for some $t_0 \in Q$? Let's suppose that this is possible. Notice since $m \in \mathcal{C}$

$$\liminf_n (x_n - |x_n|)(t_0) \geq \inf\{2m(t_0), 0\} > -\infty.$$

If

$$\liminf_n (x_n - |x_n|)(t_0) < 0,$$

it's because there's a subsequence $(x_n^{'})$ of x_n so that for some $\epsilon_0 > 0$

$$2x_n^{'}(t_0) = x_n^{'}(t_0) - |x_n^{'}(t_0)| < -\epsilon_0$$

for all n. It follows that

$$x_n^{'}(t_0) < \frac{-\epsilon_0}{2}$$

for all n. But if n is BIG we can arrange

$$x_n^{'}(t_0) > \frac{-\epsilon_0}{4}$$

(that's what our opening words (10) of wisdom guarantee). Aha! While

$$x_n^{'}(t_0) > \frac{-\epsilon_0}{4},$$

we also have

$$x_n^{'}(t_0) < \frac{-\epsilon_0}{2}.$$

Drawing the conclusion from this that $\frac{\epsilon_0}{2} < \frac{\epsilon_0}{4}$ is easy and leaves us with a clear-cut contradiction to all that is right in our world. We conclude that

$$\liminf_n (x_n - |x_n|) \geq 0.$$

We know that

$$\limsup_n (x_n - |x_n|) \leq 0,$$

and so 1° follows. □

2° If $(x_n) \subseteq \mathcal{C}, m \in \mathcal{C}, x_n \geq m$, and $\liminf_n x_n \geq 0$, then

$$\liminf_n f(x_n) \geq 0.$$

Again start with a look at $x_n - |x_n|$. As before since $2m \in \mathcal{C}$,

$$-\infty < \inf\{2m, 0\} \leq x_n - |x_n| \leq 0.$$

By 1°, we know that

$$\lim_n (x_n - |x_n|)(t) = 0$$

for each $t \in Q$, and since $\mathcal{C}$ is a lattice, we can use the the (BCθ) condition on f to conclude

$$\lim_n f(x_n - |x_n|) = 0.$$

Can $\liminf_n f(x_n) < 0$? If so, then it's because there is a subsequence (x_n') of (x_n) and an $\epsilon_0 > 0$ so that

$$f(x_n') < -\epsilon_0$$

for all n. We can (and do) assume that

$$\lim_n f(x_n')$$

exists as well. But now

$$\lim_n f(x_n'), \lim_n f(x_n' - |x_n'|)$$

both exist, and so

$$\lim_n f(|x_n'|)$$

exists too with

$$\begin{aligned}\lim_n f(|x_n'|) &= \lim_n (f(|x_n'| - x_n') + f(x_n')) \\ &= \lim_n f(x_n') \leq -\epsilon_0,\end{aligned}$$

which is not possible, and so we have 2°. □

4.2. Upper and lower integrals. Let $\mathcal{L}^*$ denote the set of all real-valued functions z on Q for which there exist two sequences $(x_n), (y_n) \subseteq \mathcal{C}$ such that

$$\limsup_n y_n \leq z \leq \liminf_n x_n.$$

$\mathcal{L}^*$ is a linear space containing $\mathcal{C}$ since $\mathcal{C}$ is a vector lattice.

Given $z \in \mathcal{L}^*$, the *upper integral of* z, denoted by $\overline{\int}(z)$, is defined by

$$\inf\{\liminf_n f(x_n) : \text{there exists } m \in \mathcal{C}, (x_n) \subseteq \mathcal{C}, x_n \geq m, \liminf_n x_n \geq z\};$$

the *lower integral of* z, denoted by $\underline{\int}(z)$, is defined by

$$\sup\{\limsup_n f(x_n) : \text{there exists } M \in \mathcal{C}, (x_n) \subseteq \mathcal{C}, x_n \leq M, \limsup_n x_n \leq z\}.$$

Obviously

$$\underline{\int} z = -\overline{\int}(-z).$$

Note. In each of the above definitions we can suppose that $\lim_n f(x_n)$ exists and is real valued since lim inf's and lim sup's are taken over sequences which are eventually finite. So we replace $\liminf_n f(x_n)$ and $\limsup_n f(x_n)$ with $\lim_n f(x_n)$ throughout.

From the definitions we have

3° If $z \in \mathcal{L}^*, z \geq 0$, and $\overline{\int}(z) < P < \infty$, then we can find $(x_n) \subseteq \mathcal{C}, x_n \geq 0, \liminf x_n \geq z$ with $f(x_n) < P$ for all n.

The value of 3° is found in the accessibility it affords us to epsilonics; since (with Banach) we frequently call on computing lim sup's and lim inf's. This is a critical aid.

Lemma 4.11. *For any* $x \in \mathcal{C}$, $\bar{\int}(x) = f(x)$.

Proof. On the one hand, we can let $x_n = x$ for all n and $m = x$. This done, we plainly have

$$\liminf_n x_n \geq x, \quad \text{and} \quad x_n \geq m;$$

hence

$$\bar{\int}(x) \leq \liminf_n f(x_n) = f(x).$$

On the other hand, if $(x_n) \subseteq \mathcal{C}$ and $m \in \mathcal{C}$ with

$$\liminf_n x_n \geq x, \quad \text{and} \quad x_n \geq m,$$

then

$$\liminf_n (x_n - x) \geq 0 \quad \text{and} \quad x_n - x \geq m - x.$$

2° steps in to say

$$\text{“}0 \leq \liminf_n f(x_n - x) = \liminf_n f(x_n) - f(x)\text{”};$$

we see that

$$f(x) \leq \liminf_n f(x_n),$$

and with this we conclude

$$f(x) \leq \bar{\int}(x). \qquad \square$$

Lemma 4.12. *If* $z_1, z_2 \in \mathcal{L}^*$ *with* $\bar{\int}(z_1), \bar{\int}(z_2) < \infty$, *then* $\bar{\int}(z_1 + z_2) \leq \bar{\int}(z_1) + \bar{\int}(z_2)$.

Proof. Suppose P_1, P_2 are numbers such that

$$\bar{\int}(z_1) < P_1 \quad \text{and} \quad \bar{\int}(z_2) < P_2.$$

There are sequences $(x_n^{(1)}), (x_n^{(2)}) \subseteq \mathcal{C}$ and functions $m_1, m_2 \in \mathcal{C}$ such that

$$\liminf_n x_n^{(1)} \geq z_1, \qquad x_n^{(1)} \geq m_1 \text{ for all } n,$$
$$\liminf_n x_n^{(2)} \geq z_2, \qquad x_n^{(2)} \geq m_2 \text{ for all } n,$$

and

$$\lim_n f(x_n^{(1)}) < P_1, \ \lim_n f(x_n^{(2)}) < P_2.$$

Letting $x_n = x_n^{(1)} + x_n^{(2)}$ and $m = m_1 + m_2$, we see that

$$\liminf_n x_n \geq z_1 + z_2, \ x_n \geq m.$$

It follows that

$$\overline{\int}(z_1 + z_2) \leq \lim_n f(x_n) = \lim_n f(x_n^{(1)}) + \lim_n f(x_n^{(2)}) < P_1 + P_2.$$

Enough said. $\square$

Lemma 4.13. *For any $z \in \mathcal{L}^*$,*

$$\underline{\int}(z) \leq \overline{\int}(z).$$

Proof. There is nothing to prove if $\overline{\int}(z) = +\infty$. If $\overline{\int}(-z) = +\infty$, then $\underline{\int}(z) = -\overline{\int}(-z)$, so again there is nothing to prove. If $\overline{\int}(z), \overline{\int}(-z) < \infty$, then Lemma 4.12 kicks in to give

$$0 = f(0) = \overline{\int}(0) = \overline{\int}(z - z) \leq \overline{\int}(z) + \overline{\int}(-z)$$

so that

$$\underline{\int}(z) = -\overline{\int}(-z) \leq \overline{\int}(z).$$

$\square$

Lemma 4.14. *If $z \in \mathcal{L}^*$ and $\overline{\int}(z) < \infty$, then*

$$\overline{\int}\left(\frac{z + |z|}{2}\right) < \infty$$

and

$$\overline{\int}(z) = \overline{\int}\left(\frac{z + |z|}{2}\right) + \overline{\int}\left(\frac{z - |z|}{2}\right).$$

Proof. Suppose $\overline{\int}(z) < P < \infty$. Find $m \in \mathcal{C}$ and $(x_n) \subseteq \mathcal{C}$ so that $x_n \geq m$ for all n, and $\liminf_n x_n \geq z$, $\lim_n f(x_n) < P$. Notice that if $x_n \geq m$, then

$$\frac{x_n - |x_n|}{2} \geq \frac{m - |m|}{2};$$

this can be seen by a simple analysis of cases. Hence

$$\begin{aligned} f\left(\frac{x_n + |x_n|}{2}\right) &= f\left(x_n - \left(\frac{x_n - |x_n|}{2}\right)\right) \\ &= f(x_n) - f\left(\frac{x_n - |x_n|}{2}\right) \\ &\leq f(x_n) - f\left(\frac{m - |m|}{2}\right) \end{aligned}$$

and

$$\begin{aligned}\bar{\int}\left(\frac{z+|z|}{2}\right) &\leq \liminf_n f\left(\frac{x_n+|x_n|}{2}\right)\\ &\leq \lim_n f(x_n) - f\left(\frac{m+|m|}{2}\right)\\ &< \infty.\end{aligned}$$

Now

$$\begin{aligned}P &> \lim_n f(x_n)\\ &\geq \liminf_n f\left(\frac{x_n+|x_n|}{2}\right) + \liminf_n f\left(\frac{x_n-|x_n|}{2}\right)\\ &\geq \bar{\int}\left(\frac{z+|z|}{2}\right) + \bar{\int}\left(\frac{z-|z|}{2}\right);\end{aligned}$$

it follows from P's arbitrary nature among members $> \bar{\int}(z)$ that

$$\bar{\int}(z) \geq \bar{\int}\left(\frac{z+|z|}{2}\right) + \bar{\int}\left(\frac{z-|z|}{2}\right).$$

Lemma 4.12 tells the rest of this tale. □

Two more lemmas are plain and worth mentioning.

Lemma 4.15. *If $z_1, z_2 \in \mathcal{L}^*$ satisfy $z_1 < z_2$, then $\bar{\int}(z_1) \leq \bar{\int}(z_2)$; in particular if $z \in \mathcal{L}^*$ and $z \geq 0$, then $\bar{\int}(z) \geq 0$.*

Lemma 4.16. *If $z \in \mathcal{L}^*$, then $\bar{\int}(\lambda z) = \lambda \bar{\int}(z)$ for any real number $\lambda \geq 0$.*

4.3. The Integral. Let $\mathcal{L}$ be the set

$$\mathcal{L} = \{z \in \mathcal{L}^* : \bar{\int}(z) = \underline{\int}(z), \text{ with both finite}\}.$$

Lemma 4.17. *$\mathcal{L}$ is a linear space and $\bar{\int}$ is a linear functional on $\mathcal{L}$. Moreover $\mathcal{C} \subseteq \mathcal{L}$ and $\bar{\int}$ extends f.*

Lemma 4.18. *If $z \in \mathcal{L}$, then $|z| \in \mathcal{L}$, that is, $\mathcal{L}$ is a vector lattice.*

Proof. Since

$$|z| = \left(\frac{z+|z|}{2}\right) + \left(\frac{|z|-z}{2}\right),$$

it's enough (thanks to $\mathcal{L}$'s linearity) to show that $\frac{z+|z|}{2}, \frac{z-|z|}{2}$ both belong to $\mathcal{L}$ if $z \in \mathcal{L}$.

We recall that Lemma 4.14 ensures that if $z \in \mathcal{L}$, then $\bar{\int}\left(\frac{z+|z|}{2}\right) < \infty$ and $\bar{\int}\left(\frac{z-|z|}{2}\right) > -\infty$, as well as

$$\bar{\int}(z) = \bar{\int}\left(\frac{z+|z|}{2}\right) + \bar{\int}\left(\frac{z-|z|}{2}\right).$$

Symmetry (applying Lemma 4.14 to $-z$ and using $\bar{\int} - z = \underline{\int} z)$ shows that $\underline{\int}\left(\frac{z+|z|}{2}\right) < \infty$ and $\underline{\int}\left(\frac{z-|z|}{2}\right) > -\infty$ as well as

$$\underline{\int}(z) = \underline{\int}\left(\frac{z+|z|}{2}\right) + \underline{\int}\left(\frac{z-|z|}{2}\right).$$

But $z \in \mathcal{L}$ says that $\bar{\int} z = \underline{\int} z$, and so

$$\begin{aligned}
&\left[\bar{\int}\left(\frac{z+|z|}{2}\right) - \underline{\int}\left(\frac{z+|z|}{2}\right)\right] + \left[\bar{\int}\left(\frac{z-|z|}{2}\right) - \underline{\int}\left(\frac{z-|z|}{2}\right)\right] \\
&\quad = \left[\bar{\int}\left(\frac{z+|z|}{2}\right) + \bar{\int}\left(\frac{z-|z|}{2}\right)\right] - \left[\underline{\int}\left(\frac{z+|z|}{2}\right) + \underline{\int}\left(\frac{z-|z|}{2}\right)\right] \\
&\quad = \bar{\int} z - \underline{\int} z = 0.
\end{aligned}$$

Now Lemma 4.13 kicks in to say that the *finite* quantities

$$\bar{\int}\left(\frac{z+|z|}{2}\right), \underline{\int}\left(\frac{z+|z|}{2}\right)$$

must, in fact, be equal, and the finite quantities

$$\bar{\int}\left(\frac{z-|z|}{2}\right), \underline{\int}\left(\frac{z-|z|}{2}\right)$$

follow suit. □

An old friend is next on the agenda.

Lemma 4.19 (Monotone Convergence Theorem).

$$(\text{MC}\theta)\ \begin{cases} \textit{if } (z_n) \subseteq \mathcal{L}, z_n \le z_{n+1} \textit{ for all } n \textit{ and } z = \lim_n z_n \textit{ with} \\ \lim_n \bar{\int}(z_n) < \infty, \textit{ then } z \in \mathcal{L} \textit{ and } \bar{\int}(z) = \lim_n \bar{\int}(z_n). \end{cases}$$

Proof. We can, and do, assume $z_1 = 0$; otherwise subtract z_1 from each of the z_n's. Next note that $z \ge z_n$ for all n and so since $\bar{\int}(z_n) = \underline{\int}(z_n)$, we have

$$\underline{\int}(z) \ge \lim_n \underline{\int}(z_n) = \lim_n \bar{\int}(z_n). \tag{11}$$

Let $\epsilon > 0$. Let $w_n = z_{n+1} - z_n \geq 0$. For each n find $(w_k^{(n)}) \subseteq \mathcal{C}$ with $w_k^{(n)} \geq 0, \lim_k w_k^{(n)} \geq w_n$ and

$$f(w_k^{(n)}) \leq \overline{\int}(w_n) + \frac{\epsilon}{2^n},$$

say. Write

$$y_n = w_n^{(1))} + \cdots + w_n^{(n)};$$

so

$$\begin{aligned} y_1 &= w_1^{(1)} \in \mathcal{C} \\ y_2 &= w_2^{(1)} + w_2^{(2)} \in \mathcal{C} \\ &\vdots \\ y_n &= w_n^{(1)} + w_n^{(2)} + \cdots + w_n^{(n)} \in \mathcal{C}. \end{aligned}$$

Since $y_n = w_n^{(1)} + w_n^{(2)} + \cdots + w_n^{(n)} \geq w_1 + w_2 + \cdots w_n$,

$$\liminf_n y_n \geq \liminf_n (w_1 + w_2 + \cdots + w_n) = \sum_n w_n = z.$$

At the same time,

$$\begin{aligned} f(y_n) &= f(w_n^{(1)}) + \cdots + f(w_n^{(n)}) \\ &\leq \overline{\int}(w_1) + \frac{\epsilon}{2} + \overline{\int}(w_2) + \frac{\epsilon}{2^2} + \cdots + \overline{\int}(w_n) + \frac{\epsilon}{2^n} \\ &< \overline{\int}(w_1) + \cdots + \overline{\int}(w_n) + \epsilon \\ &= \overline{\int}(z_2) + \overline{\int}(z_3 - z_2) + \cdots + \overline{\int}(z_{n+1} - z_n) + \epsilon \\ &= \overline{\int}(z_2 + (z_3 - z_2) + \cdots + (z_{n+1} - z_n)) + \epsilon \\ &= \overline{\int}(z_{n+1}) + \epsilon \leq \lim_n \overline{\int}(z_n) + \epsilon \quad \text{(by Lemma 4.15)}. \end{aligned}$$

It follows that

$$\overline{\int}(z) \leq \liminf_n f(y_n) \leq \lim_n \overline{\int}(z_n) + \epsilon,$$

and by ϵ's arbitrary nature,

$$\begin{aligned} 0 &\leq \overline{\int}(z) \leq \lim_n \overline{\int}(z_n) \\ &\leq \underline{\int}(z) \text{ by (11)} \\ &\leq \overline{\int}(z) < \infty. \end{aligned}$$

Therefore
$$\bar{\int}(z) = \lim_n \bar{\int}(z_n). \qquad \square$$

Another old friend:

Lemma 4.20 (Dominated Convergence Theorem). *Suppose $(z_n) \subseteq \mathcal{L}, M \in \mathcal{L}$ and $|z_n| \leq M$. Then*
$$g = \liminf_n z_n, \; h = \limsup_n z_n \in \mathcal{L}$$
with
$$\bar{\int}(g) \leq \liminf_n \bar{\int}(z_n) \leq \limsup_n \bar{\int}(z_n) \leq \bar{\int}(h).$$
Consequently,
$$(DC\theta) \begin{cases} \text{Suppose } (z_n) \subseteq \mathcal{L}, M \in \mathcal{L} \text{ satisfy } |z_n| \leq M. \\ \text{If } z(t) = \lim_n z_n(t) \text{ for each } t \in Q, \text{ then } z \in \mathcal{L} \text{ and} \\ \bar{\int}(z) = \lim_n \bar{\int}(z_n). \end{cases}$$

Proof. For each i and for each $j \geq i$, write
$$g_{ij} = \min\{z_i, z_{i+1}, \ldots, z_j\}.$$
Then the sequence $(g_{ij})_{j=i}^{\infty}$ is decreasing, each member belongs to $\mathcal{L}$, and so the sequence $(M - g_{ij})_{j=i}^{\infty}$ is an increasing sequence of members of $\mathcal{L}$. (MCθ) guarantees that if
$$g_i = \lim_j g_{ij},$$
then
$$M - g_i \in \mathcal{L}$$
and
$$\bar{\int}(M - g_i) = \lim_j \bar{\int}(M - g_{ij});$$
that is, $g_i \in \mathcal{L}$ and
$$\bar{\int}(g_i) = \lim_j \bar{\int}(g_{ij}).$$
Applying (MCθ) again reveals
$$g = \liminf_n z_n = \lim_j g_{ij} \in \mathcal{L}$$
with
$$\bar{\int}(g) = \lim_i \bar{\int}(g_i) \leq \liminf_n \bar{\int}(z_n). \qquad \square$$

Lemma 4.21. *If $z \in \mathcal{L}, z \geq 0$ and $\bar{\int} z = 0$, then whenever the function x satisfies $|x| \leq z$, we have that $x \in \mathcal{L}$ and $\bar{\int} x = 0$.*

This is an immediate consequence of Lemma 4.15.

4.4. We have the integral. Now we turn to Banach's treatment of $C(Q)^*$, Q a compact metric space. We begin by recalling a few results from his classic book *Theorie des Operations Lineaires* [**7**].

Let X be a normed linear space. A linear functional on X is *bounded* if there is a $c > 0$ so that $|f(x) \leq c||x||$ for all $x \in X$. The appellation "bounded" pertains to f's boundedness on the closed unit ball

$$B_X = \{x \in X : ||x|| \leq 1\}$$

of X. It is easy to show that a linear functional's boundedness is tantamount to its continuity, and this, in turn, is assured by f's continuity at the origin. Then

$$||f|| = \sup_{x \in B_X} |f(x)|$$

is a norm on X^* and is in fact "complete". So Cauchy sequences in X^*, with respect to this just-defined norm, converge.

The Hahn–Banach theorem provides any normed linear space with lots of linear functionals in its dual.

- Let S be a linear subspace of the real normed linear space X, and let s^* be a continuous linear functional on S. Then there is an extension x^* of s^* to all of X so that $||s^*|| = ||x^*||$. We simply let $p(x) = ||s^*||\,||x||$ and apply the Hahn–Banach theorem to find a linear functional F defined on all of X with $F(x) \leq p(x)$ for all $x \in X$. Because
$$F(-x) \leq p(-x) = p(x),$$
we see that
$$F(x) \leq p(x) = ||s^*||\,||x||$$
for all $x \in X$, and so $F = x^*$ *is* in X^* and $||F|| \leq ||s^*||$. Since F extends s^*, $||s^*|| \leq ||F||$ also.

- Let x be a member of the (real) normed linear space X. Then there is an $x^* \in X^*$ with $||x^*|| = 1$ so that $x^*(x) = ||x||$. Indeed, let
$$S = \{\alpha x : \alpha \in \mathbb{R}\},$$
define $s^*(\alpha x) = \alpha||x||$, apply the Hahn–Banach theorem, and get x^*, the norm-preserving extension of s^*.

It is frequently the case that for a particular normed linear space the norm-topology is too coarse to uncover special and delicate phenomena peculiar to that space. Oft times such phenomena are best described using the weak topology.

Definition 4.22. A base for the *weak topology* of the normed linear space X is given by sets of the form

$$U(x; x_1^*, x_2^*, \ldots, x_n^*) = \{y \in X : |x_i^*(x-y)| < \epsilon, i = 1, \ldots, n\},$$

where $x \in X, x_1^*, \ldots, x_n^* \in X^*$, and $\epsilon > 0$.

The topology generated by this base is easily seen to be a Hausdorff linear topology so the operations of $(x,y) \to x+y$ and $(\lambda, x) \to \lambda x$ of $X \times X$ to X and $\mathbb{R} \times X$ to X, respectively, are continuous.

This topology is *never* metrizable and *never* complete for the infinite-dimensional normed linear spaces! Usually closures are *not* determined sequentially. Nevertheless, weak sequential convergence in special spaces often holds secrets regarding the inner nature of such spaces.

To be sure, if X is a normed linear space and (x_n) is a sequence of vectors in X, then we say that (x_n) *converges weakly* to $x \in X$ (sometimes denoted by $x = \text{weaklim}_n x_n$) if for each $x^* \in X^*$,

$$\lim_n x^*(x_n) = x^*(x).$$

It is an integral part of basic functional analysis to compute the duals of special normed linear spaces and using the specific character of the spaces involved to characterize when sequences are *weakly null* (tend to zero weakly).

We now turn our attention to life inside spaces of the form $l^\infty(Q)$, where Q is a set and $l^\infty(Q)$ denotes the normed linear space of all bounded real-valued functions x defined on Q, where x's norm is given by

$$||x||_\infty = \sup\{|x(q)| : q \in Q\}.$$

Now to bring tools such as the Hahn–Banach theorem to bear on the study of $l^\infty(Q)$, we need to know something about $l^\infty(Q)^*$. Banach knew a great deal about this (as did F. Riesz before him); he didn't formulate an *exact* description of $l^\infty(Q)^*$ but, nevertheless, understood the basics. To begin, if $x^* \in l^\infty(Q)^*$, then x^* is entirely determined by its values at members of $l^\infty(Q)$ of the form χ_E where $E \subseteq Q$; after all, simple functions are dense in $l^\infty(Q)$. Now if $F(E) = x^*(\chi_E)$, then F is a bounded finitely additive real-valued measure on 2^Q, the collection of all subsets of Q. If we define $|F|$ by

$$|F|(E) = \sup\{F(S) : S \subseteq E\}$$

for $E \subseteq Q$, then $|F|$ is a nonnegative real-valued function defined on 2^Q and $|F(E)| \leq |F|(E)$ for each $E \subseteq Q$. (Remember: $\chi_\emptyset = 0$ so $F(\emptyset) = x^*(\chi_\emptyset) = x^*(0) = 0$.) What's more, $|F|$ is also finitely additive! If $G \subseteq E_1 \cup E_2$ where

E_1 and E_2 are disjoint subsets of Q, then $G = (G \cap E_1) \cup (G \cap E_2)$, and so

$$\begin{aligned} F(G) &= F((G \cap E_1) \cup (G \cap E_2)) \\ &= F(G \cap E_1) + F(F \cap E_2) \subseteq |F|(E_1) + |F|(E_2). \end{aligned}$$

It follows that

$$|F|(E_1 \cup E_2) \le |F|(E_1) + |F|(E_2).$$

On the other hand, if E_1 and E_2 are disjoint subsets of Q, then for any $\epsilon > 0$ we can pick $G_1 \subseteq E_1$ and $G_2 \subseteq E_2$ so

$$|F|(E_1) \le F(G_1) + \frac{\epsilon}{2}, \ |F|(E_2) \le F(G_2) + \frac{\epsilon}{2}.$$

But now

$$\begin{aligned} |F|(E_1) + |F|(E_2) &\le F(G_1) + \frac{\epsilon}{2} + F(G_2) + \frac{\epsilon}{2} \\ &= F(G_1) + F(G_2) + \epsilon \\ &= F(G_1 \cup G_2) + \epsilon \ \text{(since } F \text{ is finitely additive)} \\ &\le |F|(E_1 \cup E_2) + \epsilon; \end{aligned}$$

since $\epsilon > 0$ was arbitrary, we see that

$$|F|(E_1) + |F|(E_2) \le |F|(E_1 \cup E_2)$$

whenever E_1 and E_2 are disjoint subsets of Q.

Here's the punch line: if $x^* \in l^\infty(Q)^*$, then x^* defines a bounded finitely additive measure on 2^Q—call this measure F. From F we generate $|F|$, all of whose values are nonnegative; $|F| - F$ is also a nonnegative bounded real-valued finitely additive map on 2^Q and $F = |F| - (|F| - F)$. So F is the difference of *nonnegative* bounded finitely additive maps on 2^Q. In turn, such nonnegative additive bounded maps on 2^Q define positive linear functionals on $l^\infty(Q)$, functionals that are necessarily bounded linear functionals.

Why is this last statement so? Well suppose $G : 2^Q \to [0, \infty)$ is finitely additive. If $A \subseteq B \subseteq Q$, then $B = A \cup (B \backslash A)$ so

$$G(B) = G(A \cup (B \backslash A)) = G(A) + G(B \backslash A) \ge G(A);$$

it follows that for any $E \subseteq Q$, $G(E) \le G(Q)$, and G is bounded by $G(Q)$. Moreover if $E_1, \ldots, E_k$ are pairwise disjoint subsets of Q and $a_1, \ldots, a_n \in \mathbb{R}$,

then

$$\begin{aligned}\left|\sum_{i\le n} a_i G(E_i)\right| &\le \sum_{i\le n} |a_i| G(E_i) \le \sup_{1\le i\le n} |a_i| \sum_{i\le n} G(E_i)\\ &= \left\|\sum_{i\le n} a_i \chi_{E_i}\right\|_\infty G\Big(\bigcup_{i\le n} E_i\Big)\\ &\le G(Q)\left\|\sum_{i\le n} a_i\chi_{E_i}\right\|_\infty,\end{aligned}$$

and so G determines a linear functional

$$\sum a_i \chi_{E_i} \to \sum a_i G(E_i)$$

on the simple functions which is "bounded" there on. This bounded linear functional extends to $l^\infty(Q)$ in a bounded linear fashion, a positive functional to be sure.

Now $x \in B_{l^\infty(Q)}$ means $|x(q)| \le 1$ for all $q \in Q$; it follows that $-1 \le x(q) \le 1$ for all $q \in Q$. If f is a positive linear functional defined on $l^\infty(Q)$, so $f(x) \ge 0$ whenever $x(q) \ge 0$ for all $q \in Q$, then $f(-1) \le f(x) \le f(1)$, or $|f(x)| \le f(1)$, whenever $x \in B_{l^\infty(Q)}$. Positive linear functionals on $l^\infty(Q)$ are *bounded* linear functionals.

Whenever Q is infinite, $\ell_\infty(Q)$ is (highly) nonseparable. Nevertheless, using his mastery of lim sup's and lim inf's, Banach was able to establish the following stunning characterization of weakly convergent sequences in $\ell_\infty(Q)$, a characterization which is sequential in all regards. Banach limits play a huge role.

Theorem 4.23 (Banach). *Let Q be a (nonempty) set, and let (x_n) be a (uniformly) bounded sequence in $l^\infty(Q)$. Then (x_n) is weakly null if and only if*

$$\lim_n \liminf_k |x_n(q_k)| = 0 \tag{12}$$

for each sequence (q_k) of points in Q.

Proof. *Necessity.* Suppose to the contrary that there is a sequence of points (q_k) in Q such that

$$\limsup_n \liminf_k |x_n(q_k)| > \alpha > 0$$

for some α. Then unraveling the meaning of lim sup's, we can find a strictly increasing sequence (n_j) of positive integers such that

$$\liminf_k |x_{n_j}(q_k)| > \alpha > 0$$

for each j. Now turning to the meaning of lim inf, we find a subsequence (q_{k_m}) of (q_k) such that

$$|\lim_m x_{n_j}(q_{k_m})| > \alpha > 0$$

for each j. Let $x^* \in l^\infty(Q)$ be given by

$$x^*(x) = \mathrm{LIM}((x(q_{k_m}))_m)$$

where $\mathrm{LIM} \in l^\infty(\mathbb{N})^*$ is a Banach limit. Then for each j,

$$|x^*(x_{n_j})| > \alpha$$

and so

$$\limsup_n |x^*(x_n)| > \alpha > 0.$$

It follows that (x_n) is *not* weakly null in $l^\infty(Q)$.

Sufficiency. By remarks preceding the statement of this theorem, to test (x_n)'s weak nullity, it suffices to check the action of $x^* \in l^\infty(Q)^*$, for x^* a positive linear functional of norm 1, on the sequence (x_n). So suppose x^* is such a functional, with

$$\limsup_n x^*(x_n) > \alpha > 0,$$

where (12) holds

$$\lim_n \liminf_k |x_n(q_k)| = 0$$

for each sequence (q_k) of points in Q. Let s_n be the sequence

$$s_n(q) = \begin{cases} x_n(q) & \text{if } x_n(q) \geq 0, \\ 0 & \text{otherwise,} \end{cases}$$

and let $t_n = x_n - s_n$.

One of $\limsup_n x^*(s_n)$ and $\limsup_n x^*(t_n)$ must exceed $\frac{\alpha}{2}$; after all,

$$x_n = s_n + t_n,$$

so

$$x^*(x_n) = x^*(s_n) + x^*(t_n),$$

ensuring that

$$\limsup_n x^*(x_n) = \limsup_n(x^*(q_n) + x^*(t_n)) \leq \limsup_n x^*(s_n) + \limsup_n x^*(t_n).$$

If we clip off the "bottoms" of s_n by defining

$$y_n(q) = \begin{cases} s_n(q) & \text{if } s_n(q) \geq \frac{\alpha}{6}, \\ 0 & \text{otherwise,} \end{cases}$$

then

$$||s_n - y_n||_\infty \leq \frac{\alpha}{6}.$$

What's more,

$$\begin{aligned}\limsup_n x^*(y_n) &= \limsup_n x^*(s_n - (s_n - y_n)) \\ &= \limsup_n (x^*(s_n) - x^*(s_n - y_n))\end{aligned}$$

so that

$$\limsup_n x^*(y_n) \geq \frac{\alpha}{2} - \frac{\alpha}{6} = \frac{\alpha}{3}.$$

Let

$$S_n = \left\{ q \in Q : |x_n(q)| \geq \frac{\alpha}{6} \right\},$$

and look at χ_{S_n}. Since

$$||y_n||_\infty \leq ||s_n||_\infty \leq ||x_n||_\infty \leq M,$$

say, we see that for any $q \in Q$

$$\chi_{S_n}(q) \geq \frac{y_n(q)}{M},$$

so that

$$x^*(M \cdot \chi_{S_n}) \geq x^*(y_n).$$

From this we see that

$$\limsup_n x^*(\chi_{S_n}) \geq \frac{\alpha}{3M} =: \beta > 0.$$

For $E \subseteq Q$, let $F(E) = x^*(\chi_E)$; of course $F \in l^\infty(Q)^\star$ and

$$\limsup_n F(S_n) > \beta.$$

Let n_1 be the smallest positive integer such that

$$\limsup_n F(S_n \cap S_{n_1}) > 0.$$

Such an n exists by the way! This is the crux of the matter! In fact, otherwise,

$$\lim_n F(S_n \cap S_k) = 0$$

for each k so (because F is additive)

$$\lim_n F\left(\bigcup_{j=1}^{k} (S_n \cap S_j)\right) = 0,$$

for each k as well.

Let $k_1 = 1$. Pick $m_1 > k_1$ so large that

$$F(S_{m_1}) > \beta$$

and

$$F(S_{m_1} \cap S_{k_1}) < \frac{\beta}{2}.$$

Let $k_2 > m_1$. Pick m_2 so large that

$$F(S_{m_2}) > \beta$$

and

$$F(S_{m_2} \cap (S_1 \cup \cdots \cup S_{k_2})) < \frac{\beta}{2}.$$

Continuing in this fashion, produce $k_1 < m_1 < k_2 < m_2 < \cdots$, with

$$F(S_{m_j}) > \beta$$

and

$$F(S_{m_j} \cap (S_1 \cup \cdots \cup S_{k_j})) < \frac{\beta}{2}.$$

Now disjointivity: let T_j be given by

$$T_j = S_{m_j} \setminus \left[S_{m_j} \cap \left(\bigcup_{i=1}^{k_j} S_i \right) \right].$$

By construction

$$F(T_j) > \frac{\beta}{2}$$

and this is a "no-no" since F takes disjoint sequences to 0.

So n_1 does indeed exist such that

$$\limsup_n F(S_n \cap S_{n_1}) > 0.$$

Believe it or not.

Once faith has been established for n_1 we see that there are $n_2 < n_3 < \cdots < n_k < \cdots$ so that

$$\limsup_n F(S_n \cap S_{n_1} \cap \cdots \cap S_{n_k}) > 0$$

for each k. The all-important point here is that for each k there is at least one point q_k so

$$q_k \in S_{n_1} \cap S_{n_2} \cap \cdots \cap S_{n_k}.$$

Of course if $j \geq k$, then $q_j \in S_{n_k}$, and so by how the S_n's were defined,

$$|x_{n_k}(q_j)| \geq \frac{\alpha}{6}.$$

It soon follows that

$$\limsup_n \liminf_j |x_n(q_j)| \geq \frac{\alpha}{6}.$$

This contradicts (12) and thus, the sufficiency is proven. □

This result is remarkable because it characterizes weak convergence in a highly nonseparable, nonmetrizable situation in terms of *sequences* (q_k) in Q.

Suppose Q is a nonvoid compact metric space. Then the space $C(Q)$ of all continuous real-valued functions defined on Q, equipped with the norm

$$||x||_\infty = \sup\{|x(q)| : q \in Q\}$$

is a closed linear subspace of $l^\infty(Q)$ (the space of all bounded real-valued functions on Q). We can give a much more succinct characterization of when a bounded sequence (x_n) is weakly null in $C(Q)$ than just what Banach has above. Indeed, and again this was observed by Banach, the following is so.

Theorem 4.24 (Banach). *A (uniformly) bounded sequence (x_n) in $C(Q)$ is weakly null if and only if*

$$\lim_n x_n(q) = 0$$

for each $q \in Q$.

Proof. If $(x_n) \subseteq C(Q)$ is weakly null and $q \in Q$, then using the point evaluation

$$\delta_q \in C(K)^*, \text{where } \delta_q(x) = x(q)$$

we see that

$$0 = \lim_n \delta_q(x_n) = \lim_n x_n(q).$$

Now assume that (x_n) is a (uniformly) bounded sequence in $C(Q)$ for which $\lim_n x_n(q) = 0$ for each $q \in Q$ and imagine that (x_n) is *not* weakly null in $C(Q)$. Of course the Hahn–Banach theorem tells us that (x_n) is not weakly null in $l^\infty(Q)$ either. So by Theorem 4.23 there must be a subsequence $(x_n^{'})$ of (x_n) and a sequence of points (q_k) in Q and an $\alpha > 0$ so that for each n

$$\liminf_k |x_n^{'}(q_k)| \geq \alpha > 0. \tag{13}$$

But (q_k) is a sequence in the compact (hence sequentially compact) metric space Q and so (q_n) has a subsequence $(q_k^{'})$ that converges to some $q_0 \in Q$. By (13) it must be that for all n

$$|x_n^{'}(q_0)| \geq \alpha > 0.$$

OOPS. □

This theorem of Banach actually applies to any compact Hausdorff space Ω; that is to say, *if Ω is a compact Hausdorff space and (x_n) is a uniformly bounded sequence in $C(\Omega)$ such that $\lim_n x_n(\omega) = 0$ for each $\omega \in \Omega$, then (x_n) is weakly null in $C(\Omega)$*. The usual proof of this relies on the extension to general Ω's of the representation of $C(\Omega)^*$ as the space of regular Borel measures on Ω.

However using a device by S. Eilenberg, this result can be deduced directly from Banach's result. How? Well, suppose Ω is a compact Hausdorff space and (x_n) is a uniformly bounded sequence in $C(\Omega)$ for which $\lim_n x_n(\omega) = 0$ for each $\omega \in \Omega$. Introduce an equivalence relation on points in Ω as follows: $\omega \sim \omega'$ if $x_n(\omega) = x_n(\omega')$ for all n. If we define the distance between the equivalence classes $[\omega]$ and $[\omega']$ containing ω and ω', respectively, by

$$D([\omega], [\omega']) = \sum_n |x_n(\omega) - x_n(\omega')|2^{-n},$$

then D is a metric on the space $\widetilde{\Omega}$ of equivalence classes modulo "$\sim$". What's more, the map $\omega \to [\omega]$ that takes a point ω into the equivalence class $[\omega]$ that contains ω is a continuous surjection. By the way, we defined $\sim$, each x_n lifts to an $\widetilde{x_n} \in C(\widetilde{\Omega})$; moreover, the range of $\widetilde{x_n}$ is precisely that of x_n. If we call q the map taking $\omega \in \Omega$ to $[\omega] \in \widetilde{\Omega}$, then q is a continuous surjection of Ω onto $\widetilde{\Omega}$; consequently the linear operation $q^o : C(\widetilde{\Omega}) \to C(\Omega)$ given by

$$q^o(f)(\omega) = f(q(\omega))$$

is an isometric embedding. The point is just this: for any $\omega \in \Omega$,

$$\lim_n \widetilde{x_n}([\omega]) = \lim_n x_n(\omega) = 0.$$

Hence by Banach's theorem, $(\widetilde{x_n})$ is weakly null in $C(\widetilde{\Omega})$. Since $x_n = q^o(\widetilde{x_n})$, the sequence (x_n) is also weakly null.

Now let's turn to Banach's approach to integration. Start with $\mathcal{C} = C(Q)$, Q a compact metric space. If f is a positive linear functional on $\mathcal{C}$, then $|f(x)|$ is bounded by $f(1)$ so long as $x \in B_{\mathcal{C}}$. Why? Well if $x \in B_{\mathcal{C}}$ then $|x| \leq 1$ and $|x(q)| \leq 1$ for all $q \in Q$. So $x \in B_{\mathcal{C}}$ means $-1 \leq x(q) \leq 1$; from this it follows that $-f(1) = f(-1) \leq f(x) \leq f(1)$. Okay? *Any positive linear functional f on $C(Q)$ is a bounded linear functional with norm $f(1)$.* By Banach's theorem (Theorem 4.24), if $(x_n) \subseteq B_{\mathcal{C}}$ and $|x_n| \leq M \in \mathcal{C}$ with $\lim_n x_n(q) = 0$ for all $q \in Q$, then (x_n) is weakly null in $C(Q)$, hence $\lim_n f(x_n) = 0$. *This is $(BC\theta)$ in Banach's integration theory!*

So we can extend f to a vector lattice $\mathcal{L}$ of real-valued functions defined on Q in a linear fashion so that the extension, denoted by $\int df$, enjoys the fruits of (MCθ) and (DCθ):

(MCθ) If $(z_n) \subseteq \mathcal{L}, (z_n)$ an ascending sequence with $z = \lim_n z_n$, then $z \in \mathcal{L}$ whenever $\lim_n \int z_n df < \infty$. In this case $\int z = \lim_n \int z_n$.

(DCθ) Suppose $(z_n) \subseteq \mathcal{L}$ and $M \in \mathcal{L}$ satisfy $|z_n| \leq M$. Then $\liminf z_n$, $\limsup z_n \in \mathcal{L}$, and

$$\int \liminf z_n \leq \liminf \int z_n \leq \limsup \int z_n \leq \int \limsup z_n.$$

So should $(z_n) \subseteq \mathcal{L}, \mathcal{M} \in \mathcal{L}$ satisfy $|z_n| \leq \mathcal{M}$, if $z(q) = \lim_n z_n(q)$ for each $q \in Q$, then $z \in \mathcal{L}$ and $\int z = \lim_n \int z_n$.

What does this provide us with? To start, let F be a closed subset of Q. Consider the continuous function $\phi_F(q) = d(q; F)$; notice that $q \in F$ precisely when $\phi_F(q) = 0$. If we consider $\phi_n \in C(Q)$ to be

$$\phi_n(q) = \inf \left\{ \phi_F(q), \frac{1}{n} \right\},$$

then $\lim_n \phi_n(q) = \chi_F(q)$ for each $q \in Q$. Moreover, $|\phi_n| \leq 1 \in \mathcal{C} \subseteq \mathcal{L}$; hence $\chi_F \in \mathcal{L}$ and $\int \chi_F df = \lim_n \int \phi_n df$ by either (MCθ) or (DCθ), take your pick.

Suppose $E \subseteq Q$ and $\chi_E \in \mathcal{L}$. Then $1 = \chi_E + \chi_{E^c}$ so $\chi_{E^c} \subseteq \mathcal{L}$, also—after all $1 \in \mathcal{C}$ and $\chi_E \in \mathcal{L}$. If $A, B \subseteq Q$ and $\chi_A, \chi_B \in \mathcal{L}$, then since $\chi_{A\cap B} = \inf\{\chi_A, \chi_B\}$, we see that $\chi_{A\cap B} \in \mathcal{L}$. Because

$$\chi_A + \chi_B = \chi_{A\cup B} - \chi_{A\cap B},$$

we see that should $\chi_A, \chi_B \in \mathcal{L}$, then $\chi_{A\cup B} \in \mathcal{L}$ as well. Therefore

$$\{E \subseteq Q : \chi_E \in \mathcal{L}\}$$

is an algebra of sets containing all the closed subsets of Q.

Suppose (E_n) is an ascending sequence of subsets of Q such that $\chi_{E_n} \in \mathcal{L}$ for each n. Then

$$\chi_{\bigcup_n E_n} = \lim_n \chi_{E_n}$$

and

$$|\chi_{E_n}| \leq 1 \in \mathcal{C} \subseteq \mathcal{L}$$

so $\chi_{\bigcup_n E_n} \in \mathcal{L}$ also—here you can appeal to (MCθ) or (DCθ), take your choice.

Therefore $\{E \in Q : \chi_E \in \mathcal{L}\}$ is a σ-algebra of subsets of Q containing each and every closed subset of Q. Therefore for any Borel set $B \subseteq Q, \chi_B \in \mathcal{L}$. Thus we have proved

Theorem 4.25. *$\mathcal{L}$ contains all of the indicator functions on Borel sets.*

Suppose (B_n) is a sequence of pairwise disjoint Borel subsets of Q. Then $\chi_{B_n}, \chi_{\bigcup_n B_n} \in \mathcal{L}$ and

$$\chi_{\bigcup_n B_n} = \sum_n \chi_{B_n};$$

an appeal to (MCθ) or (DCθ) soon reveals that

$$\int \chi_{\bigcup_n B_n} df = \sum_n \int \chi_{\phi_n} df.$$

$\int df$ acts in a countably additive fashion on $\mathcal{B}o(Q)$! It is a measure on $\mathcal{B}o(Q)$; i.e.,

$$\mu_f = \int df, \text{ where } \mu_f(E) = \int \chi_E df, \text{ whenever } \chi_E \in \mathcal{L}$$

is a countably additive measure defined on some σ-field that contains the Borel σ-field. The total mass of μ_f is $f(1)$.

We initially thought to include the above construction as an appendix; that's how it appeared originally, as an appendix to Saks' book [**111**] *Theory of the Integral.* But it is too beautiful to be perpetually relegated to appendices.

5. Notes and Remarks

5.1. The Problem of Measure. Classically there were two such problems: the difficult problem of measure and the easy problem of measure. The reader should be aware that these adjectives are arbitrary and somewhat inaccurate!

The "difficult" problem of measure asks if one can assign to every set E in $\mathbb{R}^n$ a nonnegative real number $\mu(E)$, its measure, so that

(i) if $E = [0,1]^n$, then $\mu(E) = 1$;

(ii) if $\rho : \mathbb{R}^n \to \mathbb{R}^n$ is a surjective isometry, then $\mu(E) = \mu(\rho E)$;

(iii) if E is the union of any sequence (E_k) of pairwise disjoint sets, then

$$\mu(E) = \sum_k \mu(E_k).$$

We have seen that Vitali's construction of a nonmeasurable set answers this question in the negative in $\mathbb{R}$; in the Notes and Remarks to Chapter 7, Vitali's construction is modified to show that in any nondiscrete locally compact topological group, if Σ is a σ-field of sets containing the Borel sets and μ is a translation-invariant nonnegative, countably additive extension of Haar measure defined on Σ, then there is a non-μ-measurable set.

The "easy" problem of measure asks if one can assign to every set E in $\mathbb{R}^n$ a nonnegative real number $\mu(E)$, its measure, so that

(i) if $E = [0,1]^n$, then $\mu(E) = 1$;

(ii) if $\rho : \mathbb{R}^n \to \mathbb{R}^n$ is a (surjective) isometry, then $\mu(E) = \mu(\rho E)$;

(iii) if E is the union of the pairwise disjoint sets $E_1, \ldots, E_k$, then

$$\mu(E) = \mu(E_1) + \cdots + \mu(E_k).$$

The lessening of the third condition seriously changes the nature of the problem, and the solution to this problem now depends on the dimension n of the space $\mathbb{R}^n$ on which the measures reside.

For $n = 1, 2$, there is a measure μ solving the easy problem of measure; this is due to S. Banach, and it is *not* easy. For $n \geq 3$, there is no such measure and, again, the solution due to F, Hausdorff is anything but easy.

Since it was proved first, we take up Hausdorff's negative solution for $\mathbb{R}^3$ as a start. This solution revolves around the paradoxical theorem known as *Hausdorff's paradox*. For a beautiful exposition of this and related topics, we refer the reader to the paper of Karl Stromberg in the American Mathematical Monthly [**123**].

Theorem 4.26 (Hausdorff's Paradox). *The unit sphere S^2 of $\mathbb{R}^3$ can be decomposed into the disjoint union*

$$S^2 = Q \cup R \cup S \cup T$$

of four sets: Q is countable, R, S, T are pairwise congruent, and R is congruent to $S \cup T$.

Now suppose that μ is a solution to the easy problem of measure in $\mathbb{R}^3$. Normalize μ so $\mu(B_{\mathbb{R}^3}) = 1$. With Hausdorff's paradox in hand, write $B_0 = B_{\mathbb{R}^3} \setminus \{0\}$ in the form

$$B_0 = Q_0 \cup R_0 \cup S_0 \cup T_0,$$

where the subscript "0" indicates that the set is the union of rays emanating from the origin and ending on the sphere at a point of the subscript set (e.g., $Q_0 = \bigcup_{q \in Q} (0, q]$).

We first show that $\mu(Q_0) = 0$. Introduce spherical coordinates and assume (a rotation of S^2 may be needed) that $0z$ passes through the none of the points of Q. A point in S^2 is determined by two angles: "latitude" which measures $N - S$ and "longitude" which measures $E - W$.

The set of longitudes of points of Q is countable as is the set of differences thereof. So we can choose an angle λ distinct from all the angles of longitude of Q and different from all the differences. Rotate $0z$ through the angle λ; Q moves to a (countable) set Q' which is disjoint from Q. Again we can find an angle λ' distinct from all the points of $Q \cup Q'$ and from all the differences thereof. Rotate $0z$ through the angle λ' to get a set Q'' which is the image of Q'; Q, Q' and Q'' are congruent by construction. Continue in this vein.

For any n, we get pairwise disjoint sets

$$Q, Q', Q^1, \ldots, Q^{(n-1)},$$

and accompanying them are the pairwise disjoint sets

$$Q_0, Q_0', Q_0^1, \ldots, Q_0^{(n-1)}.$$

Of course all these sets are pairwise congruent as well. Then

$$\begin{aligned} \mu(B_0) &\geq \mu(Q_0 \cup \cdots \cup Q_0^{(n-1)}) \\ &= \mu(Q_0) + \mu(Q_0') + \mu(Q_0'') + \cdots + \mu(Q_0^{(n-1)}) \\ &= n\mu(Q_0). \end{aligned}$$

But $\mu(B_0)$ is finite, and we can do the above for each $n \in \mathbb{N}$. It must be that $\mu(Q_0) = 0$.

We also have

$$\begin{aligned} \mu(B_0) &= \mu(Q_0 \cup R_0 \cup S_0 \cup T_0) = \mu(Q_0) + \mu(R_0) + \mu(S_0) + \mu(T_0) \\ &\quad \text{(because } Q_0, R_0, S_0, T_0 \text{ are pairwise disjoint)} \\ &= \mu(R_0) + \mu(S_0) + \mu(T_0) \ \text{ (because } \mu(Q_0) = 0) \\ &= 3\mu(R_0) \ \text{ (because } R_0, S_0, T_0 \text{ are congruent).} \end{aligned}$$

At the same time

$$\mu(B_0) = \mu(R_0) + \mu(S_0 \cup T_0) = 2\mu(R_0)$$

because R_0 and $S_0 \cup T_0$ are congruent.

Hence

$$2\mu(R_0) = 3\mu(R_0) = \mu(B_0).$$

Since $\mu(B_0) > 0$, chaos is the only explanation.

One can easily piggyback this negative solution of the easy problem of measure in $\mathbb{R}^3$ to a negative solution in $\mathbb{R}^n$ for any $n > 3$. After all, if ν were to be a positive solution to the easy problem of measure for $\mathbb{R}^4$, then

$$\mu(E) = \nu([0,1] \times E)$$

would be a positive solution for $\mathbb{R}^3$, etc.

Banach's positive solution for $\mathbb{R}$ (and $\mathbb{R}^2$) is a delicious mix of his natural cleverness and ideas from functional analysis—ideas similar to what goes into the Hahn–Banach theorem. A few words on Banach's proof seem in order.

In what follows we will be considering bounded real-valued functions defined on $\mathbb{R}$ which are periodic of period 1. If f is such a function, then f can be viewed as a bounded real-valued function defined on the circle C of radius $\frac{1}{2\pi}$ having center at the origin. The value $f(x)$ corresponds to f being evaluated at an arc x on C where we proceed x units in a counterclockwise fashion starting at the point $\left(\frac{1}{2\pi}, 0\right)$. Let f be such a function.

Definition 4.27. We say that $f \sim 0$ if given an $\epsilon > 0$, there exist $a_1, \ldots, a_n \in \mathbb{R}$ such that for all real x

$$\left| \frac{1}{n} \sum_{k=1}^{n} f(x + a_k) \right| < \epsilon.$$

We say that $f \sim g$ (for two bounded functions defined on C) if $(f - g) \sim 0$, that is, if given an $\epsilon > 0$ there are $a_1, \ldots, a_n \in \mathbb{R}$ such that for all real x

$$\left| \frac{1}{n} \sum_{k=1}^{n} f(x + a_k) - g(x + a_k) \right| < \epsilon.$$

Proposition 4.28. *Some basic facts:*

(i) *If $f \sim 0$ and $g \sim 0$, then $f + g \sim 0$.*

(ii) *If $f \sim 0$ and $C \in \mathbb{R}$, then $C \cdot f \sim 0$.*

(iii) *If $f(x) \geq c > 0$ for all x, then it is* not *so that $f \sim 0$.*

(iv) *If $f \sim 0$ and $a \in \mathbb{R}$, then $f_a(x) := f(x + a)$ also has $f_a \sim 0$.*

(v) *If $f \sim g$ and $g \sim h$, then $f \sim h$.*

Hence "$\sim$" is an equivalence relationship among the bounded real-valued functions on C. If f is a bounded real-valued function defined on C, then the equivalence class $[f]$ of f will be denoted by F, and we shall refer to F as a *hyperfunction*, keeping Banach's terminology.

Some natural conventions will be followed:

- If F is the hyperfunction $[f]$ and a is a real-valued constant function, then $A \cdot F$ will be the hyperfunction $[a \cdot f]$, where $A = [a]$.
- If $F = [f]$ and $G = [g]$, then $F + G$ will be $[f + g]$.
- If $F = [f]$, then $-F$ will be $[-f]$.

One easily realizes much of the usual relationships hold in the space of hyperfunctions. A fundamental question to be entertained is, What about order?

It is easiest to discuss the order by working with (bounded real-valued) functions f on C.

Definition 4.29. We say $f \succ 0$ if there exists $c > 0$ and $a_1, \ldots, a_n \in \mathbb{R}$ so that for all real x,

$$\frac{1}{n}\sum_{k=1}^{n} f(x + a_k) \geq c.$$

If $-f \succ 0$, then we write $f \prec 0$. We say that $f \succ g$ if $(f - g) \succ 0$.

If $F_1 = [f_1]$ and $F_2 = [f_2]$ are hyperfunctions, we'll say that $F_1 > F_2$ if for any $f \in [f_1 - f_2]$ we have $f \succ 0$.

Proposition 4.30. *More basic facts:*

(i) *If $f \succ 0$ and $a \in \mathbb{R}$, then $f_a \succ 0$ where $f_a(x) := f(x + a)$.*

(ii) *If $f \succ 0$ and $g \succ 0$, then $(f + g) \succ 0$.*

(iii) *If $f(x) > 0$ for all x, then it is* not *so that $f \prec 0$.*

(iv) At most one *of the following holds for any f:*

$$f \succ 0, \quad f \sim 0, \quad f \prec 0.$$

(v) *If $f \succ 0$ and $g \sim 0$, then $(f + g) \succ 0$.*

(vi) *If $f \succ g, f \sim f_1$, and $g \sim g_1$, then $f_1 \succ g_1$.*

(vii) *Let $F = [f]$ be a given hyperfunction. Then there are hyperfunctions $C_1 = [c_1]$ and $C_2 = [c_2]$ where $c_1, c_2 \in \mathbb{R}$ are constant functions such that $C_1 > F > C_2$.*

We turn now to the study of linear spaces of hyperfunctions with special attention being paid to $\Omega(\mathcal{R})$ and $\Omega(\mathcal{L})$. Here $\Omega(\mathcal{R})$ is the linear space of all hyperfunctions $F = [f]$ where some $f \in F$ is Riemann integrable (keep in mind that Riemann integrable functions are necessarily bounded), and $\Omega(\mathrm{L})$ is the linear space of hyperfunctions each of whose members contain a bounded Lebesgue integrable function.

Generally if Ω is a linear space of hyperfunctions and $\Phi = [\phi]$ is a hyperfunction that does not belong to Ω, then $\Omega(\Phi)$ denotes the smallest linear space of hyperfunctions containing Ω and Φ. Now $\Omega(\Phi)$ consists of all hyperfunctions of the form $F + C \cdot \Phi$ where c is a real constant and $F \in \Omega$. Naturally $C = [c]$.

We'll attempt to avoid being too pedantic by using c_1 for both the constant real number and the hyperfunction $[c_1]$. With this in mind, notice that for any $\Psi \in \Omega(\Phi)$, Ψ can be represented in *one and only one way* in the form

$$\Psi = F + c\Phi.$$

In fact, if we also have

$$\psi = F^{'} + c^{'}\Phi,$$

then

$$(c - c')\Phi = F' - F.$$

If $c = c'$, then $c - c' = 0$ and $F' = F$. If $c \neq c'$, then

$$\Phi = \frac{1}{c - c'}F' - \frac{1}{c - c'}F,$$

and so $\Phi \in \Omega$, which is assumed not to be. Either way the uniqueness of the representation $\Psi = F + c\phi$ is established.

We further take note that if $\Psi_1, \Psi_2, \Psi_3 \in \Omega(\Phi)$ and $\Psi_3 = a_1\Psi_1 + a_2\Psi_2$ $(a_1, a_2 \in \mathbb{R})$, then on representing Ψ_1, Ψ_2, Ψ_3 as

$$\Psi_1 = F_1 + c_1\Phi, \ \Psi_2 = F_2 + c_2\Phi, \ \Psi_3 = F_3 + c_3\Phi,$$

we know that

$$F_3 = a_1F_1 + a_2F_2 \text{ and } c_3 = a_1c_1 + a_2c_2,$$

thanks to the uniqueness of the representation of Ψ_1, Ψ_2, and Ψ_3.

Now let Ω be a linear space of hyperfunctions.

Definition 4.31. A map $A : \Omega \to \mathbb{R}$ is called *linear* if for any constants c_1, c_2 and any $F_1, F_2 \in \Omega$ we have

$$A(c_1F_1 + c_2F_2) = c_1A(F_1) + c_2A(F_2).$$

We say that A is a *positive linear functional* if in addition to being linear, $A(F) \geq 0$ whenever $F \in \Omega$ satisfies $F \geq 0$.

Theorem 4.32. *Let Φ be a hyperfunction not belonging to the linear space Ω of hyperfunctions, and suppose that there exist $F_1, F_2 \in \Omega$ such that*

$$F_1 > \Phi > F_2.$$

Let $A : \Omega \to \mathbb{R}$ be a positive linear functional. Then there exists a positive linear functional $\bar{A} : \Omega(\Phi) \to \mathbb{R}$ that extends A to $\Omega(\Phi)$.

The hypothesis of Theorem 4.32 tells us that

$$\alpha = \inf\{A(F) : F \in \Omega, F > \Phi\}$$

is real. Now if $\Psi \in \Omega(\Phi)$ is written in the form

$$\Psi = F + c\Phi$$

where $c \in \mathbb{R}, F \in \Omega$, then

$$\bar{A}(\Psi) = A(F) + c\alpha$$

is a positive linear functional on $\Omega(\Phi)$ that extends A.

Banach now calls on transfinite induction and with this firmly under control proves the basic:

Proposition 4.33. *Let $(\Omega_\alpha)_{\alpha<\beta}$ be a well ordered (ascending) family of linear spaces of hyperfunctions (so $\Omega_\alpha \subseteq \Omega_\gamma$ if $\alpha < \gamma < \beta$), and suppose that for each $\alpha < \beta$, A_α is a positive linear functional on Ω_α such that $A_\gamma|_{\Omega_\alpha} = A_\alpha$ whenever $\alpha < \gamma < \beta$. Let*

$$\Omega = \bigcup_{\alpha<\beta} \Omega_\alpha,$$

and for $F \in \Omega$ set $A(F) = A_{\alpha(F)}(F)$, where $\alpha(F)$ denotes the first ordinal so that $F \in \Omega_\alpha$. Then A is a positive linear functional on Ω.

Proposition 4.34. *Suppose Ω is a linear space of hyperfunctions containing the hyperfunction 1. Assume A is a positive linear functional defined on Ω. Then A can be extended to a positive linear functional $\bar{A}$ defined on the space of* all *hyperfunctions.*

Now we're prepared for the main course of Banach's offerings.

First an appetizer.

Theorem 4.35. *Suppose f is Riemann integrable. Then $f \sim c$ where $c = \int_0^1 f(x)\,dx$.*

Proof. Suppose f is Riemann integrable, and define f_n on C by

$$f_n(x) = \frac{1}{n}\sum_{k=1}^{n} f\left(x + \frac{k}{n}\right) = \frac{1}{n}\sum_{k=1}^{n} f_{\frac{k}{n}}(x).$$

Then (f_n) converges uniformly to $\int_0^1 f(x)\,dx$. Hence if we choose n very big

$$\left|\frac{1}{n}\sum_{k=1}^{n} f\left(x + \frac{k}{n}\right) - c\right| < \epsilon$$

for all x; i.e., for n big

$$\left|\frac{1}{n}\sum_{k=1}^{n} \left[f\left(x + \frac{k}{n}\right) - c\right]\right| < \epsilon$$

for all x. This is tantamount to $f - c \sim 0$ or $f \sim c$. □

Before the next delightful tidbit we recall the following feature of $\mathbb{R}$:

Fact 4.36. The real line $\mathbb{R}$ can be written as the disjoint union of two sets N and F, where N is a set of Lebesgue measure zero, and F is a set of the first Baire category.

Proof. To see this, let (q_n) be a listing of the rationals, and let I_{ij} be an open interval with center q_i of length 2^{-i-j}. Set $G_j = \bigcup_{ij} I_{ij}$ and $N = \bigcap_j G_j$. If $\epsilon > 0$ is given and $2^{-j} < \epsilon$, then $N \subseteq \bigcap_i I_{ij}$, and

$$\sum_i l(I_{ij}) = \sum_i \frac{1}{2^{i+j}} = \frac{1}{2^j} < \epsilon.$$

Therefore N is a set of Lebesgue measure zero.

Of course each G_j is open and contains Q so N is a dense $\mathcal{G}_\delta$ set. On the other hand, each G_j^c is nowhere dense, and so $N^c = \left(\bigcap_j G_j\right)^c = \bigcup_j (G_j^c)$ is of the first category. □

Proposition 4.37. *There exists a bounded Lebesgue integrable function ρ with $\int_0^1 \rho(x)\,dx = 0$ such that if ϕ is Riemann integrable and $\phi \succ \rho$, then $\int_0^1 \phi(x)\,dx > 1$.*

If N is a set of Lebesgue measure zero with N^c a set of the first Baire category, then $\rho = \chi_N$, the indicator function of N, is bounded, Lebesgue integrable, and has $\int_0^1 \rho(x)\,dx = 0$. This ρ fits the bill.

Remark 4.38. If ρ is the function of Proposition 4.37, then

$$\inf\left\{\int_0^1 \phi(x)\,dx : \phi \succ \rho \text{ and } \phi \text{ is Riemann integrable}\right\} = 1.$$

After all, the constant functions $\phi(x) = 1 + \epsilon$ are Riemann integrable and satisfy $\phi \succ \rho$.

Moreover the ρ in Proposition 4.37 is *not* equivalent to any Riemann integrable function in the sense of "$\sim$". Indeed suppose $\rho \sim \psi$ where ψ is Riemann integrable. Then $\rho \sim c = \int_0^1 \psi(x)\,dx$ so $\rho - c \sim 0$. But $\rho - c \sim 0$ means that given $\epsilon > 0$ we can find $a_1, \ldots, a_n \in \mathbb{R}$ so that for all real x

$$\left|\frac{1}{n}\sum_{k=1}^{n}[\rho(x + a_k) - c]\right| < \epsilon. \tag{14}$$

But we've chosen ρ so that there are points where $\frac{1}{n}\sum_{k=1}^{n} \rho(x + a_k)$ takes the value one and points where $\frac{1}{n}\sum_{k=1}^{n} \rho(x+a_k)$ is zero. There is absolutely no way for (14) to hold in light of this.

Theorem 4.39. *There is a positive linear functional H defined on the vector space of all bounded real-valued functions on the circle C satisfying the following:*

(i) *$H(c) = c$ for each real constant c;*

(ii) *for any* $a \in \mathbb{R}$ *and* $\sigma = \pm 1$, *if* $f_a^\sigma(x) := f(\sigma x + a)$, *then*

$$H(f) = H(f_a^\sigma);$$

(iii) *if* f *is a bounded Lebesgue integrable function, then*

$$H(f) = \int_0^1 f(x)\,dx.$$

Proof. Denote by $\Omega(\mathcal{L})$ the linear space of hyperfunctions, each containing a bounded Lebesgue integrable function on C. For any $X \in \Omega(\mathcal{L})$, we let A be the operation

$$A(X) = L - \int_0^1 \phi(x)\,dx,$$

where ϕ is a bounded Lebesgue integrable function belonging to X.

The operator $A(X)$ is well defined:

$$L - \int_0^1 f(x)\,dx = 0$$

for each and every f that's Lebesgue integrable and satisfies $f \sim 0$. Why is this so?

Suppose $f \sim 0$, and let $\epsilon > 0$ be given. Then one can find reals $a_1, \ldots, a_n$ so that for all real x

$$\left|\frac{1}{n}\sum_{k=1}^{n} f(x + a_k)\right| < \epsilon.$$

It follows that

$$L\int_0^1 \left|\frac{1}{n}\sum_{k=1}^{n} f(x + a_k)\right|\,dx < \epsilon,$$

and so

$$\left|L - \int_0^1 \frac{1}{n}\sum_{k=1}^{n} f(x + a_k)\,dx\right| < \epsilon.$$

But

$$L\int_0^1 f(x)\,dx = L - \int_0^1 f(x + a_k)\,dx = \frac{1}{n}\sum_{k=1}^{n} L - \int_0^1 f(x + a_k)\,dx.$$

So

$$|L - \int_0^1 f(x)\,dx| < \epsilon.$$

Since $\epsilon > 0$ was arbitrary, we're left with the conclusion that

$$L - \int_0^1 f(x)\,dx = 0,$$

whenever $f \sim 0$.

The operation A is linear: if X_1, X_2 and X are in $\Omega(\mathcal{L})$ with $X = c_1X_1 + c_2X_2$ ($c_1, c_2 \in \mathbb{R}$), and if ϕ_1, ϕ_2 are bounded Lebesgue integrable functions in the hyperfunctions X_1, X_2, respectively, then $\phi = c_1\phi_1 + c_2\phi_2 \in X$, and

$$\begin{aligned} A(X) &= L - \int_0^1 c_1\phi_1(x) + c_2\phi_2(x)\, dx \\ &= c_1 L - \int_a^b \phi_1(x)\, dx + c_2 L \int_a^b \phi_2(x)\, dx \\ &= c_1 A(X_1) + c_2 A(X_2). \end{aligned}$$

The operation A is positive: Suppose $X \geq 0$, and let ϕ be a bounded Lebesgue integrable function in X such that $\phi \succ 0$ or $\phi \sim 0$. We've already seen that if $\phi \sim 0$, then

$$L - \int_0^1 \phi(x)\, dx = 0,$$

and so

$$A(X) = L - \int_0^1 \phi(x)\, dx = 0.$$

What if $\phi \succ 0$? Well, in this case there are reals $a_1, \ldots, a_n$ and $c > 0$ so that for all real x,

$$\frac{1}{n} \sum_{k=1}^{n} \phi(x + a_k) > c.$$

It follows that

$$\frac{1}{n} \sum_{k=1}^{n} L - \int_0^1 \phi(x + a_k)\, dx > c.$$

But as we've seen (using the invariance of the integral under translation(!)), this is tantamount to

$$L - \int_0^1 \phi(x)\, dx > c,$$

and so $A(X) > c > 0$.

Regardless, $A(X) \geq 0$ if $X \geq 0$.

Because $\Omega(\mathcal{L})$ contains the constant hyperfunction 1, Proposition 4.34 is applicable. On using Proposition 4.34 we find a positive linear functional $\bar{A}$ defined on the linear space of all hyperfunctions such that $\bar{A}|_{\Omega(\mathcal{L})} = A$.

Now we define an operation G with domain the linear space of all bounded real-valued functions that are periodic of period 1 (hence bounded functionals defined on C). Let f be such a function, and suppose f is a member of the hyperfunction F. As you might expect, set $G(f) = \bar{A}(f)$. Then G is well defined because F is the only hyperfunction containing f, G is linear (this is straightforward), and should $f \geq 0$ (i.e., $f(x) \geq 0$ for all x) then

$G(f) \geq 0$ as well. To see that this last assertion is so, for any $\epsilon > 0$, let $\phi(x) := f(x) + c$, and realize that $\phi \succ 0$ so the hyperfunction Φ determined by ϕ verifies $\Phi > 0$ and so $\bar{A}(\Phi) \geq 0$. In turn, this tells us that $G(\phi) \geq 0$ so $G(\epsilon + f) \geq 0$. But G is additive so $G(f) \geq -G(\epsilon) = -\epsilon$, which because $\epsilon > 0$ was arbitrary tells us that $G(f) \geq 0$.

Since $f_a \sim f$ (recall $f_a(x) := f(x+a)$), f and f_a determine the same hyperfunction F, and so

$$G(f_a) = G(f) = \bar{A}(F).$$

It is plain that

$$G(f) = L - \int_0^1 f(x)\,dx$$

if f is Lebesgue integrable—one need only recall the opening arguments of this argument.

Now we'll define H.

Let f be a bounded real-valued function, periodic with period 1. Denote by $f^{(+1)}$ the function $f(x)$ and $f^{(-1)}$ the function $f^{(-1)}(x) := f(-x)$. Then

$$Hf = \frac{1}{2}(Gf^{(+1)} + Gf^{(-1)}).$$

The functional H does all that we have claimed: it's a positive, linear functional. If $f(x) = c$ is a constant function, then

$$A(f) = \frac{1}{2}(G(c) + G(c)) = c,$$

$$H(f_a^{(\sigma)}) = \frac{1}{2}(Gf_a^{\sigma} + Gf_a^{(-\sigma)}) = \frac{1}{2}(Gf^{\sigma} + Gf^{(-\sigma)}) = H(f),$$

and finally if f is Lebesgue integrable, then (since it's assumed that f is periodic with period one),

$$\begin{aligned} H(f) &= \frac{1}{2}(Gf^{(+1)} + Gf^{(-1)}) = \frac{1}{2}(Gf^{(+1)} + G\tilde{f}) \\ &= \frac{1}{2}(L - \int_0^1 f(x)\,dx + L - \int_0^1 \tilde{f}(x)\,dx) = L - \int_0^1 f(x)\,dx, \end{aligned}$$

where $\tilde{f}(x) = f(1-x)$.

Banach makes special note that the above extension of Lebesgue measure is noteworthy in that it satisfies conditions (i), (ii), *and* (iii). □

In fact there is a positive linear functional defined on C that satisfies conditions (i) and (ii) of Theorem 4.39 but is *not* identical to the Lebesgue integral for all bounded Lebesgue integrable functions.

Indeed, consider the linear space $\Omega(\mathcal{R})$ consisting of hyperfunctions that contain a bounded Riemann integrable function, and let ρ be the bounded Lebesgue integrable function of Proposition 4.37. Suppose ρ generates the hyperfunction $\mathcal{R}$.

Define the functional $A : \Omega(\mathcal{R}) \to \mathbb{R}$ by

$$A(\Phi) = \int_0^1 \phi(x)\, dx,$$

where $\phi \in \Phi$ is a Riemann integrable function. Then A is a positive linear functional on $\Omega(\mathcal{R})$.

We extend A to a positive linear functional $\bar{A}_1$ on the linear span of $\Omega(\mathcal{R}) \cup \{R\}$ by

$$\bar{A}_1(\Phi + cR) = A(\Phi) + c.$$

By Theorem 4.32 and by Remark 4.38, if

$$\alpha = \inf\{A(\Phi) : \Phi \in \Omega(\mathbb{R}); \Phi > R\},$$

then $\alpha = 1$. In particular $\bar{A}_1(R) = 1$.

Now we can apply Proposition 4.34 to extend $\bar{A}_1$ to a positive linear functional $\bar{A}$ on the space of all hyperfunctions.

If we now proceed as in the proof of Theorem 4.39, we find a positive linear functional H defined on the space of all bounded real-valued functions on C such that

$$H(\rho) = 1,$$

where we keep in mind that

$$L \int_0^1 \rho(x)\, dx = 0.$$

We make special mention of the fact that $H(f) = H(g)$ whenever $f \sim g$. To see this, it suffices to show that $H(f) = 0$ whenever $f \sim 0$ (this, thanks to H's being a positive linear functional). Suppose $f \sim 0$. Then regardless of $\epsilon > 0$ we can find reals $a_1, \dots, a_n$ so that for all x

$$-\epsilon \leq \frac{1}{n} \sum_{k=1}^{n} f(x + a_i) \leq \epsilon.$$

It follows that

$$H\left(\frac{1}{n} \sum_{i=1}^{n} f_{a_i}\right) = \frac{1}{n} \sum_{i=1}^{n} H(f_{a_i}) = H(f),$$

and so

$$-\epsilon \leq H(f) \leq \epsilon.$$

Now $\epsilon > 0$ being arbitrary gives rise to the realization that $H(f) = 0$. Of course, if f is Riemann integrable, then

$$f \sim c = \int_0^1 f(x)\,dx,$$

and so

$$H(f) = c = \int_0^1 f(x)\,dx.$$

One consequence of Theorem 4.39 worthy of note is the following solution of the problem of measure for $\mathbb{R}$ (we use Banach's notation):

Theorem 4.40. *One can assign to each subset E of points of the unit circle C a number $H(E)$ in such a way that*

(i) $H(E) \geq 0$;

(ii) $H(C) = 1$;

(iii) $H(E_1 \cup E_2) = H(E_1) + H(E_2)$ *whenever* $E_1 \cap E_2 = \emptyset$;

(iv) *if E_1 and E_2 are congruent (there is an isometry of $\mathbb{R}$ into $\mathbb{R}$ that takes E_1 onto E_2), then $H(E_1) = H(E_2)$;*

(v) $H(E) =$ *Lebesgue measure of E if E is Lebesgue measurable.*

Banach turns his attention to $\mathbb{R}^2$. Here's what he does.

Again, C will denote the (true unit) circle with center the origin of the plane and the radius $1/2\pi$.

Let $f(x, y)$ be a bounded real-valued function defined on a (nonempty) bounded domain $E \subseteq \mathbb{R}^2$, and let H be the functional whose existence was the point of Theorem 4.39.

Let ξ be an angle (point on C).

Rotate the OX, OY axes system through the angle ξ (in a counterclockwise manner, if you please), and let (x', y') be the new coordinates of (x, y) in the OX', OY' system. Recall from your days of analytic geometry that

$$x = x' \cos\xi - y' \sin\xi \quad \text{and} \quad y = x' \sin\xi + y' \cos\xi$$

(so the transformations $(x, y) \to (x', y')$ and $(x', y') \to (x, y)$ are linear isometries).

Look at

$$\tilde{f}(x', y') = f(x' \cos\xi - y' \sin\xi, x' \sin\xi + y' \cos\xi)$$

defined on the set $\tilde{E}$ of points (x', y') which one obtains by rotating E through the angle $-\xi$.

He showed that solvable groups are amenable and noted that Hausdorff's paradox leads to the existence of a free subgroup of G_n $(n \geq 3)$ of rank 2, something forbidden for solvable groups.

If one proceeds like what was done following the statement of Hausdorff's paradox, then it is easy to see that the amenability (or nonamenability) of G_n holds the answer to the easy problem of measure in $\mathbb{R}^n$.

Results related to the problem of measure often lead to paradoxical statements, and so attract the attention of many from different areas of mathematical endeavors. I. P. Natanson's *Theory of Functions of a Real Variable* [**94**] relates the problem to classical real variables. K. Stromberg's paper [**123**] *The Banach–Tarski Paradox* presents the Hausdorff paradox, and its more famous companion, the Banach–Tarski paradox, in a clear and complete way. And S. Wagon's volume "The Banach–Tarski Paradox" in Cambridge University's *Encyclopedia of Mathematics* series gives an overview of this fascinating area of mathematics.

Banach's work, as outlined above, solves (negatively) the case for $n = 1$ of the Banach–Ruziewicz problem, which asks if the (normalized) Lebesgue measure is the unique finitely additive rotation-invariant probability defined on the Lebesgue measurable subsets of the n-sphere. For $n > 1$, the answers to this problem are all of a positive nature.

For $n = 2$ and $n = 3$, this uniqueness of normalized Lebesgue measure among finitely additive rotation invariant probabilities was established by V. G. Drinfeld [**33**]; while for all $n > 3$ this was established independently by G. Margulis [**77**] and D. Sullivan [**125**], based in part on earlier work by J. Rosenblatt [**106**] and D. Každan [**65**].

For the Euclidean spaces, Banach [**4**] showed that there was more than one rotation invariant finitely additive measure, normalized so that the measure of the unit ball is one, on n-space in case $n = 1$ or $n = 2$. For $n > 2$, G. Margulis [**78**] has shown that Lebesgue measure is the unique finitely additive, suitably normalized, rotation invariant measure on the σ-field of all Lebesgue measurable sets.

A remarkable construction due to S. Kakutani and J. Oxtoby [**64**] shows that Lebesgue measure in n-space can be extended in a countable additive, nonnegative, translation-invariant manner to very large σ-fields. The interested reader will find a discussion of this work in their Appendix 2.

Chapter 5

Compact Groups Have a Haar Measure

1. The Arzelá–Ascoli Theorem

When we first study analysis, one of the basic results encountered is the fact that if f is a continuous real-valued function defined on a closed, bounded interval, then f is uniformly continuous. This result has a natural extension to the setting of topological groups. The proof of this extension is easy but nevertheless worthy of close attention.

Theorem 5.1. *Let G be a topological group, and let M be a nonempty compact subset of G. Then any continuous function $f : G \to \mathbb{R}$ is left uniformly continuous on M; i.e., given $\epsilon > 0$, there is an open set V containing the identity of G so that if $x, y \in M$ and $x \in yV$, then $|f(x) - f(y)| \leq \epsilon$.*

Proof. Let $\epsilon > 0$ be given. For each $a \in M$, there is an open set V_a that contains the identity such that if $x \in M$ and $x \in aV_a$, then $|f(x) - f(a)| \leq \frac{\epsilon}{2}$. Since $e \cdot e = e$, there is an open set W_a that contains the identity e and satisfies $W_a \cdot W_a \subseteq V_a$. Now if $a \in M$, then $a \in a \cdot W_a$ so $\{a \cdot W_a : u \in M\}$ covers M; we can find $a_1, \ldots, a_n \in M$ so that

$$M \subseteq (a_1 \cdot W_{a_1}) \cup \cdots \cup (a_n \cdot W_{a_n}).$$

Look at

$$V = W_{a_1} \cap \cdots \cap W_{u_n}.$$

Then V is open and contains e. Let $x, y \in M$ with $x \in yV$. To see that $|f(x) - f(y)| \leq \epsilon$, note that if $y \in M$, then $y \in a_i \cdot W_{a_i}$ for some $i = 1, \ldots, n$.

It follows from this and $W_a \subseteq V_a$ that $|f(y) - f(a_i)| \leq \frac{\epsilon}{2}$. Also

$$x \in yV \subseteq a_i \cdot W_{a_i} \cdot V \subseteq a_i \cdot W_{a_i} \cdot W_{a_i} \subseteq a_i \cdot V_{a_i},$$

and so $|f(x) - f(a_i)| \leq \frac{\epsilon}{2}$ too.

In sum, if $x, y \in M$ with $x \in yV$, then

$$|f(x) - f(y)| \leq |f(x) - f(a_i)| + |f(a_i) - f(y)| \leq \frac{\epsilon}{2} + \frac{\epsilon}{2} = \epsilon. \qquad \square$$

Note. Of course, we can also show that the continuous function of Theorem 5.1 is right uniformly continuous.

If S is a compact Hausdorff space, then we denote by $C(S)$ the Banach space of all continuous real-valued functions on S equipped with the sup norm $||\cdot||_\infty$:

$$||f||_\infty = \sup\{|f(s)| : s \in S\}.$$

The task before us is to identify the subsets of $C(S)$ that are relatively norm compact.

Let $\mathcal{K}$ be a nonempty collection of members of $C(S)$. We say $\mathcal{K}$ is *equicontinuous at a point* $s \in S$ if, given $\epsilon > 0$, there is an open set U which contains s so that whenever $t \in U$, then

$$|f(t) - f(s)| \leq \epsilon$$

for all $f \in \mathcal{K}$. We say that $\mathcal{K}$ is *equicontinuous* if $\mathcal{K}$ is equicontinuous at each point of S.

Our first result shows that equicontinuity allows us to arrive at uniform conclusions from pointwise hypotheses.

Fact 5.2. If $\mathcal{K}$ is equicontinuous, then the topology of pointwise convergence on $\mathcal{K}$ and the norm topology coincide.

Proof. We need to show that if $f_0 \in \mathcal{K}$ and $\epsilon > 0$, then there is a finite collection $s_1, \ldots, s_n \in S$ so that should

$$|f(s_i) - f_0(s_i)| \leq \frac{\epsilon}{2}$$

for $i = 1, \ldots, n$, $f \in \mathcal{K}$, then

$$||f - f_0|| \leq \epsilon.$$

Since $\mathcal{K}$ is equicontinuous, for each point $s \in S$ we can find an open set U_s in S containing s so that if $t \in U_s$, then

$$|f(t) - f(s)| \leq \frac{\epsilon}{4}$$

for all $f \in \mathcal{K}$. Now

$$S = \bigcup_{s \in S} U_s$$

is compact so there are points $s_1, \ldots, s_n \in S$ so that each $t \in S$ belongs to a $U_{s_j}, 1 \leq j \leq n_j$; i.e., there are points $s_1, \ldots, s_n \in S$ so that for any $t \in S$ there is an $s_j, 1 \leq j \leq n$, for which

$$|f(t) - f(s_j)| \leq \frac{\epsilon}{4} \tag{15}$$

for all $f \in \mathcal{K}$.

We have our sought after points $s_1, \ldots, s_n$. Assume $f \in \mathcal{K}$ and

$$|f(s_j) - f_0(s_j)| \leq \frac{\epsilon}{2} \tag{16}$$

for $j = 1, \ldots, n$. Then given $t \in S$ there is j, $1 \leq j \leq n$, so that (15) applies.

It follows that

$$\begin{aligned} |f(t) - f_0(t)| &\leq |f(t) - f(s_j)| + |f(s_j) - f_0(s_j)| + |f_0(s_j) - f_0(t)| \\ &\leq \frac{\epsilon}{4} + \frac{\epsilon}{2} + \frac{\epsilon}{4} = \epsilon \end{aligned}$$

by (15), (16), and (15), if you please. □

Equicontinuity is contagious.

Fact 5.3. If $\mathcal{K} \subseteq C(S)$ is equicontinuous, then $\mathcal{K}$'s pointwise closure (in $\mathbb{R}^S$) is also equicontinuous.

Proof. Let $s \in S$ and $\epsilon > 0$. There is an open set U_s containing s so that if $t \in U_s$, then

$$|f(t) - f(s)| < \frac{\epsilon}{3}$$

for all $f \in \mathcal{K}$. Let $g \in \overline{\mathcal{K}^p}$, $\mathcal{K}$'s pointwise closure in $\mathbb{R}^S$. Then given $s, t \in S$ there is $f \in \mathcal{K}$ so that

$$|f(s) - g(s)|, |f(t) - g(t)| < \frac{\epsilon}{3},$$

since this is what it means for g to be in $\overline{\mathcal{K}^p}$. If $t \in U_s$, then

$$\begin{aligned} |g(s) - g(t)| &\leq |g(s) - f(s)| + |f(s) - f(t)| + |f(t) - g(t)| \\ &< \frac{\epsilon}{3} + \frac{\epsilon}{3} + \frac{\epsilon}{3} = \epsilon. \end{aligned}$$

□

Now we are ready for the centerpiece of this section.

Theorem 5.4 (Arzelá and Ascoli). *A nonempty subset $\mathcal{K}$ of $C(S)$ is relatively norm compact if and only if $\mathcal{K}$ is uniformly bounded and equicontinuous.*

Proof. Assume $\mathcal{K}$ is uniformly bounded and equicontinuous. By Fact 5.3 the pointwise closure, $\overline{\mathcal{K}^p}$ of $\mathcal{K}$ is also equicontinuous. But $\mathcal{K}$ is bounded in each coordinate by its uniform bound, and so $\overline{\mathcal{K}^p}$ is also bounded in each coordinate by the same uniform bound. Tychonoff's theorem tells us that $\overline{\mathcal{K}^p}$ is compact in $\mathbb{R}^S$, the topology of pointwise convergence. But now Fact 5.2 tells us that $\overline{\mathcal{K}^p}$ is norm compact so $\mathcal{K}$ is relatively norm compact.

On the other hand, if $\mathcal{K}$ is relatively norm compact in $C(S)$, then $\mathcal{K}$ is uniformly bounded. Further, if $\epsilon > 0$ is given then there are $f_1, \dots, f_n \in \mathcal{K}$ so that each $f \in \mathcal{K}$ is within $\frac{\epsilon}{3}$ of one of the f_j's, $1 \le j \le n$. Suppose $s \in S$. For each j, $1 \le j \le n$, there is an open set U_j in S containing s so that

$$|f_j(t) - f_j(s)| \le \frac{\epsilon}{3}$$

for all $t \in U_j$. It follows that if $f \in \mathcal{K}$ and $t \in U_1 \cap \cdots \cap U_n$, then judicious choice of j allows us to estimate:

$$\begin{aligned} |f(s) - f(t)| &\le |f(s) - f_j(s)| + |f_j(s) - f_j(t)| + |f_j(t) - f(t)| \\ &\le \frac{\epsilon}{3} + \frac{\epsilon}{3} + \frac{\epsilon}{3} = \epsilon, \end{aligned}$$

and so $\mathcal{K}$ is indeed equicontinuous. □

An added bonus.

Corollary 5.5. *Let G be a topological group, and let M be a nonempty compact subset of G. Suppose $\mathcal{K} \subseteq C(G)$ is equicontinuous on M. Then given $\epsilon > 0$ there is an open set V containing the identity of G so that if $x, y \in M$ and $x \in yV$, then $|f(x) - f(y)| \le \epsilon$ for all $f \in \mathcal{K}$.*

Proof. (It is useful to take a close look at the proof of Theorem 5.1 before looking at this proof.)

Let $\epsilon > 0$ be given. For each $a \in M$ there is an open set V_a that contains the identity e of G such that if $x \in M$ and $x \in aV_a$, then $|f(x) - f(a)| \le \frac{\epsilon}{2}$ for all $f \in \mathcal{K}$. Since $e \cdot e = e$ there is an open set W_a that contains e and satisfies

$$W_a \cdot W_a \subseteq V_a.$$

Now if $a \in M$, then $a \in a \cdot W_a$. Hence $\{a \cdot W_a : a \in M\}$ forms an open cover of M. Therefore we can find $a_1, \dots, a_n \in M$ so

$$M \subseteq (a_1 \cdot W_{a_1}) \cup \cdots \cup (a_n \cdot W_{a_n}).$$

Clearly,

$$V = W_{a_1} \cap \cdots \cap W_{a_n}$$

is open and contains e. Let $x, y \in M$ and suppose $x \in yV$. Since $y \in M$, $y \in a_i \cdot W_{a_i}$ for some $i, 1 \leq i \leq n$. It follows that

$$|f(y) - f(a_i)| \leq \frac{\epsilon}{2}$$

for all $f \in \mathcal{K}$. After all, $W_a \subseteq V_a$. Further,

$$x \in yV \subseteq a_i \cdot W_{a_i} \cdot V \subseteq a_i \cdot W_{a_i} \cdot W_{a_i} \subseteq a_i V_{a_i},$$

and so

$$|f(x) - f(a_i)| \leq \frac{\epsilon}{2}$$

for all $f \in \mathcal{K}$ as well.

In sum, if $x, y \in M$ and $x \in yV$, then, regardless of $f \in \mathcal{K}$,

$$|f(x) - f(y)| \leq |f(x) - f(a_i)| + |f(a_i) - f(y)| \leq \frac{\epsilon}{2} + \frac{\epsilon}{2} = \epsilon. \qquad \square$$

2. Existence and Uniqueness of an Invariant Mean

In this section, we'll show how to ascribe to each $f \in C(G)$ a mean $M(f)$, which is at one and the same time linear in f, nonnegative when f is, and is a true average with the values at f and any right translate of f, identical.

Let G be a compact topological group. Denote by $\mathcal{F}(G)$ the collection of nonempty finite subsets of G and by $C(G)$ the Banach space of all continuous real-valued functions defined on G, equipped as usual with the supremum norm.

Throughout this section, if $F_1, F_2 \in \mathcal{F}(G)$, then by $F_1 \cdot F_2$, we mean all words $a \cdot b$, where $a \in F_1$ and $b \in F_2$; in particular, if $a_1 \cdot b_1 = a_2 \cdot b_2$ but $a_1 \neq a_2$, then we distinguish $a_1 \cdot b_1$ and $a_2 \cdot b_2$.

Definition 5.6. If $F \in \mathcal{F}(G)$ and $f \in C(G)$, then

$$\mathrm{RAve}_F f(x) := \frac{1}{|F|} \sum_{a \in F} f(xa), \quad x \in G.$$

Naturally $\mathrm{RAve}_F f \in C(G)$.

Lemma 5.7. (i) *If* $f \in C(G)$, *then* $\min f, \max f$, *and* $\operatorname{Osc} f = \max f - \min f$ *all exist.*

(ii) *If* $f \in C(G)$ *and* $F \in \mathcal{F}(G)$, *then*

$$\operatorname{Osc} \mathrm{RAve}_F f \leq \operatorname{Osc} f.$$

In fact,

$$\min f \leq \min \mathrm{RAve}_F f \leq \max \mathrm{RAve}_F f \leq \max f.$$

Note that $\mathcal{K}$ is uniformly bounded since

$$\begin{aligned} |\mathrm{RAve}_F f(x)| &= \frac{1}{|F|}\left|\sum_{a\in F} f(xa)\right| \\ &\leq \frac{1}{|F|}\sum_{a\in F}|f(xa)| \\ &\leq \frac{1}{|F|}|F|\cdot\|f\| = \|f\|_\infty. \end{aligned}$$

Thus $\mathcal{K}$ is uniformly bounded, equicontinuous family in $C(G)$. □

Lemma 5.9 brings the Arzelá–Ascoli theorem into play.

With Lemmas 5.8 and 5.9 in hand, the plan of attack is clear. We want an averaging technique which will give a true average, assigning values in a uniformly distributed manner. If the function f is constant, then we will plainly want to assign that value of constancy to f. With the aforementioned lemmas in hand, we handle nonconstant functions thusly: If f is not constant, then we can find $F_1 \in \mathcal{F}(G)$ so that

$$\mathrm{Osc}\mathrm{RAve}_{F_1} < \mathrm{Osc} f.$$

If $\mathrm{RAve}_{F_1}(f)$ is constant, then its value of constancy is the natural value to ascribe to f. If $\mathrm{RAve}_{F_1}(f)$ is not constant, then we appeal to Lemma 5.9 again to find $F_2 \in \mathcal{F}(G)$ so that

$$\mathrm{Osc}\mathrm{RAve}_{F_2}\mathrm{RAve}_{F_1} < \mathrm{Osc}\mathrm{RAve}_{F_1}(f).$$

Continuing in this vein, we see that in the worst case, we can find a sequence $(F_n) \subseteq \mathcal{F}(G)$ so that for each n

$$\mathrm{Osc}\mathrm{RAve}_{F_{n+1}}\mathrm{RAve}_{F_n}(f) < \mathrm{Osc}\mathrm{RAve}_{F_n}(f).$$

Appealing to Messrs. Arzelá and Ascoli, we can pass to a sequence $(F_n^{'}) \subset \mathcal{F}(G)$ so $(\mathrm{RAve}_{F_n^{'}}(f))$ is uniformly convergent.

The point is that because our averages were taken with respect to right translates, in the long run, judicious choices of the F_n's ought to produce an average that is right invariant. Remarkably enough the wisdom needed has already been provided by von Neumann.

Lemma 5.10. *Let $f \in C(G)$ and $\mathcal{K} = \{\mathrm{RAve}_F f : F \in \mathcal{F}(G)\}$. Then*

$$\inf_{g\in\mathcal{K}} \mathrm{Osc}\, g = 0.$$

Proof. Let

$$s = \inf_{g\in\mathcal{K}} \mathrm{Osc}\, g = \inf\{\mathrm{Osc}\,\mathrm{RAve}_F f : F \in \mathcal{F}(G)\}.$$

Therefore there exists (F_n) in $\mathcal{F}(G)$ such that $\mathrm{Osc}(\mathrm{RAve}_{F_n})(f) \searrow s$. Thanks to Arzelá and Ascoli we can, by passing to subsequences if necessary, assume that

$$\mathrm{RAve}_{F_n} f \to g \in C(G)$$

uniformly. It's plain that on assuming the uniform convergence of (RAve_{F_n}) that

$$\min \mathrm{RAve}_{F_n} f \to \min g \quad \text{and} \quad \max \mathrm{RAve}_{F_n} f \to \max g,$$

and so

$$\mathrm{Osc}\mathrm{RAve}_{F_n} f \to \mathrm{Osc} g.$$

Thus $\mathrm{Osc} g = s$.

Here's the point: g is constant! Indeed if g were not constant there would be an $F_0 \in \mathcal{F}(G)$ such that

$$s_0 = \mathrm{Osc}\mathrm{RAve}_{F_0} g < \mathrm{Osc} g = s$$

thanks to Lemma 5.8. Since $(\mathrm{RAve}_{F_n} f)$ is uniformly convergent, there exists N such that

$$||\mathrm{RAve}_{F_N} f - g||_\infty < \frac{s - s_0}{3};$$

i.e., for any $x \in G$,

$$|\mathrm{RAve}_{F_N} f(x) - g(x)| \le \frac{s - s_0}{3}.$$

But this is quickly seen to mean

$$|\mathrm{RAve}_{F_0}\mathrm{RAve}_{F_N} f(x) - \mathrm{RAve}_{F_0} g(x)| \le \frac{s - s_0}{3}$$

for all $x \in G$ as well. It follows that

$$|\mathrm{Osc}\mathrm{RAve}_{F_0}\mathrm{RAve}_{F_N} f - \mathrm{Osc}\mathrm{RAve}_{F_0} g| < 2\left(\frac{s - s_0}{3}\right);$$

i.e.,

$$|\mathrm{Osc}\mathrm{RAve}_{F_0}\mathrm{RAve}_{F_N} f - s_0| < 2\left(\frac{s - s_0}{3}\right).$$

But this in turn means that

$$\mathrm{Osc}\mathrm{RAve}_{F_0}\mathrm{RAve}_{F_N} f < s_0 + 2\left(\frac{s - s_0}{3}\right) = \frac{2}{3}s + \frac{1}{3}s_0 < s.$$

But

$$\mathrm{Osc}\mathrm{RAve}_{F_0}\mathrm{RAve}_{F_N} f = \mathrm{Osc}\mathrm{RAve}_{F_0 F_N} f,$$

and

$$s = \inf_{F \in \mathcal{F}(G)} \mathrm{Osc}\mathrm{RAve}_F f.$$

This should elicit an "OOPS" because

$$\mathrm{RAve}_{F_0}\mathrm{RAve}_{F_N} f = \mathrm{RAve}_{F_0 \cdot F_N} f \in \mathcal{K}.$$

Therefore g is constant and $s = 0$; i.e.,

$$\inf_{g\in\mathcal{K}} \operatorname{Osc} g = 0. \qquad \square$$

We say the real number p is a *right mean* of f if for each $\epsilon > 0$, there is an $F \in \mathcal{F}(G)$ such that

$$|\mathrm{RAve}_F f(x) - p| < \epsilon$$

for all $x \in G$; i.e.,

$$||\mathrm{RAve}_F f - p||_\infty < \epsilon.$$

Theorem 5.11. *Every $f \in C(G)$ has a right mean.*

Proof. By the techniques used in Lemma 5.10, there is a constant function h (say $h(x) \equiv p$) and a sequence $(F_n) \subseteq \mathcal{F}(G)$ such that

$$\lim_n ||\mathrm{RAve}_{F_n} f - h||_\infty = 0;$$

i.e.,

$$||\mathrm{RAve}_{F_n} f - p||_\infty \to 0$$

as $n \to \infty$. Plainly p is a right mean of f. $\square$

It's plain that each $f \in C(G)$ has a *left mean* as well, that is, there is a $q \in \mathbb{R}$ so that if $\epsilon > 0$ is given, there exists an $F \in \mathcal{F}(G)$ so that

$$\left| \frac{1}{|F|} \sum_{a\in F} f(ax) - q \right| < \epsilon$$

for all $x \in G$. For obvious reasons, we define

$$\mathrm{LAve}_F f(x) = \frac{1}{|F|} \sum_{a\in F} f(ax).$$

Theorem 5.12. *Let $f \in C(G)$. Let p be a right mean of f, and let q be a left mean of f. Then $p = q$.*

Proof. Let $\epsilon > 0$. Find $A, B \in \mathcal{F}(G)$ so that

$$||\mathrm{RAve}_A f - p||_\infty \le \frac{\epsilon}{2}, \qquad ||\mathrm{LAve}_B f - q||_\infty \le \frac{\epsilon}{2}.$$

Now

$$\begin{aligned}
\mathrm{RAve}_A\mathrm{LAve}_B f(x) &= \mathrm{RAve}_A \frac{1}{|B|}\sum_{b\in B} f(bx) \\
&= \frac{1}{|A|}\frac{1}{|B|}\sum_{a\in A} S(xa) \ \ (\text{where } \ S(x) = \sum_{b\in B} f(bx)) \\
&= \frac{1}{|A|}\frac{1}{|B|}\sum_{a\in A}\sum_{b\in B} f(bxa) \\
&= \frac{1}{|B|}\frac{1}{|A|}\sum_{b\in B}\sum_{a\in A} f(bxa) \\
&= \frac{1}{|B|}\sum_{b\in B}\frac{1}{|A|}\sum_{a\in A} f(bxa) \\
&= \frac{1}{|B|}\sum_{b\in B}\mathrm{RAve}_A f(bx) \\
&= \mathrm{LAve}_B\mathrm{RAve}_A f.
\end{aligned}$$

Further,

$$\mathrm{RAve}_A(\mathrm{LAve}_B f - q) = \mathrm{RAve}_A\mathrm{LAve}_B f - q$$

and

$$\mathrm{LAve}_B(\mathrm{RAve}_A f - p) = \mathrm{LAve}_B\mathrm{RAve}_A f - p.$$

So for any $x \in G$,

$$\begin{aligned}
|p-q| &= |p - \mathrm{RAve}_A\mathrm{LAve}_B f(x) + \mathrm{RAve}_A\mathrm{LAve}_B f(x) - q| \\
&\leq |p - \mathrm{RAve}_A\mathrm{LAve}_B f(x)| + |\mathrm{RAve}_A\mathrm{LAve}_B f(x) - q| \\
&= |p - \mathrm{LAve}_B\mathrm{RAve}_A f(x)| + |\mathrm{RAve}_A\mathrm{LAve}_B f(x) - q| \\
&= |\mathrm{LAve}_B(p - \mathrm{RAve}_A f(x))| + |\mathrm{RAve}_A(\mathrm{LAve}_B f(x) - q)| \\
&\leq |p - \mathrm{RAve}_A f(x)| + |\mathrm{LAve}_B f(x) - q| \\
&\quad (\text{since } |\mathrm{LAve}_B f| \leq |f| \text{ and } \mathrm{RAve}_B f| \leq |f|) \\
&< \frac{\epsilon}{2} + \frac{\epsilon}{2} = \epsilon,
\end{aligned}$$

and $p = q$. Go figure. □

Corollary 5.13. *For any $f \in C(G)$ there is a unique number $M(f)$ that is both a right and left mean.*

Theorem 5.14. *The functional M on $C(G)$satisfies the following:*

(i) *M is linear.*

(ii) *$Mf \geq 0$ if $f \geq 0$.*

(iii) $M(1) = 1$.

(iv) $M({}_af) = M(f) = M(f_a)$ *for each* $a \in G$, *where* ${}_af(x) = f(ax)$ *and* $f_a(x) = f(xa)$.

(v) $M(f) > 0$ *if* $f \geq 0$ *but* $f \neq 0$.

(vi) $M(\hat{f}) = M(f)$ *where* $\hat{f}(x) = f(x^{-1})$ *for each* $x \in G$.

Proof. We start by showing

$$M(\text{RAve}_F f) = M(f) \tag{17}$$

for each $f \in C(G)$ and each $F \in \mathcal{F}(G)$. Suppose that $M(f) = p$. If $\epsilon > 0$ is given to us, then we can find $F_0 \in \mathcal{F}(G)$ such that

$$||\text{LAve}_{F_0} f - p||_\infty \leq \epsilon;$$

i.e.,

$$\left| \frac{1}{|F_0|} \sum_{b \in F_0} f(bx) - p \right| \leq \epsilon$$

for all $x \in G$. It follows that for any $x \in G$,

$$|\text{RAve}_F \text{LAve}_{F_0} f(x) - p| \leq \epsilon.$$

Since

$$\text{RAve}_F \text{LAve}_{F_0} f = \text{LAve}_{F_0} \text{RAve}_F f,$$

p is a left mean of $\text{RAve}_F f$. Hence by our previous result,

$$M(\text{RAve}_F f) = p$$

and

$$M(\text{RAve}_F f) = M(f).$$

To see that M is linear, let $M(f) = p$ and $M(h) = q$. Pick $H_0 \in \mathcal{F}(F)$ so that

$$||\text{RAve}_{H_0} h - q||_\infty \leq \epsilon;$$

i.e., for all $x \in G$,

$$\left| \frac{1}{|H_0|} \sum_{b \in H_0} h(xb) - q \right| \leq \epsilon;$$

i.e., if $E \in \mathcal{F}(G)$ and $x \in G$, then

$$|\text{RAve}_{E \cdot H_0} h(x) - q| = |\text{RAve}_E \text{RAve}_{H_0} h(x) - q| < \epsilon.$$

Now

$$p = M(f) = M(\text{RAve}_F f)$$

for any $F \in \mathcal{F}$. Therefore p is the right mean of $\text{RAve}_{H_0} f$. Hence there exists $F_0 \in \mathcal{F}(G)$ so that

$$||\text{RAve}_{F_0} \text{RAve}_{H_0} f - p||_\infty \leq \epsilon;$$

i.e., for all $x \in G$,

$$|\text{RAve}_{F_0 \cdot H_0} f(x) - p| = |\text{RAve}_{F_0} \text{RAve}_{H_0} f(x) - p| \leq \epsilon.$$

Since we already know that for all $x \in G$ and each $E \in \mathcal{F}(G)$,

$$\left| \text{RAve}_{E \cdot H_0} h(x) - q \right| \leq \epsilon,$$

it follows that by taking $E = F_0$ we get for each $x \in G$,

$$|\text{RAve}_{F_0 \cdot H_0}(f + h)(x) - (p + q)| \leq 2\epsilon.$$

Thus

$$M(f + h) = M(f) + M(h).$$

It follows from this and the easily established fact that $M(kf) = kM(f)$ that M is linear, and we have shown (i).

Parts (ii) and (iii) are clear. To see (iv), since

$$\text{RAve}_F f(xa) = \text{RAve}_{a \cdot F} f(x), \tag{18}$$

$$\begin{aligned} M(f_a) &= M(\text{RAve}_F f_a(x)) \text{ (by (17))} \\ &= M(\text{RAve}_F f(xa)) \\ &= M(\text{RAve}_{a \cdot F} f(x)) \text{ (by (18))} \\ &= M(f) \text{ (by (17))}. \end{aligned}$$

Similarly,

$$\text{LAve}_F f(ax) = \text{LAve}_{F \cdot a} f(x), \tag{19}$$

and so

$$\begin{aligned} M({}_a f) &= M(\text{LAve}_F({}_a f(x))) \\ &\quad \text{(by (17)) actually it's equivalent with left averages)} \\ &= M(\text{LAve}_F f(ax)) \\ &= M(\text{LAve}_{F \cdot a} f(x)) \text{ (by (19))} \\ &= M(f) \text{ (by (17) actually it's equivalent with left averages)}, \end{aligned}$$

and thus

$$M({}_a f) = M(f) = M(f_a).$$

For (v), suppose that $f \in C(G), f \geq 0, f \not\equiv 0$. Then there is $\alpha > 0$ such that $U = \lfloor f > \alpha \rfloor$ is nonempty and open. It's easy to see that $\{Ua^{-1} : a \in G\}$ is an open cover of the compact G. It follows that for some $a_1, \ldots, a_m \in G$

$$G = Ua_1^{-1} \cup Ua_2^{-1} \cup \cdots \cup Ua_m^{-1}.$$

Let's check to see how this plays out.

defined on K. This extension, whose value at $f \in B(\mathcal{A})$ will still be denoted by $\int f\ \lambda$, satisfies

$$\left|\int f\, d\lambda\right| \leq \int |f|\, d\lambda \leq ||f||_\infty.$$

Moreover, if $f \geq 0$, then $\int f\, d\lambda \geq 0$.

It is worth noting that $C(K) \subseteq B(\mathcal{A})$. In fact, if $\epsilon > 0$ and $x \in C(K)$, then we can approximate x by an $\mathcal{A}$-simple function x_ϵ as follows. Cover $x(K)$ by open sets $G_1, \ldots, G_n$ each of diameter less than ϵ. Let $A_1 = G_1, A_2 = G_2 \setminus G_1, \ldots, A_j = G_j \setminus (G_1 \cup \cdots \cup G_{j-1}), \ldots$ If $A_j \neq \emptyset$, let $x_\epsilon(k) = x(k)$ for all $k \in A_j$, and $x_\epsilon(k) = 0$ if $k \notin \bigcup_{j=1}^n G_j$. It's plain that

$$|x(k) - x_\epsilon(k)| \leq \epsilon$$

for all $k \in K$ and x_ϵ is $\mathcal{A}$-simple.

Of course, $\int d\lambda$ acts as continuous positive linear functional on $C(K)$ with norm (as a functional) less than or equal to 1, which is $\lambda(K)$. In order to assume more about λ, namely, we need to know that λ is *weakly regular* in the sense that given $\epsilon > 0$ and $A \in \mathcal{A}$ there is a closed set $F \subseteq A$ and an open set $U \supseteq A$ so that $\lambda(U \setminus F) < \epsilon$. The result follows:

Fact 5.16. If λ is a weakly regular finitely additive probability on $\mathcal{A}$, then $f \to \int f\, d\lambda$ is a norm one functional on $C(K)$.

Proof. Let $\epsilon > 0$ be given. Let $E_1, \ldots, E_n$ be pairwise disjoint member $\mathcal{A}$ so that $K = E_1 \cup \cdots \cup E_n$. For each j, $1 \leq j \leq n$, pick F_j closed, $F_j \subseteq E_j$ with $\lambda(E_j \setminus F_j) \leq \frac{\epsilon}{n}$. Using K's normality, let $G_1, \ldots, G_n$ be open sets that are disjoint and so that $F_j \subseteq G_j$. Shrink the G_j's if necessary so that $\alpha(G_j \setminus F_j) < \frac{\epsilon}{n}$–$\lambda$'s weak regularity comes in handy again.

Let $x_1, \ldots, x_n \in C(K)$ be chosen so that $0 \leq x_j \leq 1$ and $x_j = 0$ outside of G_j but $x_j = 1$ inside of F_j. Set

$$x_0 = \sum_{j=1}^n x_j \in C(K).$$

Then $||x_0||_\infty = 1$ and

$$\left|\int x_0\, d\lambda - 1\right| \leq 3\epsilon.$$

It follows from this and the usual epsilonics that

$$\left|\left|f \to \int f\, d\lambda\right|\right|_{C(K)^*} = 1. \qquad \square$$

Now let x^* be a positive linear functional on $C(K)$. Because $x \in C(K)$ has $||x||_\infty \leq 1$ means $-1 \leq x \leq +1$ (as functions on K),

$$-x^*(1) = x^*(-1) \leq x^*(x) \leq x^*(1),$$

and so $||x^*|| = x^*(1)$. Applying Kantorovich's theorem to extend x^* to a positive linear functional y^* on $l^\infty(K)$ we see that $||y^*|| = y^*(1)$ as well. Now for any $A \in \mathcal{A}, y^*(\chi_A) = \lambda(A)$ is a nonnegative finitely additive measures on $\mathcal{A}$. If we assume $||y^*|| = 1 (= ||x^*||)$, then λ is a finitely additive probability and for any $f \in B(\mathcal{A})$,

$$y^*(f) = \int f \, d\lambda.$$

If λ is weakly regular, then the functional $x \to \int x \, d\lambda$ already achieves x^*'s norm on $C(K)$. Otherwise? Otherwise, we will show how to modify λ to get a finer representing measure.

Generically, F is a closed subset of K, G is open subset of K, and E is any subset of K. Set

$$\mu_1(F) = \inf\{\lambda(G) : F \subseteq G\}, \qquad \mu_2(E) = \sup\{\mu_1(F) : F \subseteq E\}.$$

Both μ_1, μ_2 take nonnegative values and each is ascending on its domain.

Suppose $F_1 \backslash G_1 \subseteq G$. Then $F_1 \subseteq G_1 \cup G$, and so

$$\lambda(G_1 \cup G) \leq \lambda(G_1) + \lambda(G)$$

implies

$$\mu_1(F_1) \leq \lambda(G_1) + \lambda(G).$$

Therefore

$$\mu_1(F_1) \leq \lambda(G_1) + \mu(F_1 \backslash G_1),$$

after all, $F_1 \backslash G_1$ is a closed set contained in G. Since G_1 is any open set, we let G_1 range over the open sets that contain a given $F \cap F_1$, and we see

$$\mu_1(F_1) \leq \mu_1(F \cap F_1) + \mu_1(F_1 \backslash G_1).$$

But $F_1 \backslash F = F_1 \backslash (F_1 \cap F)$ which contains $F_1 \backslash G_1$ so

$$\mu_1(F_1) \leq \mu_1(F \cap F_1) + \mu_2(F_1 \backslash F).$$

If we now let $E \subseteq K$ and F_1 range over the closed sets contained in E the result is

$$\mu_2(E) \leq \mu_2(F \cap E) + \mu_2(E \backslash F), \tag{20}$$

where E and F are unrestricted within their generic classes.

We plan to show the reverse inequality stated in (20) holds as well.

Suppose F_1 and F_2 are *disjoint* closed sets. Envelop F_1 and F_2 in *disjoint* open sets G_1 and G_2, respectively. Let G contain $F_1 \cup F_2$. Then

$$\lambda(G) \geq \lambda(G \cap G_1) + \lambda(G \cap G_2) \geq \mu_1(F_1) + \mu_1(F_2);$$

hence

$$\mu_1(F_1 \cup F_2) \geq \mu(F_1) + \mu(F_2).$$

Now look to generic E, F, and let F_1 range over the closed subsets of $E \cap F$, and let F_2 range over the closed sets contained in $E \backslash F$. The result is

$$\mu_2(E) \geq \mu_1(F_1 \cup F_2) \geq \mu_2(E \cap F) + \mu_2(E \backslash F),$$

and so

$$\mu_2(E) = \mu_2(E \cap F) + \mu_2(E \backslash F). \tag{21}$$

Look familiar? It should. Were μ_2 an outer measure, then (21) would say that each closed set F is μ_2-measurable.

Now μ_2 is *not* necessarily an outer measure but it's not far from being such. We say that a set $E \subseteq K$ is a μ_2-set if

$$\mu_2(A) = \mu_2(A \cap E) + \mu_2(A \cap E^c)$$

for each $A \in \mathcal{A}$. What we've just seen is that each closed set is a μ_2-set. Moreover, following the reasoning encountered in the Carathéodory procedure, the collection of μ_2-sets is an algebra of sets. Let's see that this is so.

Of course, $\emptyset, K$ are both μ_2-sets and E is a μ_2-set if and only if E^c is. Suppose E_1 and E_2 are μ_2-sets. Take $A \in \mathcal{A}$. Since E_1 is a μ_2-set

$$\mu_2(A \cap E_2) = \mu(A \cap E_2 \cap E_1) + \mu_2(A \cap E_2 \cap E_1^c). \tag{22}$$

Since E_2 is a μ_2-set,

$$\mu_2(A) = \mu(A \cap E_2) + \mu_2(A \cap E_2^c). \tag{23}$$

Also

$$\mu_2(A \cap (E_1 \cap E_2)^c) = \mu_2(A \cap (E_1 \cap E_2)^c \cap E_2) + \mu_2(A \cap (E_1 \cap E_2)^c \cap E_2^c).$$

But

$$(E_1 \cap E_2)^c \cap E_2 = E_1^c \cap E_2 \text{ and } (E_1 \cap E_2)^c \cap E_2^c = E_2^c,$$

and so

$$\mu_2(A \cap (E_1 \cap E_2)^c) = \mu(A \cap E_2 \cap E_1^c) + \mu_2(A \cap E_2^c). \tag{24}$$

It follows that

$$\begin{aligned}
\mu_2(A) &= \mu_2(A \cap E_2) + \mu_2(A \cap E_2^c) \text{ (by (23))} \\
&= \mu_2(A \cap (E_1 \cap E_2)) + \mu_2(A \cap E_2 \cap E_1^c) + \mu_2(A \cap E_2^c) \text{ (by (22))} \\
&= \mu_2(A \cap (E_1 \cap E_2)) + \mu_2(A \cap (E_1 \cap E_2)^c) \text{ (by (24))},
\end{aligned}$$

and $E_1 \cap E_2$ is also a μ_2-set. The collection of μ_2-sets is an algebra of sets containing the topology of K.

If E_1 and E_2 are disjoint μ_2-sets and $A \in \mathcal{A}$, then

$$\begin{aligned} \mu_2(A \cap (E_1 \cup E_2)) &= \mu_2(A \cap (E_1 \cup E_2) \cap E_1) + \mu_2(A \cap (E_1 \cup E_2) \cap E_1^c) \\ &\quad \text{(since } E \text{ is a } \mu_2\text{-set)} \\ &= \mu_2(A \cap E_1) + \mu(A \cap E_2), \end{aligned}$$

and so μ_2 is finitely additive on the algebra of μ_2-sets.

If we restrict μ_2 to $\mathcal{A}$ we get a finitely additive probability μ. Moreover for any closed set $F \subseteq K$,

$$\mu(F) = \mu_2(F) = \mu_1(F),$$

as careful inspection of the respective definitions will reveal.

Let $A \in \mathcal{A}$. By μ_2's very definition

$$\begin{aligned} \mu(A) &= \mu_2(A) = \sup\{\mu_1(F) : F \text{ is closed and } F \subseteq A\} \\ &= \sup\{\mu(F) : F \text{ is closed, } F \subseteq A\}; \end{aligned}$$

further,

$$\begin{aligned} \mu(A) &= \mu(K) - \mu(A^c) = \mu(K) - \sup\{\mu(F) : F \text{ is closed, } F \subseteq A\} \\ &= \mu(K) + \inf\{\mu(F^c) - \mu(K) : F \text{ is closed, } F \subseteq A^c\} \\ &= \inf\{\mu(F^c) : F \text{ is closed, } F \subseteq A\} \\ &= \sup\{\mu(G) : G \text{ is open } A \subseteq G\}. \end{aligned}$$

It remains to show that

$$\int x \, d\lambda = \int x \, d\mu$$

for each and every $x \in C(K)$. But again we need only do this for nonnegative real-valued x's. Since each $x \in C(K)$ is bounded, we'll turn our attention to $x \in C(K)$ so that $0 \leq x \leq 1$.

Let $\epsilon > 0$. Partition K into *disjoint* members $E_1, \ldots, E_n$ of $\mathcal{A}$ in such a way that

$$\int x \, d\mu \leq \sum_{j=1}^{n} \left(\inf_{k \in E_j} f(k) \right) \mu(E_j) + \epsilon.$$

Since μ is regular, we can find closed sets $F_1, \ldots, F_n$ with $F_j \subseteq E_j$ such that

$$\int x \, d\mu \leq \sum_{j=1}^{n} \left(\inf_{k \in E_j} f(k) \right) \mu(F_j) + 2\epsilon.$$

But K is normal and $x \in C(K)$ so we can envelop the F_j's within *disjoint* open sets G_j where $F_j \subseteq G_j$ for $j = 1, \ldots, n$ with

$$\inf_{k \in G_j} x(k) \geq \inf_{k \in E_j} x(k) - \frac{\epsilon}{n(\mu(K))}.$$

This being done,

$$\int x \, d\mu \leq \sum_{j=1}^{n} \left(\inf_{k \in G_j} x(k) \right) \mu(G_j) + 3\epsilon.$$

But remember that

$$\mu_1(F) = \mu_2(F) = \mu(F)$$

for closed sets $F \subseteq K$, and if G is open with $F \subseteq G$, then $\mu(F) \leq \lambda(G)$. It follows that $\mu(G) \leq \lambda(G)$ and

$$\sum_{j=1}^{n} \left(\inf_{k \in G_j} x(k) \right) \mu(G_j) \leq \sum_{j=1}^{n} \left(\inf_{k \in G_j} x(k) \right) \lambda(G_j) \leq \int x \, d\lambda.$$

So

$$\begin{aligned}
\int x \, d\mu &\leq \sum_{j=1}^{n} \left(\inf_{k \in G_j} x(k) \right) \mu(G_j) + 3\epsilon \\
&\leq \sum_{j=1}^{n} \left(\inf_{k \in G_j} x(k) \right) \lambda(G_j) + 3\epsilon \\
&\leq \int x \, d\lambda + 3\epsilon.
\end{aligned}$$

Since $\epsilon > 0$ was arbitrary, we have that for any $x \in C(K)$, with $0 \leq x \leq 1$,

$$\int x \, d\mu \leq \int x \, d\lambda.$$

If we apply this to $(1 - x)$ and keep in mind that

$$\mu(K) = \lambda(K),$$

then the reverse inequality follows from

$$\int (1 - x) d\mu \leq \int (1 - x) d\lambda,$$

and that is all she wrote!

Our next task is to relate positive linear functionals on $C(K)$ to *countably additive* regular Borel measures on the Borel σ-field $\mathcal{B}_o(K)$. Here's a safe first step.

Theorem 5.17 (Alexandroff). *Let μ be a regular finitely additive nonnegative measure defined on the Borel field $\mathcal{A}$ (the field generated by the open sets) of the compact Hausdorff space K. Then μ is countably additive on $\mathcal{A}$.*

Proof. Let $\epsilon > 0$. Suppose (E_n) is a sequence of pairwise disjoint members of $\mathcal{A}$ with $E = \bigcup_n E_n \in \mathcal{A}$. There is an $F \in \mathcal{A}$, F closed, $F \subseteq E$, with $\mu(E \setminus F) < \epsilon$.

Moreover for each n we can find a G_n that is open with $E_n \subseteq G_n$ and

$$\mu(G_n \backslash E_n) < \frac{\epsilon}{2^n}.$$

Now

$$F \subseteq E = \bigcup_n E_n \subseteq \bigcup_n G_n;$$

so K's compactness, which is inherited by the closed set F ensures us that there is an N so

$$F \subseteq G_1 \cup \cdots \cup G_N.$$

It follows that

$$\begin{aligned} \sum_n \mu(E_n) &\geq \sum_n |\mu|(G_n) - \epsilon \\ &\geq \sum_{n \leq N} |\mu|(G_n) - \epsilon \\ &\geq \mu(F) - \epsilon \\ &\geq \mu(E) - 2\epsilon. \end{aligned}$$

So $\epsilon > 0$ being arbitrary soon reveals that

$$\sum_n \mu(E_n) \geq \mu(E) = \mu\left(\bigcup_n E_n\right).$$

On the other hand, for any $M \in \mathbb{N}$,

$$\mu(E) = \mu\left(\bigcup_n E_n\right) \geq \mu\left(\bigcup_{n \leq M} E_n\right) = \sum_{n \leq M} \mu(E_n),$$

so

$$\mu\left(\bigcup_n (E_n)\right) \geq \sum_n \mu(E_n)$$

too. So μ is countably additive. □

Now we note that having a nonnegative real-valued countably additive measure μ on the field $\mathcal{A}$, if $A \subseteq K$ and we define $\mu^*(A)$ by

$$\mu^*(A) = \inf\left\{\sum_n \mu(E_n)\right\},$$

where the infimum is taken over all sequence (E_n) of members of $\mathcal{A}$ such that $A \subseteq \bigcup_n E_n$, then the result that μ^* is an outer measure on K. Two points are worth making here: First, Carathéodory's Theorem (see Chapter 2) tells us that the μ^*-measurable sets constitute a σ-field of subsets of K on which μ^* is countably additive. Second, if $E \in A$, then $\mu^*(E) = \mu(E)$.

We need to establish this second feature of μ^*, so let $E \in \mathcal{A}$. Of course $E \subseteq E$ tells us immediately that

$$\mu^*(E) \leq \mu(E).$$

On the other hand, if $E \in \mathcal{A}$ and $E \subseteq \bigcup_n E_n$ where each $E_n \in \mathcal{A}$, then the disjointification of E_n

$$D_1 = E_1, D_2 = E_2 \backslash E_1, \ldots, D_j = E_j \backslash \bigcup_{k=1}^{j-1} E_k, \ldots,$$

is a disjoint sequence of members of $\mathcal{A}$ with $\bigcup_n E_n = \bigcup_n D_n$. It follows that

$$\begin{aligned}
\mu(E) &= \mu\left(E \cap \left(\bigcup_n D_n\right)\right) \\
&= \mu\left(\bigcup_n (E \cap D_n)\right) \\
&= \sum_n \mu(E \cap D_n) \quad (\mu \text{ is countably additive on } \mathcal{A}) \\
&\leq \sum_n \mu(D_n) \\
&\leq \sum_n \mu(E_n).
\end{aligned}$$

It follows that $\mu(E) \leq \mu^*(E)$ too. So if $E \in \mathcal{A}$, then $\mu(E) = \mu^*(E)$.

Finally, we note that each $E \in \mathcal{A}$ is μ^*-measurable. To show this, we need to test *vis-à-vis* arbitrary $A \subseteq K$. Since μ^* is an outer measure and

$$A = (A \cap E) \cup (A \cap E^c),$$

then

$$\mu^*(A \cap E) + \mu^*(A \cap E^c) \leq \mu^*(A).$$

Now let $\epsilon > 0$ be given and find $(E_n) \subseteq \mathcal{A}$ such that

$$A \subseteq \bigcup_n E_n$$

and

$$\sum_n \mu(E_n) \leq \mu^*(A) + \epsilon.$$

Notice that

$$A \cap E \subseteq \bigcup_n (E_n \cap E) \qquad \text{and} \qquad A \cap E^c \subseteq \bigcup_n (E_n \cap E^c).$$

Therefore

$$\begin{aligned} \mu^*(A \cap E) + \mu^* A \cap E^c) &\leq \sum_n \mu(E_n \cap E) + \sum_n \mu(E_n \cap E^c) \\ &= \sum_n \mu(E_n \cap E) + \mu(E_n \cap E^c) \\ &= \sum_n \mu(E_n) \\ &\leq \mu^*(A) + \epsilon. \end{aligned}$$

But $\epsilon > 0$ was arbitrary so we have

$$\mu^*(A \cap E) + \mu(A \cap E^c) \leq \mu^*(A).$$

We have shown that for every $E \in \mathcal{A}$ and for every $A \subseteq K$,

$$\mu^*(A) = \mu^*(A \cap E) + \mu^*(A \cap E^c),$$

and so each member of $\mathcal{A}$ is μ^*-measurable.

Alas when we consider the extensive μ^* of μ to the Borel σ-field $\mathcal{B}o(K)$, we get a countably additive *regular* Borel probability. Let's see why μ^* is regular.

Let $E \in \mathcal{B}o(K)$, and let $\epsilon > 0$ be given. We can find a sequence (E_n) in $\mathcal{A}$ such that $E \subseteq \bigcup_n E_n$ and

$$\mu^*\left(\bigcup_n E_n \setminus E\right) < \frac{\epsilon}{2}.$$

But μ is a regular on $\mathcal{A}$ so for every n there exists an open set G_n such that

$$E_n \subseteq G_n \text{ and } \mu(G_n \setminus E_n) < \frac{\epsilon}{2^n}.$$

Now $E \subseteq \bigcup E_n \subseteq \bigcup G_n = G$, an open set. Further

$$\begin{aligned} \mu^*\left(G \setminus E\right) &\leq \mu^*\left(G \setminus \left(\bigcup_n E_n\right)\right) + \mu^*\left(\left(\bigcup_n E_n\right) \setminus E\right) \\ &= \mu^*\left(\bigcup_n \left(G_n \setminus E_n\right)\right) + \mu^*\left(\left(\bigcup_n E_n\right) \setminus E\right) \\ &= \sum_n \mu^*\left(G_n \setminus E_n\right) + \mu^*\left(\left(\bigcup_n E_n\right) \setminus E\right) < \epsilon, \end{aligned}$$

and so μ^* is outer regular on $\mathcal{B}o(K)$ and inner regularity follows.

4. Notes and Remarks

We have followed Pontrjagin's [**104**] classic in our discussion of von Neumann's proof of the existence of Haar measure (or invariant mean) for compact groups. A careful look at von Neumann's proof will indicate why he was so quick to see the possibility of an abstract presentation of solutions to the easy problem of measure; of course, as any good Aristotelian understands, it's a long way between possibility and execution. Our presentation of the computation of $C(K)^*$ follows that of Dunford and Schwartz, as found scattered throughout Part I of their three-part bible [**34**].

4.1. Other Proofs of the Existence of Haar Measure in Compact Groups.

4.1.1. *Marriage and Haar Measure.* Can one produce an invariant measure "abstractly"? There are several ways of doing so; presently we will follow a path indicated by Maak, as expressed in the notes of V. Milman and G. Schechtman [**83**]. It calls on basic tools of functional analysis and a combinatorial solution to the "marriage problem". It is well motivated by a simple example.

Let Π denote the unit circle,

$$\Pi = \{z \in \mathbb{C} : |z| = 1\}.$$

Then Π is a compact metric topological group, a prototype of such, if you please. It also possesses a natural invariant probability Borel measure, normalized Lebesgue measure λ, a measure that requires construction, to be sure.

However, λ can be realized by an abstract limiting procedure if we but think in a Riemannian manner. For each n, let $\{z_1, \ldots, z_n\}$ be the n-primitive roots of unity, and let μ_n be

$$\mu_n = \frac{1}{n}\sum_{k=1}^{n} \delta_{z_k},$$

where δ_z is the (continuous linear) functional on $C(\Pi)$ of evaluation at z,

$$\delta_z(f) = f(z).$$

The sequence (μ_n) lies in the dual unit ball $B_{C(\Pi)^*}$ of $C(\Pi)^*$ and so has a weak* limit point. (This is Banach's proof of the Banach–Alaoglu theorem.) Actually, more is so. The sequence (μ_n) is itself weak*-convergent, that is, for any $f \in C(\Pi)$, $\lim_n \mu_n(f)$ exists and

$$\lim_n \mu_n(f) := \lim_n \frac{1}{n}\sum_{k=1}^{n} f(z_k) = \int_0^{2\pi} f(t)d\lambda(t).$$

After all, the n primitive roots of unity are uniformly distributed about Π, and so an appeal to the basic theory of the Riemann integral tells us the limiting procedure above is valid.

We want to modify this procedure to a more general setting and so call on functional analysis for tools to do so. The basic tool will be the Banach–Alaoglu theorem. Recall that if X is a Banach space with dual X^*, then the weak*-topology of X^* is the topology of pointwise convergence on members of X. Salient features of this topology include

- the closed unit ball B_{X^*} of X^* is always weak*-compact;
- B_{X^*} is metrizable in the weak*-topology if and only if X is separable.

Our setting will be $C(S)$ where S is a compact metric space; in this case, $C(S)$ is a separable Banach space and so $B_{C(S)^*}$ is weak*-compact and metrizable. In particular, any sequence (μ_n) in $B_{C(S)^*}$ has a weak*-convergent subsequence.

We now present a combinatorial lemma, sometimes referred to as the *marriage problem.* Its proof goes back to P. Halmos and H. E. Vaughan [**51**].

Suppose for each boy b_i in the set B of n boys,

$$B = \{b_1, \ldots, b_n\},$$

there is a collection $G(b_i)$ of girls with whom b_i is acquainted.

Theorem 5.18. *With bigamy an anathema, for each boy in B to be able to marry a girl with whom he is acquainted it is both necessary and sufficient that regardless of $C \subseteq B$*

$$\left|\bigcup_{b\in C} G(b)\right| \geq \left|C\right| \tag{25}$$

holds.

Proof. Since it is plain that (25) is necessary, we'll concentrate on proving sufficiency of (25). So we suppose (25) to be in effect and prove the possibility of a wise matchmaker by an induction on the number of boys in B. Though not universally associated with marriage, we nevertheless introduce the notion of perfect cliques.

Definition 5.19. A clique $C \subseteq B$ is *perfect* if

$$\left|\bigcup_{b\in C} G(b)\right| = \left|C\right|.$$

Lemma 5.20. *If C and C' are perfect, then so is $C \cup C'$.*

Proof. By (25) we know that

$$\left| \bigcup_{b \in C \cup C'} G(b) \right| \geq \left| C \cup C' \right|.$$

Further

$$|C \cup C'| = |C| + |C'| - |C \cap C'|,$$

and, since it's too much to expect all the $G(b)$'s to be singletons,

$$\begin{aligned} \left| \bigcup_{b \in C \cup C'} G(b) \right| &\leq \left| \bigcup_{b \in C} G(b) \right| + \left| \bigcup_{b \in C'} G(b) \right| - \left| \bigcup_{b \in C \cap C'} G(b) \right| \\ &= |C| + |C'| - \left| \bigcup_{b \in C \cap C'} G(b) \right| \\ &\leq |C| + |C'| - |C \cap C'| \ (\text{by } (25)) \\ &= |C \cup C'|. \end{aligned}$$

So $C \cup C'$ is perfect. ☐

So for now we suppose that for any collection of fewer than n boys, (25) is sufficient to warm the cockles of the matchmaker's cold heart. Let B be a collection of n boys which given any $C \subseteq B$, we have

$$\left| \bigcup_{b \in C} G(b) \right| \geq |C|.$$

We'll take a good look at $G(b_n)$.

If $g \in G(B_n)$, then looking at

$$G(b_1) \setminus \{g\}, G(b_2) \setminus \{g\}, \ldots, G(b_{n-1}) \setminus \{g\},$$

we see two possibilities:

MAYBE for some $g \in G(b_n)$

$$G(b_1) \setminus \{g\}, G(b_2) \setminus \{g\}, \ldots, G(b_{n-1}) \setminus \{g\}$$

satisfies (25) and so we can effect a marriage of each $b_1, b_2, \ldots, b_{n-1}$ to a girl of his acquaintances and still have the lovely g left to marry b_n when $g \in G(b_n)$. Of course in this case we're happy onlookers and the proof is done.

MAYBE for each $g \in G(b_n)$, the collection

$$G(b_1) \setminus \{g\}, G(b_2) \setminus \{g\}, \ldots, G(b_{n-1}) \setminus \{g\}$$

fails (25); i.e., for each $g \in G(b_n)$ there is a $C_g \subseteq \{b_1, b_2, \ldots, b_{n-1}\}$ such that

$$\left| \bigcup_{b_i \in C_g} G(b_i) \setminus \{g\} \right| < |C_g|.$$

But deleting g from $G(b_i)$ can change the number of girls in this collection by at most 1 and

$$\{G(b_i) : b_i \in C_g\}$$

satisfies (25) by our inductive hypothesis so each of the $G(b_i)$'s, with $b_i \in C_g$ must contain g! It follows that

$$\left| \bigcup_{b_i \in C_g} G(b_i) \setminus \{g\} \right| = \left| \bigcup_{b_i \in C_g} G(b_i) \right| - 1 < |C_g|.$$

Since $\{G(b_i) : b_i \in C_g\}$ satisfies (25), it must be that

$$\left| \bigcup_{b_i \in C_g} G(b_i) \right| = |C_g|;$$

i.e., $\{G(b_i) : b_i \in C_g\}$ is perfect for each $g \in G(b_n)$.

Our lemma now assures us that $\{G(b_i) : b_i \in \bigcup_{g \in G(b_n)} C_g\}$ is also perfect and so

$$\left| \bigcup \{G(b_i) : b_i \in \bigcup_{g \in G(b_n)} C_g\} \right| = \left| \bigcup_{g \in G(b_n)} C_g \right|.$$

Keep in mind that

$$G(b_n) \subseteq \bigcup_{b_i \in C_g} G(b_i).$$

So if we add b_n to $\bigcup\{C_g : g \in G(b_n)\}$, we have

$$\begin{aligned} \left| G(b_n) \cup \{G(b_i) : b_i \in \bigcup_{g \in G(b_n)} C_g\} \right| &= \left| \bigcup \{G(b_i) : b_i \in \bigcup_{g \in G(b_n)} C_g\} \right| \\ &= \left| \bigcup \{C_g : g \in G(b_n)\} \right| \\ &< \left| \{b_n\} \cup \left\{ \bigcup_{g \in G(b_n)} C_g \right\} \right|, \end{aligned}$$

and we've found a collection $C \subseteq B$,

$$C = \{b_n\} \cup \left\{ \bigcup_{g \in G(b_n)} C_g \right\},$$

which violates (25).

We see then that the second of the possible outcomes is defined to fail, and so we deduce that the first "MAYBE" is so. $\square$

We will apply our result to proving the existence of a probability measure μ on a compact metric space S that is G-invariant, where G is the group of isometries of S onto itself. The use of the solution to the marriage problem in this context is due to Maak.

Proof. Let S be a compact metric space, and let G be the group of isometries of S onto S. Let $\epsilon > 0$. Let N_ϵ be a minimal ϵ-net; i.e,

$$S = \bigcup_{x \in N_\epsilon} \{y \in S : d(x, y) < \epsilon\},$$

and $|N_\epsilon|$ is minimal.

For $f \in C(S)$, define

$$\mu_{N_\epsilon}(f) = \frac{1}{|N_\epsilon|} \sum_{x \in N_\epsilon} f(x).$$

Clearly, μ_{N_ϵ} is a probability measure on S and as such belongs to $B_{C(S)^*}$, the unit ball of $C(S)^*$. Since $C(S)$ is a separable Banach space, $B_{C(S)^*}$ is compact and metrizable in the weak*-topology. Hence for a suitably chosen sequence $\epsilon_n \searrow 0$, we have that $(\mu_{N_{\epsilon_n}})$ is weak*-convergent to a probability $\mu \in B_{C(S)^*}$.

Our first task will be to show that μ is independent of the choices of N_ϵ made, ϵ-by-ϵ. To this end, we will show that if $\widetilde{N_\epsilon}$ is another minimal ϵ-net, then not only are N_ϵ and $\widetilde{N_\epsilon}$ equinumerous but we can find a bijection ϕ of N_ϵ onto $\widetilde{N_\epsilon}$ such that for any $x \in N_\epsilon$,

$$d(x, \phi(x)) < 2\epsilon.$$

We'll say that $x \in N_\epsilon$ and $y \in \widetilde{N_\epsilon}$ are *acquainted* if

$$\{s \in S : d(x, s) < \epsilon\} \cap \{s \in S : d(y, s) < \epsilon\} \neq \emptyset.$$

We claim that if K is any subset of N_ϵ and L is the subset of $\widetilde{N_\epsilon}$ consisting of s's that are acquainted with some member of K, then L has at least as many members as K does. Why is this so? Well, L consists of precisely those s's in $\widetilde{N_\epsilon}$ such that

$$\{s' \in S : d(s, s') < \epsilon\} \cap \left(\bigcup_{t \in N_\epsilon} \{s' \in S; d(t, s') < \epsilon\} \right) \neq \emptyset.$$

Were L to have fewer members than K, then the set

$$L \cup (N_\epsilon \setminus K)$$

would have cardinality

$$\begin{aligned} |L \cup (N_\epsilon \setminus K)| &\leq |L| + |(N_\epsilon \setminus K)| \\ &< |K| + |(N_\epsilon \setminus K)| = |N_\epsilon|. \end{aligned}$$

Yet for any $x \in S$, x within ϵ of some point of N_ϵ, maybe x is within ϵ of a point of L and maybe not! But if x is *not* within ϵ of a point of L, it is *not* within ϵ of a point of K—after all, that's how L was defined! So x is within ϵ of a point $(N_\epsilon \setminus K)$. Then $L \cup (N_\epsilon \setminus K)$ is an ϵ-net for S with fewer members than N_ϵ. Oops! Our claim is verified.

Now our marriage counselor steps in to advise us that there is a bijection ϕ of N_ϵ onto $\widetilde{N_\epsilon}$ such that for each $t \in N_\epsilon$, $\phi(t)$ is acquainted with t; i.e., there is a bijection $\phi : N_\epsilon \to \widetilde{N_\epsilon}$ such that for each $t \in N_\epsilon$,

$$d(t, \phi(t)) < 2\epsilon.$$

But this entails the following: if $f \in C(S)$, then

$$\begin{aligned} |\mu_{N_\epsilon}(f) - \mu_{\widetilde{N_\epsilon}}(f)| &\leq \frac{1}{|N_\epsilon|} \sum_{t \in N_\epsilon} |f(t) - f(\phi(t))| \\ &\leq \omega_f(2\epsilon), \end{aligned}$$

where

$$\omega_f(\eta) := \sup\{|f(t) - f(s)| : d(s,t) \leq \eta\}$$

is the modulus of (uniform) continuity of $f \in C(S)$. It follows from this that

$$\text{weak} * \lim_n \mu_{N_{\epsilon_n}}, \text{weak} * \lim_n \mu_{\widetilde{N_{\epsilon_n}}},$$

coexist and coincide.

What of it? Well if $g \in G$, then

$$\widetilde{N_\epsilon} = \{g(t) : t \in N_\epsilon\}$$

is a minimal ϵ-net if N_ϵ is, so for any $f \in C(S)$

$$\begin{aligned} \mu(f \circ g) &= \lim_n \frac{1}{|N_\epsilon|} \sum_{t \in N_\epsilon} f(g(t)) \\ &= \lim_n \frac{1}{|N_\epsilon|} \sum_{s \in \widetilde{N_\epsilon}} f(s) \\ &= \lim_n \mu_{\widetilde{N_{\epsilon_n}}}(f) = \mu(f), \end{aligned}$$

and μ is G-invariant. □

4.1.2. *Fixed Points in Haar Measure.* We start with the following classical result.

Theorem 5.21 (Markov–Kakutani Fixed Point Theorem). *Suppose K is a compact convex subset of the locally convex (Hausdorff) linear topological space E, and suppose that T is a continuous linear operator $T : E \to E$ such that $T(K) \subseteq K$. Then there is an element $k \in K$ so that*

$$T(k) = k.$$

Proof. Fix $k_0 \in K$, and set

$$k_n = \frac{1}{n+1}(k_0 + T(k_0) + \cdots + T^n(k_0)).$$

Each $k_n \in K$. Because K is compact the sequence (k_n) has a limit point k. Let $\phi \in E^*$. Then

$$T^*(\phi) = \phi \circ T \in E^*,$$

and so given an $\epsilon > 0$ and $n_0 \in \mathbb{N}$ there is an $n > n_0$ so that

$$|(\phi(k_n - k)| < \epsilon$$

and

$$|(T^*(\phi)(k_n - k)| = |(\phi \circ T)(k_n - k)| < \epsilon.$$

So

$$\begin{aligned}
|\phi(k - T(k))| &= |\phi(k) - \phi(T(k))| \\
&= |\phi(k) - \phi(k_n) + \phi(k_n) - \phi(T(k_n)) + \phi(T(k_n)) - \phi(T(k))| \\
&\le |\phi(k) - \phi(k_n)| + |\phi(k_n) - \phi(T(k_n))| \\
&+ |\phi(T(k_n)) - \phi(T(k))| \\
&\le \epsilon + \phi(k_n - T(k_n)) + \epsilon \\
&= \epsilon + \frac{\phi(k_0 - T^{n+1}(k_0))}{n+1} + 2\epsilon \\
&\le \frac{2\sup\{\phi(x)| : x \in K\}}{n_0+1} + 2\epsilon.
\end{aligned}$$

Since $\epsilon > 0$ and $n_0 \in \mathbb{N}$ are arbitrary, it follows that

$$\phi(k - T(k)) = 0.$$

Since $\phi \in E^*$ is arbitrary,

$$k = T(k)$$

follows. □

Here's an easy consequence of the Markov–Kakutani Fixed Point Theorem using some transfinite hijinks.

Corollary 5.22. *Suppose $\mathcal{A}$ is a commuting family of continuous linear operators on the locally convex (Hausdorff) linear topological space, K is a compact convex subset of E, and $T(K) \subseteq K$ for each $T \in \mathcal{A}$. Then there is a $k \in K$ such that*

$$T(k) = k$$

for each $T \in \mathcal{A}$.

Proof. First we notice that if $T_1, \ldots, T_n \in \mathcal{A}$, then there is a $k \in K$ so that

$$T_1(k) = k, T_2(k) = k, \ldots, T_n(k) = k.$$

This follows by induction on n. The Markov–Kakutani theorem shows it so for $n = 1$. Suppose we know it to be so for $n > 1$, and we let

$$K_n = \{k \in K : T_1(k) = k, T_2(k) = k, \ldots, T_n(k) = k\}.$$

Then K_n is a compact, convex subset of E, and if $k \in K_n$, then for $j = 1, \ldots, n$ we see that

$$T_{n+1}k = T_{n+1}(T_j(k)) = T_j(T_{n+1}(k))$$

since members of $\mathcal{A}$ commute. So $T_{n+1}(k) \in K_n$ too. Applying the Markov–Kakutani theorem again tells us that T_{n+1} has a fixed point in K_n.

So for any finite set $\mathcal{F} \subseteq \mathcal{A}$, if

$$K_{\mathcal{F}} = \{k \in K : T(k) = k \text{ for } T \in \mathcal{F}\},$$

then $K_{\mathcal{F}}$ is a nonempty closed convex subset of K and the family

$$\{K_{\mathcal{F}} : \mathcal{F} \subseteq \mathcal{A}, \mathcal{F} \text{ finite}\}$$

has the finite intersection property. By K's compactness

$$\bigcap \{K_{\mathcal{F}} : \mathcal{F} \subseteq \mathcal{A}, \mathcal{F} \text{ finite}\}$$

is nonempty. Of course any point in the intersection is a common fixed point for all $T \in \mathcal{A}$. □

A consequence of the above corollary is the existence of a Haar measure on a compact abelian group. Since the main idea of this proof will soon be revisited to deduce the existence of a Haar measure on arbitrary compact group topological groups, we present the proof in outline only.

The key is the fact the dual $C(G)^*$ of the Banach space $C(G)$ of continuous scalar-valued functions on G is the space $M(G)$ of regular Borel measures on G. Inside $M(G)$ we concentrate on the weak*-compact convex set $\mathcal{P}(G)$ of probabilities in $M(G)$.

For $f \in C(G)$ and $g \in G$, define

$$T_g : C(G) \to C(G)$$

by

$$T_g(f)(h) = f(gh)$$

for $h \in G$. The collection $\{T_g : g \in G\}$ is a *commuting* family of linear isometries of $C(G)$ onto itself. What's more, each T_g^* takes $P(G)$ into itself:

$$(T_g^*(\mu)(f)) = \int (T_g f)\, d\mu = \int f(gh)\, d\mu(h) = \int f(h)\, d\mu_g(h),$$

where $\mu_g(A) = \mu(g^{-1}A)$. This shows

$$T_g^* \mu = \mu_g.$$

The map T_g^* is weak*-weak* continuous on $M(G)$ to $M(G)$. By the Markov–Kakutani theorem and its corollary, there is a common fixed point in $\mathcal{P}(G)$. It's plain that the common fixed point is a Haar measure.

For general compact topological groups we need the following.

Theorem 5.23 (Ryll-Nardzewski). *Suppose $\mathcal{S}$ is a semigroup of continuous linear operators on a Hausdorff locally convex topological linear topological space E, and let K be a compact convex subset of E. Assume that $S(K) \subseteq K$ for each $S \in \mathcal{S}$. Suppose further that whenever k_1 and k_2 are distinct points of K, then*

$$0 \notin \overline{\{S(k_1 - k_2) : S \in \mathcal{S}\}}.$$

Then there exists $k \in K$ such that for each $S \in \mathcal{S}$

$$Sk = k.$$

Remark 5.24. It must be mentioned that this result of Ryll-Nardzewski [**109, 110**] is a consequence of the Markov–Kakutani theorem and martingale argumentation.

Corollary 5.25. *Let G be a compact topological group. Then there exists a regular Borel probability μ on G such that*

$$\mu(g^{-1}A) = \mu(A)$$

for each $g \in G$ and each Borel set $A \subseteq G$.

Proof. The proof proceeds along the same path as that outlined in the case when G is abelian. This time we see that the collection $\{T_g : g \in G\}$ is a group of isometries of $C(G)$ onto $C(G)$ and, as before,

$$T_g^*(\mu) = \mu_g,$$

ensuring that each T_g^* is a weak*-weak* continuous linear operator taking $\mathcal{P}(G)$ into $\mathcal{P}(G)$. What remains to be shown is that the added condition of

Ryll-Nardzewski is met. To this end, suppose p_1, p_2 are distinct members of $\mathcal{P}(G)$. There must be an $f \in C(G)$, $||f||_\infty = 1$ so

$$\int f\, dp_1 - \int f\, dp_2 = \epsilon > 0.$$

Then $\{T_g f : g \in G\}$ is equicontinuous and norm-bounded, and so (thanks to Arzelá and Ascoli) it is relatively norm compact in $C(G)$. Hence there are $g_1, \ldots, g_n \in G$ such that if $g \in G$ there is an i, $1 \leq i \leq n$ so that

$$||T_g^{-1}(f) - T_{g_i}(f)||_\infty < \frac{\epsilon}{4}. \tag{26}$$

If we let

$$V = \left\{\mu \in \mathcal{P}(G) : \left|\int T_{g_1} f\, d\mu\right|, \ldots, \left|\int T_{g_n} f\, d\mu\right| < \frac{\epsilon}{4}\right\},$$

then V is a weak* open set containing zero. Then if $g \in G$ and g_i satisfies (26),

$$\begin{aligned} |(T_g^* p_1 - T_g^* p_2)(T_{g_i} f)| &\geq |(T_g^* p_1 - T_g^* p_2)(T_g^{-1} f)| \\ &\quad - |(T_g^* p_1(T_g^{-1} f - T_{g_i} f))| - |(T_g^* p_2(T_g^{-1} f - T_{g_i} f))| \\ &\geq \epsilon - \frac{\epsilon}{4} - \frac{\epsilon}{4} = \frac{\epsilon}{2}. \end{aligned}$$

It follows that $T_g^*(p_1 - p_2) \notin V$.

The hypotheses of the Ryll-Nardzewski theorem are fulfilled.

The common fixed point μ of all the T_g^*'s satisfies

$$\mu(A) = \mu(g^{-1}(A)),$$

for all $g \in G$ and all Borel sets $A \subseteq G$. □

The elder of the authors was shown this proof by D. J. H. "Ben" Garling, whose Cambridge University notes, *The Martingale Convergence Theorem and its Applications*, are a treasure trove of fascinating proofs and mathematical connections.

Of course we specially direct the ambitious reader to the original papers of Ryll-Nardzewski ([**109**], [**110**]), which use ergodic theoretic ideas to prove his fixed-point result.

1. Homogeneous Spaces

Let G be a compact topological group, and let K be a compact Hausdorff space. We say that G *acts transitively* on K if there is a continuous map $G \times K \to K : (g, k) \mapsto g(k)$ such that

(i) $e(k) = k$ for all $k \in K$ (e is the identity of G);

(ii) $(g_1 g_2)(k) = g_1(g_2(k))$ for all $g_1, g_2 \in G, h \in K$;

(iii) given $k_1, k_2 \in K$, there is a $g \in G$ so that $g(k_1) = k_2$.

It is noteworthy that each $g \in G$ may be viewed as a homeomorphism of K onto itself; after all, the map $k \to g(k)$ is continuous and has $k \to g^{-1}(k)$ as an inverse.

Condition (iii) says, in particular, that the space K is *homogeneous*; i.e., we can move points of K around K via homeomorphisms (members of G, in fact) of K onto itself.

If μ is the unique translation invariant Borel probability on G, then μ induces a G-invariant Borel probability on K. This is an important construction, one worth understanding in general as well as in special cases.

Suppose H is a closed subgroup of the compact topological group G. Consider the set G/H (of left cosets of H in G) with the so-called "quotient topology", that is, the strongest topology that makes the natural map $q_H : G \to G/H$ (taking $g \in G$ to $gH \in G/H$) continuous. So $U \subseteq G/H$ is open precisely when $q_H^{\leftarrow}(U)$ is open in G. In other words, a typical open set in G/H is of the form $\{xH : x \in V\}$ when V is open in G. Because H is supposed to be closed, this topology is Hausdorff; because q_H is continuous and surjective, G/H is compact.

More is so. G acts transitively on G/H. The map $(g, g'H) \to gg'H$ fits the bill in the definition.

In fact, any transitive action of a compact group on a compact space is of the sort just described. To be sure we need to tell when seemingly different spaces are the same under G's action. Let G act transitively on each of the compact Hausdorff spaces K_1, K_2. We say that *K_1 and K_2 are isomorphic under G's action* if there is a homeomorphism $\phi : K_1 \to K_2$ such that

$$\phi(g(k_1)) = g(\phi(k_1))$$

for each $k_1 \in K_1$.

Theorem 6.1 (Weil). *Let the compact group G act transitively on the compact Hausdorff space K. Then there is a closed subgroup H of G such that K and G/H are isomorphic under G's action.*

Proof. Fix $k_0 \in K$. Look at

$$H = \{g \in G : g(k_0) = k_0\}.$$

H is called an *isotopy subgroup*. It is plain that H is a closed subgroup of G. A natural candidate for the isomorphism of G/H and K is at hand: $\phi : G/H \to K$ given by

$$\phi(gH) = g(k_0).$$

For $g_1, g_2 \in G$, $g_1(k_0) = g_2(k_0)$ precisely when

$$g_1^{-1}(g_2(k_0)) = g_1^{-1}(g_1(k_0)) = e(k_0) = k_0,$$

or $g_1^{-1}g_2 \in H$, which is tantamount to $g_1H = g_2H$. This assures us that ϕ is well defined and injective.

The transitivity of G's action ensures ϕ's surjectivity. To see that ϕ is also continuous, fix $g \in G$ and let V be an open set in K containing $g(k_0)$. By the continuity of the map

$$(g, k) \to g(k)$$

on $G \times K$, there is an open set U in G which contains g so that $u(k_0) \in V$ for all $u \in U$. But $q_H(U)$ is open in G/H's quotient topology and $q_H(U) \subseteq \phi^{\leftarrow}(V)$. This shows that ϕ is a continuous bijection between the compact Hausdorff spaces G/H and K; as such, ϕ is a homeomorphism.

Further if $g_1, g_2 \in G$, then

$$g_1(\phi(g_2H)) = g_1(g_2(k_0)) = (g_1g_2)(k_0) = \phi(g_1(g_2H)).$$

Thus G/H and K are isomorphic under G's action. □

Note that because of this isomorphism theorem we can consider *any* G/H where H is the isotopy subgroup associated with *any* $k_0 \in K$.

Now we're ready for the main course.

Theorem 6.2 (Weil). *Suppose the compact group G acts transitively on the compact Hausdorff space K. Then there is a unique G-invariant regular Borel probability measure on K.*

Proof. We identify K with G/H where H is an isotopy subgroup of G as in our previous theorem. Let

$$q_H : G \to G/H$$

be the natural quotient map. Suppose μ is the normalized Haar measure on G and define $\mu_{G/H}$ on G/H by

$$\mu_{G/H}(B) = \mu(q_H^{\leftarrow}(B))$$

for any Borel set $B \subseteq G/H$.

If $g \in G$ and B is Borel subset of G/H, then

$$\begin{aligned} g(q_H^{\leftarrow}(B)) &= \{gx : xH \in B\} \\ &= \{gx : gxH \in gB\} \\ &= q_H^{\leftarrow}(gB). \end{aligned}$$

Therefore

$$\begin{aligned} \mu_{G/H}(gB) &= \mu(q_H^{\leftarrow}(gB)) \\ &= \mu(g(q_H^{\leftarrow}(B)) \\ &= \mu(q_H^{\leftarrow}(B)) = \mu_{G/H}(B), \end{aligned}$$

and $\mu_{G/H}$ is G-invariant.

Uniqueness is a touchier issue, as is always the case it seems. We take a close look at how regular countably additive Borel measures, members of $\text{rca}(\mathcal{B}_o(G)) = C(K)^*$ act on $C(G)$.

Take $\phi \in C(G)$ and $g \in G$. Define $\phi_g \in C(G)$ by

$$\phi_g(x) = \phi(gx).$$

Denote by μ_H the Haar measure (normalized so $\mu_H = 1$) on H. The map $G \to C(G)$ that takes g to ϕ_g is uniformly continuous (this is an easy modification of Theorem 5.1) so that

$$\hat{\phi}(g) = \int_H \phi_g(h) d\mu_H(h), \quad g \in G$$

defines a member $\hat{\phi}$ of $C(G)$.

Suppose $g_1 H = g_2 H$. Then $g_1^{-1} g_2 \in H$,

$$\begin{aligned} \hat{\phi}(g_1) &= \int_H \phi_{g_1}(h)\, d\mu_H(h) \\ &= \int_H \phi_{g_1}(g_1^{-1} g_2 h)\, d\mu_H(h) \ \text{(by } \mu_H\text{'s invariance and } g_1^{-1} g_2 \in H) \\ &= \int_H \phi(g_2 H)\, d\mu_H(h) \\ &= \int_H \phi_{g_2}(h)\, d\mu_H(h) = \hat{\phi}(g_2). \end{aligned}$$

Therefore $\hat{\phi}$ is constant on the left cosets of H so we can lift $\hat{\phi}$ to a continuous function $\tilde{\phi}$ on G/H:

$$\tilde{\phi}(gH) = \hat{\phi}(g).$$

To summarize: if $\phi \in C(G)$, then we define $\hat{\phi} \in C(G)$ and from this we get $\tilde{\phi} \in C(G/H)$.

Remarkably, each member of $C(G/H)$ comes about from this procedure. In fact, if $f \in C(G/H)$, then $f \circ q_H \in C(G)$ and for any $g \in G$

$$\begin{aligned}
\widetilde{(f \circ q_H)}(gH) &= \widehat{(f \circ q_H)}(g) \\
&= \int_H (f \circ q_H)_g(h)\, d\mu_H(h) \\
&= \int_H (f \circ q_H)(gh) d\mu_H(h) \\
&= \int_H f(ghH) d\mu_H(h) \\
&= \int_H f(gH) d\mu_H(h) \\
&= f(gH)\mu_H(H) = f(gH).
\end{aligned}$$

In other words, $f = \widetilde{(f \circ q_H)}$.

Now we look at G's action. Take *any* G-invariant regular Borel probability measure ν on G/H. For $\phi \in C(G)$ define

$$\lambda(\phi) = \int_{G/H} \tilde{\phi}(gH)\, d\nu(gH).$$

Then λ is a probability measure in $C(G)^\star$. Moreover, λ is translation invariant. Indeed if $x \in G$

$$\begin{aligned}
\lambda(\phi_x) &= \int_{G/H} \tilde{\phi}_x(gH)\, d\nu(gH) \\
&= \int_{G/H} \hat{\phi}_x(g)\, d\nu(gH) \\
&= \int_{G/H} \phi(xg)\, d\nu(gH) \\
&= \int_{G/H} \tilde{\phi}(xgH)\, d\nu(gH) \\
&= \int_{G/H} \tilde{\phi}(gH)\, d\nu(gH) = \lambda(\phi).
\end{aligned}$$

So λ is nothing else but normalized Haar measure on G.

If ν_1 and ν_2 are G-invariant regular Borel probabilities on G/H and if $x = \widetilde{(x \circ q_H)} \in C(G/H)$, then

$$\begin{aligned} \nu_1(x) &= \int_{G/H} x(gH)\, d\nu_1(gH) \\ &= \int_{G/H} \widetilde{(x \circ q_H)}(gH)\, d\nu_1(gH) \\ &= \lambda(x \circ q_H) \\ &= \int_{G/H} \widetilde{(x \circ q_H)}(gH)\, d\nu_2(gH) \\ &= \nu_2(x); \end{aligned}$$

in other words, ν_1 and ν_2 are the same. □

The worth of an abstract construction lies, at least in part, it its applicability to concrete cases. Our first application is classical and was well known before Weil's general theorem. It is, nonetheless, interesting and important.

Consider $O(n)$ the orthogonal group of order n, a compact group, and S^{n-1}, the unit sphere in $\mathbb{R}^n$, which is our compact Hausdorff space. The action of $O(n)$ on S^{n-1} is given, naturally by

$$(u, x) \rightarrow u(x).$$

If we consider the rotations of $(n+1)$-dimensional Euclidean space, then they act as isometries on the n-sphere. Since Hausdorff measures are invariant under all such maps on any given metric space, the Hausdorff n-measure is invariant under the action of the natural rotations (that generated the homogeneous space in the first place!)

Exercise 6.3. Verify that $O(n)$ acts transitivity on S^{n-1} in a suitable fashion.

Acknowledging the descriptions of members of $O(n)$ as rotations of $\mathbb{R}^n$, a direct application of Weil's theorem says, "There is a unique rotation-invariant regular Borel probability measure on S^{n-1}."

Exercise 6.4. Let ρ denote the unique rotation invariant Borel probability on the sphere S^{n-1} of $\mathbb{R}^n$ and for $y \in \mathbb{R}^n$ denote by f_y the (continuous) function defined on S^{n-1} by

$$f_y(x) = (x, y),$$

where $(\cdot, \cdot)$ is the inner product in $\mathbb{R}^n$. Show that if $||y|| = 1$, then

$$||f_y||_{L^1(\rho)} = \int |x_n| \, d\rho(x),$$

and so the map

$$y \rightarrow M_n^{-1} f_y,$$

where

$$M_n = ||f_y||_{L^1(\mu)},$$

is a linear isometry of E into $L^1(\rho)$.

Geometry is replete with examples of compact Hausdorff spaces that are homogeneous spaces on which various compact groups act transitively.

Here are a few more.

Again our group is $O(n)$ but this time our underlying compact space is

$$\sum(n) = \{(x,y) \in S^{n-1} \times S^{n-1} : x \perp y\},$$

where $x \perp y$ means x is perpendicular to y. Note that $(x,y) \in \sum(n)$ precisely when for any real numbers a, b,

$$||ax + by||^2 = a^2 + b^2.$$

It is easy to see from this that $\sum(n)$ is a closed subset of $S^{n-1} \times S^{n-1}$, hence, is compact. The action of $O(n)$ is natural enough, too: $(u,(x,y)) \rightarrow (ux, uy)$.

Exercise 6.5. Verify that $O(n)$ acts transitively on $\sum(n)$.

Let $1 \leq m \leq n$. Denote by $\sum^{(m)}(n)$ the set

$$\sum^{(m)}(n) = \Big\{(x_1, \ldots, x_m) \in \underbrace{S^{n-1} \times \cdots \times S^{n-1}}_{m \text{ times}} : \{x_1, \ldots, x_m\} \text{ is orthonormal}\Big\}.$$

Note that $(x_1, \ldots, x_m) \in \sum^{(m)}(n)$ precisely when, regardless of the real numbers $a_1, \ldots, a_m$, we have

$$\left|\left|\sum_{j=1}^{m} a_j x_j\right|\right|^2 = \sum_{j=1}^{m} a_j^2.$$

With this in mind, $\sum^{(m)}(n)$ is a compact set of $(S^{n-1})^m$ is easy to see.

Exercise 6.6. Verify that the action of $O(n)$ on $\sum^{(m)}(n)$ given by

$$(u,(x_1, \ldots, x_m)) \rightarrow (ux_1, ux_2, \ldots, ux_m)$$

is a transitive one, establishing, with a modicum of tender love and care, that $O(n)$ acts transitively on $\sum^{(m)}(n)$.

Finally, let $\mathcal{G}_m(n)$ denote the m-dimensional Grassmanian manifold, that is, $\mathcal{G}_m(n)$ is the space of all m-dimensional linear subspaces of $\mathbb{R}^n$. There is a natural surjection of $\sum^{(m)}(n)$ onto $\mathcal{G}_m(n)$ that takes $(x_1, \ldots, x_m) \in \sum^{(m)}(n)$ to the linear span of $\{x_1, \ldots, x_m\} \in \mathcal{G}_m(n)$. If we equip $\mathcal{G}_m(n)$ with the natural quotient topology, the result is a compact Hausdorff space.

Exercise 6.7. Verify that $O(n)$ acts transitively on $\mathcal{G}_m(n)$. The map reflecting the action of $O(n)$ on $\mathcal{G}_m$ is plain: if $\{x_1, \ldots, x_m\}$ is an orthonormal set in $\mathbb{R}^n$, then

$$(u, \text{span}\{x_1, \ldots, x_m\}) = \text{span}\{ux_1, \ldots, ux_m\}.$$

Here we interject that the geometry imparted above on $\mathcal{G}_m(n)$ is such that if $E = \text{span}\{x_1, \ldots, x_m\}$ and $E' = \text{span}\{x_1', \ldots, x_m'\}$ are members of $\mathcal{G}_m(n)$ and if each x_k is close to x_k' in $\mathbb{R}^n$, then E is close to E' in $\mathcal{G}_m(n)$.

In this way we find that there is a unique rotation invariant probability Borel measure on the m-dimensional Grassmanian manifold $\mathcal{G}_m(n)$.

2. Unitary Representations: The Peter–Weyl Theorem

We enter here with a discussion of the wonderful theorem of F. Peter and H. Weyl which asserts that every compact topological group G has a complete system of irreducible unitary representations.

What's this all about? While topological groups have very regular features, they remain quite general. To pursue a deeper understanding of them, it is necessary to hypothesize some added structure, such as assuming the groups are locally compact or even compact. A worthy goal is to be able to aptly compare such groups with classically studied, well-understood groups, such as groups of matrices with complex entries and multiplication being that of matrix multiplication. The Peter–Weyl theorem gives secondary information about compact groups in terms of groups of matrices

Our approach is modeled after that presented in Pontrjagin's classic [**104**], with a bit of help from more modern mathematical technology.

A *unitary representation* of a compact group G is a homomorphism $x \to U_x$ of G into the group of unitary operators on some Hilbert space H with the added feature that for each $h \in H$,

$$x \to U_x(h)$$

is a continuous complex-valued function on G. The representation $x \to U_x$ is *irreducible* if the only closed linear subspaces of H left invariant by every one of the operators U_x are $\{0\}$ and H.

We say that G admits a *complete system of irreducible unitary representations* if given any $g \in G$ with $g \neq e$ (the identity of G), then there is an irreducible unitary representation $x \to U_x$ of G over the Hilbert space H such that $U_g \neq \mathrm{id}_H$.

We open with the classical *Schur's Lemma*, a purely algebraic tool that is indispensable to our needs. We follow Schur's Lemma with a proof that for compact groups, irreducible unitary representations are necessarily finite dimensional. We follow this with a proof of the so-called "orthogonality relations" calling on Schur's Lemma to sort out the differences in the entries of nonequivalent irreducible representations of a compact group.

We next pass to a description of integral operators on a compact group, concentrating on highly symmetric such operators. Particular attention is paid to convolution operators. The principal feature of such operators is their approximative power. It is essential, too, that the eigenspaces of these operators are translation invariant, and so give rise to unitary representations, albeit not necessarily irreducible.

Finally we find out that the entries of irreducible unitary representations are sufficiently rich and diverse so as to establish the completeness of the system of *all* irreducible unitary representations of a compact group.

To get us started, we call in a classical piece of algebra.

Lemma 6.8 (Schur's Lemma). *Let E and F be finite-dimensional complex linear spaces, and suppose $\mathcal{R}$ and $\mathcal{S}$ are irreducible families of linear transformations on E and F, respectively. Assume that $T : E \to F$ is a linear transformation such that given any $B \in \mathcal{S}$ there is an $A \in \mathcal{R}$ such that*

$$BT = TA.$$

Then precisely one of the following holds:

(i) $T = 0$ *or*

(ii) $\dim E = \dim F$, *and T is invertible.*

Proof. *Step One.* Suppose $\dim E = m$, $\dim F = n$, and $\{e_1, \ldots, e_m\}$, $\{f_1, \ldots, f_n\}$ are a bases for E and F, respectively. Let

$$b_j = Te_j, \text{ and } L = \operatorname{span}\{b_1, \ldots, b_m\} = T(E).$$

Then L is invariant under the action of $\mathcal{S}$. To see this, let $B \in \mathcal{S}$. Pick $A \in \mathcal{R}$ so that

$$BT = TA.$$

If $Ae_i = \sum_{j=1}^m a_{i_j} e_j$ then

$$Bb_i = BTe_i = TAe_i = T\Big(\sum_{j=1}^m a_{i_j} e_j\Big) = \sum_{j=1}^m a_{i_j} b_j \in L;$$

so B leaves L invariant.

Step Two. We apply the irreducibility of $\mathcal{S}$ and conclude that either $L = \{0\}$ or $L = F$.

In case $L = \{0\}$, we have $TE = L = \{0\}$, and $T = 0$.

What happens if $L = F$? Well, $TE = F$, and so

$$\dim TE = \dim F = n,$$

so

$$m = \dim E \geq \dim T(E) = n.$$

Step Three. Let E' and F' denote the duals of E and F, respectively, and $\mathcal{R}'$ and $\mathcal{S}'$ denote the collections of adjoint operators of $\mathcal{R}$ and $\mathcal{S}$, respectively.

Suppose $T \neq 0$ (so $m \geq n$). Then $T' \neq 0$ as well. Since $\mathcal{R}'$ and $\mathcal{S}'$ are also irreducible families and for any $B' \in \mathcal{S}'$ we can find an $A' \in \mathcal{R}'$ so that

$$T'B' = A'T',$$

we can mimic the argument of Step Two but armed now with $T' \neq 0$. The conclusion is

$$T'F' = E',$$

and so

$$n = \dim F = \dim F' \geq \dim T'F' = \dim E' = \dim E = m.$$

Therefore $m = n$, and T is surjective between two vector spaces of the same finite number of dimensions so T is invertible. □

Corollary 6.9. *Let E be a finite-dimensional complex linear space, and let $\mathcal{R}$ be an irreducible family of linear transformations on E. Suppose A is a linear transformation on E such that*

$$AB = BA$$

for each $B \in \mathcal{R}$. Then $A = \alpha\, \mathrm{id}_E$ for the some $\alpha \in \mathbb{C}$.

Proof. If $A \neq 0$ and α is a nonzero eigenvalue of A, then $A - \alpha\, \mathrm{id}_E$ commutes with each $B \in \mathcal{R}$. Schur's Lemma tells us either $A - \alpha\, \mathrm{id}_E = 0$ or $A - \alpha\, \mathrm{id}_E$ is invertible. The latter is not in the cards, otherwise what does an eigenvalue mean? So $A - \alpha\, \mathrm{id}_E = 0$ is what is left, and that is as it should be. □

To put Schur's Lemma into play, we establish the following stunning feature of irreducible unitary representation of compact groups.

Theorem 6.10. *Let G be a compact topological group, and let $x \to V_x$ be an irreducible unitary representation of G over the Hilbert space H. Then H is finite dimensional.*

Proof. Let μ be normalized Haar measure on G; keep in mind that μ is right-, left-, and inversion-invariant.

For $h \in H$, look at the sesquilinear form on $H \times H$ given by

$$(u, v) = \int_G \langle V_x h, v\rangle \overline{\langle V_x h, u\rangle}\, d\mu(x).$$

By the F. Riesz theorem, there is a bounded linear operator

$$B_h : H \to H$$

so that for any $u, v \in H$

$$\langle B_h u, v\rangle = (u, v) = \int_G \langle V_x h, v\rangle \overline{\langle V_x h, u\rangle} d\mu(x).$$

Notice that for any $y \in G$

$$B_h V_y = V_y B_h.$$

After all, if $u, v \in H$, then

$$\begin{aligned}
\langle B_h V_y u, v\rangle &= \int_G \langle V_x h, v\rangle \overline{\langle V_x h, V_y u\rangle} d\mu(x) \\
&= \int_G \langle V_x h, v\rangle \overline{\langle V_y^* V_x h, u\rangle}\, d\mu(x) \\
&= \int_G \langle V_x h, v\rangle \overline{\langle V_y^{-1} V_x h, u\rangle}\, d\mu(x) \\
&= \int_G \langle V_x h, v\rangle \overline{\langle V_{y^{-1}} V_x h, u\rangle}\, d\mu(x).
\end{aligned}$$

Thanks to the unitary nature of $V_{y^{-1}}$,

$$\begin{aligned}
\langle B_h V_y u, v\rangle &= \int_G \langle V_{y^{-1}} V_x h, V_{y-1} v\rangle \overline{\langle V_{y^{-1}} V_x h, u\rangle} d\mu(x) \\
&= \int_G \langle V_{y^{-1}x} h, V_{y-1} v\rangle \overline{\langle V_{y^{-1}x} h, u\rangle}\, d\mu(x).
\end{aligned}$$

Now using μ's left invariance, we have

$$\begin{aligned}\langle B_h V_y u, v\rangle &= \int_G \langle V_x h, V_{y^{-1}} v\rangle \langle \overline{V_x h, u}\rangle \, d\mu(x)\\ &= \langle B_h u, V_{y^{-1}} v\rangle = \langle B_h u, V_y^{-1} v\rangle\\ &= \langle B_h u, V_y^* v\rangle = \langle V_y B_h, v\rangle,\end{aligned}$$

and we conclude that

$$\langle B_h V_y u, v\rangle = \langle V_y B_h u, v\rangle$$

for each $u, v \in H$. It follows then that

$$B_h V_y = V_y B_h.$$

But $x \to V_x$ is irreducible so there is $\alpha(h) \in C$ so

$$B_h = \alpha(h)\mathrm{id}_H;$$

that's for Schur.

So

$$\alpha(h)\langle u, v\rangle = \langle B_h u, v\rangle = \int_G \langle V_x h, v\rangle \langle \overline{V_x h, u}\rangle \, d\mu(x)$$

for any $u, v \in H$. It follows that

$$\int_G |\langle V_x h, u\rangle|^2 d\mu(x) = \alpha(h)\langle u, u\rangle = \alpha(h)||u||^2.$$

Now we take apt advantage of the unimodular character of G: If $f \in L^1(\mu)$ and f^* is defined by $f^*(x) = f(x^{-1})$, then $\int_G f^* d\mu = \int_G f \, d\mu$. It follows that

$$\begin{aligned}\alpha(h)||u||^2 &= \int_G |\langle V_x h, u\rangle|^2 \, d\mu(x)\\ &= \int_G |\langle u, V_x h\rangle|^2 \, d\mu(x)\\ &= \int_G |\langle V_x^* u, h\rangle|^2 \, d\mu(x)\\ &= \int_G |\langle V_x^{-1} u, h\rangle|^2 \, d\mu(x)\\ &= \int_G |\langle V_{x^{-1}} u, h\rangle|^2 \, d\mu(x)\\ &= \int_G |\langle V_x u, h\rangle|^2 \, d\mu(x)\\ &= \alpha(u)||h||.\end{aligned}$$

Thus regardless of $u \in H$,

$$\frac{\alpha(u)}{||u||^2} = \frac{\alpha(h)}{||h||^2},$$

so $\alpha(h) = c||h||^2$ for some constant c. In turn this says

$$\int_G |\langle V_x h, h\rangle|^2 \, d\mu(x) = \alpha(h)||h||^2 = c||h||^4.$$

If we try h with $||h|| = 1$, it follows that $c > 0$.

Now suppose (ϕ_n) is an orthonormal sequence in H. Set $h = \phi_1$ and $u = \phi_k$ for $k > 1$. Then

$$\int_G |\langle V_x \phi_k, \phi_1\rangle|^2 \, d\mu(x) = \alpha(\phi_k)||\phi_1||^2 = c.$$

But $(V_x \phi_n)$ is also an orthonormal sequence since V_x is unitary. So for any $n \in \mathbb{N}$, the Bessel–Parseval identities assure us that

$$\begin{aligned} nc &= \sum_{k=1}^{n} \int_G |\langle V_x \phi_k, \phi_1\rangle|^2 d\mu(x) = \int_G \sum_{k=1}^{n} |\langle V_x \phi_k, \phi_1\rangle|^2 d\mu(x) \\ &\leq \int ||\phi_1||^2 d\mu(x) = 1. \end{aligned}$$

Since $c > 0$, this means $nc \leq 1$ severely limits the length of any orthonormal sequence in H, and with it the dimension of H. □

Suppose $x \to V_x$ and $x \to U_x$ are two unitary representations, the first over the Hilbert space $H(V)$ and the second over the Hilbert space $H(U)$. We say these representations are *equivalent* if there is a linear isometry T of $H(V)$ into $H(U)$ so that for any $x \in G$, $TV_x = U_x T$.

Our next result shows that inequivalent irreducible unitary representatives of a compact group carry considerable information by their very inequivalence. Again Schur's Lemma plays a central role.

Theorem 6.11. *Let G be a compact topological group with normalized Haar measure μ. Suppose $x \to U_x = (u_{i_j}(x))$ and $x \to V_x = (v_{kl}(x))$ are nonequivalent irreducible unitary representations of G on the finite-dimensional Hilbert spaces E and F, respectively. Here $\{e_1, \ldots, e_m\}$ is an orthonormal basis of E, $\{f_1, \ldots, f_n\}$ is an orthonormal basis of F and $(u_{i_j}(x))$ and $(v_{kl}(x))$ are the matrix representatives of $x \to U_x$ and $x \to V_x$ on E and F with respect to these orthonormal systems. Then*

$$\int_G u_{i_j}(x)\overline{v_{kl}(x)} \, d\mu(x) = 0.$$

Proof. Let $T : F \to E$ be a linear transformation.

For any $x \in G$, let

$$A_x = U_x T V_x^{-1} = U_x T V_{x^{-1}} = U_x T V_x^*.$$

Each A_x is a linear transformation from F to E, and as x varies through G, A_x varies continuously through the finite-dimensional space $L(F,E)$ of all (bounded) linear maps from F to E. Let $A : F \to E$ be the average of the A_x's, that is,

$$A = \int_G A_x \, d\mu(x).$$

Then A is a linear transformation from F to E.

Note for any $a \in G$

$$U_a A V_{a^{-1}} = U_a A V_a^{-1} = A.$$

Indeed,

$$\begin{aligned} U_a A V_{a^{-1}} &= U_a \int_G A_x d\mu(x) V_{a^{-1}} \\ &= \int_G U_a A_x V_{a^{-1}} \, d\mu(x) \\ &= \int_G U_a U_x T V_{x^{-1}} V_{x^{-1}} \, d\mu(x) \\ &= \int_G U_{ax} T V_{(ax)^{-1}} \, d\mu(x) \quad \text{(by left invariance of } \mu) \\ &= \int_G U_x T V_{x^{-1}} \, d\mu(x) = A. \end{aligned}$$

So for each $a \in G$

$$U_a A = A V_a.$$

But $x \to U_x$ and $x \to V_x$ are not equivalent, so Schur's Lemma tells us that $A = 0$. In other words,

$$\int_G U_x T V_{x-1} \, d\mu(x) = 0,$$

and this is so for *any* linear transformation $T : F \to E$.

If T is the linear map that takes f_l to e_j but annihilates

$$f_1, \ldots, f_{l-1}, f_{l+1}, \ldots, f_n,$$

then in terms of the entries $(u_{i_j}(x))$ of U_x and $(v_{kl}(x))$ of V_x, this says (since $V_{x^{-1}} = \overline{(v_{kl}(x))})$

$$\int_G u_{i_j}(x) \overline{v_{kl}(x)} \, d\mu(x) = 0. \qquad \square$$

A fixed irreducible unitary representation also packs a lot of information. If

$$x \to U_x = (u_{i_j}(x))$$

is such a representation over the finite-dimensional Hilbert space E, then we will follow a path similar to the one followed in the above proof to deduce the pertinent information.

Let $S : E \to E$ be any linear transformation. For any $x \in G$, let

$$B_x = U_x S U_x^{-1} = U_x S U_{x^{-1}};$$

each B_x is a linear transformation in F and, as before, $x \to B_x$ is a continuous action taking G into the space $L(E, E)$ of all (bounded) linear transformations in E. The average B, given by

$$B = \int_G B_x \, d\mu(x),$$

is a (bounded) linear transformation on E, and for each $a \in G$

$$BU_a = U_a B.$$

Irreducibility tells us (in the voice of Schur) that there is an $\alpha \in \mathbb{C}$, so $B = \alpha \, \mathrm{id}_E$, that is,

$$\alpha \, \mathrm{id}_E = \int_G U_x S U_{x^{-1}} \, d\mu(x).$$

Hence

$$\begin{aligned} m\alpha = \mathrm{tr}(\alpha \, \mathrm{id}_E) &= \mathrm{tr}\left(\int_G U_x S U_{x^{-1}} \, d\mu(x) \right) \\ &= \int_G \mathrm{tr}(U_x S U_{x^{-1}}) \, d\mu(x) \\ &= \int_G \mathrm{tr}(U_x U_{x^{-1}} S) \, d\mu(x) \\ &= \int_G \mathrm{tr} S \, d\mu(x) \\ &= \mathrm{tr}(S). \end{aligned}$$

So regardless of the linear transformation $S : E \to E$

$$\frac{\mathrm{tr}(S)}{m} = \alpha,$$

and so

$$\int_G U_x S U_{x^{-1}} \, d\mu(x) = \frac{\mathrm{tr}(s)}{m} \, \mathrm{id}_E.$$

Picking S judiciously will reveal that

$$\int_G u_{i_j}(x) \overline{u_{kl}(x)} d\mu(x) = \frac{\delta_{ik}\delta_{jl}}{m}.$$

Theorem 6.11, and the above information are often referred to as the "orthogonality relations" of irreducible unitary representations of a compact group.

To bring them to bear on the study of compact groups, we make a brief visit to the theory of integral equations on a compact group.

Throughout this next discussion, G is a fixed compact topological group and μ is the normalized Haar measure on G. We let $C(G)$ denote the Banach space of continuous complex-valued functions on G, equipped with the supremum norm

$$||f||_\infty = \sup\{|f(x)| : x \in G\},$$

and for $1 \leq p < \infty$, $L^p(G)$ denotes the Banach space of (equivalence classes of) pth power integrable functions with respect to the Haar measure μ.

Let $k : G \times G \to C$ be a continuous function, and suppose $k(s,t) = \overline{k(t,s)}$ for all $s, t \in G$. Define the integral operator K on $L^1(G)$ by

$$Kf(s) = \int_G k(s,t) f(t)\, d\mu(t).$$

Notice that K takes its values in $C(G)$, thanks to k's uniform continuity on $G \times G$. In fact, if $f \in B_{L^1(G)}$ and $\epsilon > 0$, then there is an open set V in G that contains the identity so that if $s^{-1}s^{'} \in V$, then

$$|k(s,t) - k(s^{'}, t)| \leq \epsilon.$$

It follows that

$$\begin{aligned} |Kf(s) - Kf(s')| &\leq \int_G |k(s,t) - k(s',t)||f(t)|\, d\mu(t) \\ &\leq \epsilon \int_G |f(t)|\, d\mu(t) \leq \epsilon. \end{aligned}$$

So in truth K takes the closed unit ball of $L^1(G)$ into a uniformly bounded (by $||k||_\infty$), equicontinuous set in $C(G)$. Thus K acts as a compact linear operator from $L^1(G)$ into $C(G)$.

Also K acts as a bounded linear operator from the Hilbert space $L^2(G)$ to itself. Viewing K thusly, we take note that if $\langle \cdot, \cdot \rangle$ denotes the inner product in $L^2(G)$ so

$$\langle f, g \rangle = \int_G f(t)\overline{g(t)}\, d\mu(t),$$

then

$$\begin{aligned} \langle Kf, g \rangle &= \int_G \int_G k(s,t) f(t)\, d\mu(t) \overline{g(s)}\, d\mu(s) \\ &= \int_G \int_G k(s,t) f(t) \overline{g(s)}\, d\mu(t)\, d\mu(s) \\ &= \int_G f(t) \Big(\int_G \overline{k(t,s) g(s)}\, d\mu(s) \Big)\, d\mu(t) \\ &= \int_G f(t) \overline{Kg(t)} d\mu(t) = \langle f, Kg \rangle, \end{aligned}$$

with the obligatory incantation of Fubini at the appropriate juncture. Thus K acts in $L^2(G)$ as a self-adjoint operator. Its action on $L^2(G)$ is also compact.

The spectral theorem for compact self-adjoint operators gives very precise information about K. To wit:

- $||K||_{L^2(G)\to L^2(G)}$, the operator norm of K, is given by
$$||K||_{L^2(G)\to L^2(0)} = \sup\{|\langle Kf, f\rangle| : ||f||_2 \le 1\}.$$
- K's nonzero eigenvalues (λ_n) are all real and can be listed as a null sequence (perhaps only finitely nonzero) in descending order of absolute value:
$$||K||_{L^2(G)\to L^2(G)} = |\lambda_1| \ge |\lambda_2 \ge \cdots \ge |\lambda_n| \ge \cdots \searrow 0.$$
- If λ is a nonzero eigenvalue of K, then the eigenspace
$$E(\lambda) = \{f \in L^2(G) : Kf = \lambda f\}$$
corresponding to λ is a finite-dimensional (closed) linear subspace of $L^2(G)$.
- Moreover, since K's values are actually in $C(G)$, each eigenspace $E(\lambda)$ has a (finite) orthonormal basis consisting of members of $C(G)$.

Of critical interest will be those integral operators generated by members $\kappa \in C(G)$ for which $\kappa(s) = \overline{\kappa(s^{-1})}$ for all $s \in G$. Here if $k(s,t) = \kappa(t^{-1}s)$, then the operator K is just
$$\begin{aligned} Kf(s) = \int_G k(s,t)f(t)d\mu(t) &= \int_G \kappa(t^{-1}s)f(t)d\mu(t) \\ \text{(also known as)} &= (f * \kappa)(s), \end{aligned}$$

a "convolution operator".

Convolution operators, such as described above, serve many purposes. Without further comment, they are a smoothing process; more to the present point, they help approximate. Here's how. Still in the setting of a compact group G with normalized Haar measure μ, if U is an open set in G that contains G's identity e, then let $\kappa_U \in C(G)$ be a nonnegative function with support contained in U such that

- $\kappa_U(e) = 1$,
- $\int_G \kappa_U \, d\mu = 1$,
- $\kappa_U^*(s) = \overline{\kappa_U(s^{-1})} = \kappa_U(s)$ for all s.

Order these open sets by reverse inclusion: $U \geq V$ if $U \subseteq V$.

Theorem 6.12. *The net $(\kappa_U * f)_U$ converges to f in $C(G)$ for each f.*

Proof. Let $f \in C(G)$ and $\epsilon > 0$ be given.

By f's uniform continuity, there is an open set W containing e such that regardless of $s \in G$ if $t \in W$, then

$$|f(t^{-1}s) - f(s)| \leq \epsilon.$$

Let U be any open set containing c such that $U \subseteq W$ (that is, $U \geq W$). Then

$$\begin{aligned}|\kappa_U * f(s) - f(s)| &= \left| \int_G \kappa_U(t) f(t^{-1}s)\, d\mu(t) - f(s) \right| \\ &\leq \int_G \kappa_u(t) |f(t^{-1}s) - f(s)|\, d\mu(t) \\ &= \int_U \kappa_U(t) |f(t^{-1}s) - f(s)|\, d\mu(t) \\ &\leq \int_G \kappa_U(t) \cdot \epsilon\, d\mu(t) = \epsilon,\end{aligned}$$

and, like a man falling off a steep cliff, this proof has a sudden end. □

Next, we take advantage of the symmetry possessed by the κ_U's, $\kappa_U(z) = k_U(z^{-1})$. As a result we see that

$$\begin{aligned}(f^* * \kappa_U)(s) &= \int_G f^*(t)\kappa_U(t^{-1}s)\, d\mu(t) \\ &= \int_G \overline{f(t^{-1})}\kappa_U(t^{-1}s)\, d\mu(t) \\ &= \int_G \overline{f(t^{-1})}\kappa_U(s^{-1}t)\, d\mu(t).\end{aligned}$$

Let $T = s^{-1}t$ so $t^{-1} = T^{-1}s^{-1}$ and $d\mu(t) = d\mu(T)$. Then

$$\begin{aligned}(f^* * \kappa_U)(s) &= \int_G \overline{f(T^{-1}s^{-1})}\kappa_U(T)\, d\mu(T) \\ &= \int_G \kappa_U(T)\overline{f(T^{-1}s^{-1})}\, d\mu(T) \\ &= (\kappa_U * f)^*(s).\end{aligned}$$

It follows from this, the fact that $f^{**} = f$, and the previous theorem that for any $f \in C(G)$, the net $(f * \kappa_U)_U$ converges in $C(G)$ to f.

A simple, but important, computation involving $\lambda_s f(t) := f(s^{-1}t)$:

$$\begin{aligned}\lambda_s f * \kappa(x) &= \int_G \lambda_s f(t)\kappa(t^{-1}x)\, d\mu(t)\\ &= \int_G f(s^{-1}t)\kappa(t^{-1}x)\, d\mu(t).\end{aligned}$$

Again, if we let $T = s^{-1}t$ (so $t = sT$), then the left invariance of μ gives us

$$\begin{aligned}\lambda_s f * \kappa(x) &= \int_G f(T)\kappa(T^{-1}s^{-1}x)\, d\mu(T)\\ &= \lambda_s\left(\int_g f(T)\kappa(T^{-1}x)\right)\, d\mu(T)\\ &= \lambda_s((f*\kappa)(x)).\end{aligned}$$

If f is an eigenvector of the operator $f \to f * \kappa$ with eigenvalue η, then

$$(\lambda_s f) * \kappa = \lambda_s(f * \kappa) = \lambda_s(\eta f) = \eta\lambda_s(f);$$

hence $\lambda_s f$ is also an eigenvector of $f \to f * \kappa$ with eigenvalue η.

Therefore the eigenspaces of the operator $f \to f * \kappa$ are invariant under translation by λ_s.

We have seen that the eigenspaces $E(\lambda)$ of an eigenvalue λ of a convolution operator $f \to f*\kappa$, for suitable symmetric κ, is left invariant by the operator $\lambda_s f(x) = f(s^{-1}x)$. Now

$$\lambda_s|_{E(\lambda)} : E(\lambda) \to E(\lambda)$$

is unitary for each $s \in G$. If $\{g_1, \ldots, g_{d_\lambda}\}$ is an orthonormal basis of $E(\lambda)$, then the functions

$$\chi_{ki}(s) = \langle \lambda_s g_i, g_k\rangle$$

belong to $C(G)$, and because $\lambda_s g_i \in E(\lambda)$ for each $s \in G$,

$$g_i(s^{-1}t) = \lambda_s(g_i)(t) = \sum_{k=i}^{d_\lambda} \langle \lambda_s g_i, g_k\rangle g_k(t)$$

holds for almost all $t \in G$ (denoting by d_λ the dimension of $E(\lambda)$). But all in sight is continuous, so

$$g_i(s^{-1}t) = \sum_{k=1}^{d_\lambda} \langle \lambda_s g_i, g_k\rangle g_k(t)$$

holds for all $s, t \in G$.

Let $U_s = \lambda_s|_{E(\lambda)}$, so

$$U_s = (\chi_{i_j}(s)) = \langle \lambda_s g_i, g_j\rangle.$$

Since λ_s is unitary, $\chi_{i_j}(s) = \overline{\chi_{ji}(s^{-1})}$ or $U_s^* = U_{s^{-1}}$.

Further,

$$\begin{aligned}\chi_{i_j}(st) &= \langle \lambda_t g_i, \lambda_{s^{-1}} g_j\rangle \\ &= \sum_{k=1}^{d_\lambda} \langle g_k, \lambda_{s^{-1}} g_i\rangle \langle \lambda_t g_i, g_k\rangle \\ &= \sum_{k=1}^{d\lambda} \chi_{k_i}(s)\chi_{k_j}(t)\end{aligned}$$

or $U_{st} = U_s U_t$.

Therefore $s \to \mathcal{U}_s$ is a unitary representation of G over $E(\lambda)$.

Return now to the operators of the sort $K(f) = f * \kappa$, when κ is a continuous function on G satisfying $\kappa(s) = \overline{k(s^{-1})}$ for each $s \in G$. The operator K has a sequence (λ_n) of eigenvalues which we list in order of their descending module:

$$|\lambda_1| \geq |\lambda_2| \geq \cdots \geq |\lambda_n| \geq \cdots \searrow 0;$$

for each n, the eigenspace $E(\lambda_n)$ is a finite-dimensional subspace of $L^2(G)$, and we can list the orthonormal bases of the $E(\lambda_n)$'s thusly,

$$\underbrace{g_1^{(1)}, \ldots, g_{d_1}^{(1)}}_{E(\lambda_1)}, \underbrace{g_1^{(2)}, \ldots, g_{d_2}^{(2)}}_{E(\lambda_2)}, \ldots, \underbrace{g_1^{(n)}, \ldots, g_{d_n}^{(n)}}_{E(\lambda_n)}, \ldots,$$

where $d_n = \dim E(\lambda_n)$ and $g_1^{(n)}, \ldots, g_{d_n}^{(n)}$ is an orthonormal basis of $E(\lambda_n)$, consisting of members of $C(G)$.

For any $f \in L^2(G)$, we know that

$$f = \sum_n c_n g_n + g_0,$$

where $c_n = \langle f, g_n\rangle, g_n$ in the list above, and $Kg_0 = 0$.

It follows that

$$Kf = \sum_n c_n K(g_n),$$

where the sum is taken in $L^2(G)$. But $\sum_n c_n g_n$ converges in $L^2(G)$ so given $\epsilon > 0$ there is $N_\epsilon \in \mathbb{N}$ so

$$\left\| \sum_{m > N_\epsilon} c_m g_m \right\|_2 < \epsilon.$$

This ensures that

$$\left\| K\Big(\sum_{m > N_\epsilon} c_m g_m \Big) \right\|_2 < ||K||\epsilon.$$

Of course $Kg_m = \mu_m g_m$ so

$$K\Big(\sum_{m=1}^{N_\epsilon} c_m g_m\Big) = \sum_{m=1}^{N_\epsilon} c_m \mu_m g_m$$

is a linear combination of the g_m's, $m = 1, \ldots, N_\epsilon$. Alas,

$$\Big|\Big|Kf - \sum_{m=1}^{N_\epsilon} c_m \mu_m g_m\Big|\Big|_\infty = \Big|\Big|K(\sum_{m>n_\epsilon} c_m g_m)\Big|\Big|_\infty$$

$$\leq \Big|\Big|K(\sum_{m>n_\epsilon} c_m g_m)\Big|\Big|_2 < ||K||\epsilon,$$

and Kf is, in fact, the uniform limit of linear combinations of the g_m's.

Suppose $x \to U_x$ is a unitary representation of the compact group G over the finite-dimensional Hilbert H. Then there are (closed) linear subspaces $H_1, \ldots, H_n$ of H such that

$$H = H_1 \oplus H_2 \oplus \cdots \oplus H_n,$$

and the restriction $x \to U_x|_{H_i}$ is an irreducible unitary representation of G over H_i for each $i = 1, 2, \ldots, n$. After all, if S is a (closed) linear subspace of H that is left invariant under all the U_x's ($x \in G$), then so too is $S^\perp$: suppose $s \in S, s' \in S^\perp, x \in G$; then

$$\langle U_x s', s\rangle = \langle s', U_x^* s\rangle = \langle s', U_x^{-1} s\rangle = \langle s', U_{x^{-1}} s\rangle.$$

Since $U_{x^{-1}} \in S$, the last quantity is 0. It follows that $U_x s \in S^\perp$ if $s' \in S^\perp$.

Continue this idea until you run out of room.

To summarize, we start with $f \in C(G)$. Then f can be approximated by $f * \kappa_U$ for suitable U.

If $K(f) = f * \kappa_U$, then $K(f)$ is the uniform limit of linear combinations of orthonormal bases of the (finite-dimensional) eigenspaces of K.

These eigenspaces, in turn, are the underlying Hilbert spaces of unitary representations of G.

If we look closely at an eigenspace of K, say E, and monitor what the representation $s \to U_s$ is doing on E, we see that this representation is a direct sum of irreducible unitary representations of G. So E has an orthonormal basis consisting of blocks, each an orthonormal basis of an underlying Hilbert space of an irreducible unitary representations of G.

It follows from this that the system of irreducible unitary representations of G is a complete system. Indeed, if not, then there would be a $g \in G, g \neq e$

so that for every irreducible unitary representation $x \to U_x$ of G over H,

$$U_g = \mathrm{id}_H = U_e.$$

This in turn implies that $f(g) = f(e)$. But we have already noted that G is a completely regular Hausdorff space.

3. Pietsch Measures

In 1967 in [**100**], A. Pietsch introduced the absolutely p-summing operators. Soon after his introduction of this class of operators, J. Lindenstrauss and A. Pełczyński [**71**], used these operators to clarify points of Grothendieck's mysterious *Résumé* [**45**]. In particular, they used Pietsch's p-summing operators and Grothendieck's inequality to inspire a generation of functional analysts and effect a rebirth of interest in Banach space theory. Much of the power of Pietsch's p-summing operators lies in the surprising role measure theory plays in their study.

A linear operator $u : X \to Y$ between the Banach spaces X and Y is *absolutely p-summing* (here $1 \leq p < \infty$) if there exists $K > 0$ such that for any $x_1, \ldots, x_n \in X$ we have

$$\sum_{k \leq n} ||u(x_k)||^p \leq K^p \sup \left\{ \sum_{k \leq n} |x^* x_k|^p : x^* \in B_{X^*} \right\}. \tag{27}$$

Here B_{X^*} denotes the closed unit ball of dual $X^*$of X,

$$B_{X^*} = \{x^* \in X^* : ||x^*|| \leq 1\}.$$

It is not difficult to see that $u : X \to Y$ is absolutely p-summing precisely when given a sequence (x_n) of members of X for which $\sum_n |x^*(x_n)|^p < \infty$ for each $x^* \in X^*$, $\sum ||ux_n||^p < \infty$. Thanks to the closed graph theorem of Banach, it follows that there is a $K > 0$ so that

$$\left(\sum ||x_n||^p \right)^{\frac{1}{p}} \leq K \sup \left\{ \left(\sum |x^* x_n|^p \right)^{\frac{1}{p}} : x^* \in B_{X^*} \right\} \tag{28}$$

for any sequence $(x_n) \subseteq X$. The constants K in (27) and (28) are interchangeable, and the p-summing norm $\pi_p(u)$ of u is defined to be the infimum of all such K's. The collection of all p-summing linear operators from X to Y is denoted by $\Pi_p(X;Y)$; it is a linear space and with π_p as a norm, $(\Pi_p(X;Y), \pi_p)$ is a Banach space.

Suppose $K \subseteq B_{X^*}$ is weak*-closed (closed in the topology of pointwise convergence on X, a topology in which B_{X^*} is compact, by the way) and norms X, so for any $x \in X$

$$||x|| = \sup\{|x^*(x)| : x^* \in K\}.$$

Then $u : X \to Y$ is p-summing precisely when $\sum_n ||ux_n||^p < \infty$ whenever $\sum_n |x^*(x_n)|^p < \infty$ for each $x^* \in K$. Along with this we get that

$$\pi_p(u) = \sup\left\{ \left(\sum ||ux_n||^p\right)^{\frac{1}{p}} : \left(\sum |x^*(x_n)|^p\right)^{\frac{1}{p}} \le 1, \text{ all } x^* \in K \right\}.$$

The principal example of the above phenomenon is when F is a closed linear subspace of a $C(K)$-space. In this case,

$$\widetilde{K} = \{\delta_k : k \in K\} \subseteq C(K)^*$$

is a weak*-closed, norming subset of B_{F^*}, where $\delta_k(f) = f(k)$ is the usual point change.

Now one can stare at the definition of an operator being p-summing and not realize that measure theory is in play. That this is so is a profound result of Pietsch, often called Pietsch's domination theorem. Here's what it says:

Theorem 6.13 (Pietsch)**.** *Let $u : X \to Y$ be a bounded linear operator between Banach spaces X and Y. Let $K \subseteq B_{X^*}$ be a weak*-compact, norming set. Then u is an absolutely p-summing operator if and only if there exists a regular Borel probability measure μ on $(K, weak^*)$ such that for each $x \in X$*

$$||ux|| \le \pi_p(u)\left(\int_K |x^*(x)|^p \, d\mu(x^*)\right)^{\frac{1}{p}}.$$

Pietsch's original proof was somewhat complicated, but in the early seventies a shorter proof was offered by B. Maurey [**81**]. We present an outline of Maurey's proof of the existence of the "dominating probability". For the finite set $F = \{x_1, \ldots, x_n\} \subseteq X$, we define the function $\phi_F : K \to \mathbb{R}$,

$$\phi_F(x^*) = \sum_{k \le n} (||ux_k||^p - \pi_p^p(u)|x^*(x_n)|^p).$$

The collection $\mathcal{K} = \{\phi_F : F \text{ is a finite collection in } X\}$ is a convex subset of $C(K)$. Because u is absolutely p-summing with p-summing norm $\pi_p(u)$, K is disjoint from the positive cone

$$\mathcal{P} = \{f \in C(K) : f(x^*) > 0 \text{ for all } x^* \in K\},$$

an open convex set. Mazur's separation version of the Hahn–Banach theorem tells us there is a continuous linear functional $\mu \in C(K)^*$, the space of regular Borel measures on K, such that for some real constant c

$$\int_K \phi_F \, d\mu \le c < \int_K f \, d\mu \tag{29}$$

for all finite sets $F \subseteq X$ and all $f \in \mathcal{P}$. Since $0 \in \mathcal{K}$, $c \ge 0$. Since $c < \int_K f \, d\mu$ for every $f \in \mathcal{P}$, $c \le 0$. So $c = 0$, and μ is a nonzero, nonnegative measure. The defining inequality (29) is impervious to dividing by positive constants, so we can assume that μ is a probability. Testing (29) against $\phi_{\{x\}} \in \mathcal{K}$ gives the result.

Incidentally, if there is a probability μ on K so that

$$||ux|| \leq C\Big(\int_K |x^*(x)|^p \, d\mu(x^*)\Big)^{\frac{1}{p}},$$

then

$$\begin{aligned}
\sum_{j\leq n} ||ux_j||^p &\leq C^p \sum_{j\leq n} \int_K |x^*(x_j)|^p \, d\mu(x^*) \\
&\leq C^p \int_K \sum_{j\leq n} |x^*(x_j)|^p \, d\mu(x^*) \\
&\leq C^p \int_K \sup_{x^*\in K} \sum_{j\leq n} |x^*(x_j)|^p \, d\mu \\
&= C^p \sup_{x^*\in K} \sum_{j\leq n} |x^*(x_j)|^p
\end{aligned}$$

(because μ is a probability), and so u is absolutely p-summing with $\pi_p(u) \leq C$.

Since μ comes about as a result of the tandem of the Hahn–Banach theorem and the Riesz representation of $C(K)^*$, it appears that little can be said of a personal nature about μ. This is *not* so.

Theorem 6.14. *Let G be a compact topological group, and let F be a closed translation invariant linear subspace of $C(G)$ that separates the points of G. Suppose that $u : F \to X$ is an absolutely p-summing linear operator, and suppose that for each $f \in F$ and $g \in G$*

$$||u(f)|| = ||u(f_g)||,$$

where $f_g(x) = f(gx)$. Then the normalized Haar measure on G is a Pietsch measure for u.

Proof. Let μ be a Pietsch measure for u; this says μ is a regular Borel probability measure on G such that for any $f \in F$

$$||uf||^p \leq \pi_p^p(u) \int_G |f(x)|^p \, d\mu(x).$$

For $g \in G$ let $\mu_g \in C(G)^*$ be given by

$$\mu_g(f) = \mu(f_g).$$

Here $f_g(x) = f(gx)$ for all $x \in G$. Then μ_g is a regular Borel probability measure on G and is, in fact, a Pietsch measure for u. In fact, for any $f \in F$, $f_g \in F$

$$\begin{aligned} ||u(f)||^p &= ||u(f_g)||^p \\ &\leq \pi_p^p(u)\mu(|f_g|^p) \\ &= \pi_p^p(u)\mu_g(|f|^p) \\ &= \pi_p^p(u)\int_G |f(x)|^p \, d\mu_g(x). \end{aligned}$$

The map taking $g \in G$ to $\mu_g \in C(G)^*$ is a continuous function from G onto $(B_{C(G)^*}, weak^*)$. Hence the Gelfand integral

$$\nu = \int_G \mu_g \, d\mu$$

exists and lies in $B_{C(G)^*}$. Here dg indicates we are integrating with respect to the normalized Haar measure on G. Of course, ν is the Gelfand integral

$$\nu = \int_G \mu_g \, d\mu$$

and means that for any $f \in C(G)$

$$\int_G f(x) \, d\nu(x) = \int_G \mu_g(f) \, dg.$$

It's easy to see that ν is nonnegative, since $f \geq 0$ implies $\mu_g(f) \geq 0$ for all $g \in G$. It's also plain that ν is a probability:

$$\nu(1) = \int_G \mu_g(1) \, dg = \int_G 1 \, dg = 1.$$

Therefore ν is a regular Borel probability on G.

It is also true that ν is a Pietsch measure for u since for each $f \in F, g \in G$,

$$\frac{||uf||^p}{\pi_p^p(u)} \leq \mu_g(|f|^p)$$

we have that

$$\frac{||uf||^p}{\pi_p^p(u)} \leq \int_G \mu_g(|f|^p) \, dg = \nu(|f|^p)$$

as well.

Finally ν is translation invariant: if $h_0 \in G$ and $f \in C(G)$, then

$$\begin{aligned}
\nu(f_{h_0}) &= \int_G \mu(f_{h_0})\, dg = \int_G \int_G f_{h_0}(x)\, d\mu_g(x)\, dg \\
&= \int_G \int_G f_{h_0}(gx)\, d\mu(x)\, dg \\
&= \int_G \int_G f(h_0 gx)\, d\mu(x)\, dg \\
&= \int_G \int_G f(h_0 gx)\, dg\, d\mu(x) \\
&= \int_G \int_G f(gx)\, dg\, d\mu(x) \\
&= \int_G \int_G f(gx)\, d\mu(x)\, dg \\
&= \int_G \int_G f(x)\, d\mu_g(x)\, dg \\
&= \int_G \nu(f) dg = \nu(f).
\end{aligned}$$

Therefore ν is the normalized Haar measure on G. □

4. Notes and Remarks

The material from Section 1 of this chapter traces its origin in the form presented herein to A. Weil's [**136**] treatise. Another excellent source is Nachbin's delightful book, *The Haar Integral* [**89**].

We have cited but a few of the homogeneous spaces that admit invariant probabilities. It would be hard to study modern geometry or analysis without encountering examples of invariant measures on homogeneous spaces.

The power inherent in the use of these techniques can be appreciated, for example, in the study of A. Dvoretzky's celebrated spherical sections theorem [**35**]. Roughly speaking, this theorem says that in any infinite-dimensional normed linear space, one can (theoretically) find finite-dimensional subspaces of arbitrarily high dimension that are but small perturbation of Euclidean spaces. In his original exposition of this result, Dvoretzky calls on the invariant probabilities living in the Grassman manifold $\mathcal{G}(n)$.

Other proofs (simpler in execution) have since appeared, but each makes its fair share of use of invariant measures on homogeneous spaces. We refer the bravest readers to the paper of T. Figiel, J. Lindenstrauss, and V. Milman [**39**] for a proof calling on the isoperimetric inequality and invariant measures on homogeneous spaces and to the survey papers of Lindenstrauss [**70**] and V. Milman [**82**]. Once hooked on the beauty of the subject, the V. Milman and G. Schectman volume, *Asymptotic Theory of Finite Dimensional Normed Spaces* [**83**], will be a natural.

Our presentation of the Peter–Weyl theorem follows the path laid by L. Pontrjagin set forth in Chapter IV of *Topological Groups* [**104**]. Relying as it does on the existence of a normalized Haar measure on a compact group, the Pontrjagin presentation carries over to any compact group.

Schur's Lemma (Lemma 6.8) it to due to I. Schur [**114**] and is present in virtually all presentations of representations of groups.

Theorem 6.10 was uncovered by A. Gurevič [**46**] in case the compact group satisfies the second axiom of countability. See also P. Koosis [**68**] and L. Nachbin [**88**]. K. Shiga [**119**] modified the Gurevič construction to apply it to irreducible representations of compact groups over general (complex) Banach spaces.

The notion of an absolutely p-summing operator is due to A. Pietsch [**100**] who is responsible for the magical theorem wherein "Pietsch measures" make their appearance. Though A. Grothendieck isolated the absolutely 1-summing operators and the absolutely 2-summing operators in different guises, it was the identification by Pietsch of the norm π_p that was to play a critical role in the development and application of these operator ideals.

Theorem 6.14 appears in J. Diestel, H. Jarchow, and A. Tonge [**25**], where it is admittedly inspired by earlier works of Y. Gordon, S Kwapien and A. Pełczyński.

Pietsch's isolation of the p-summing operators (and his definition of their norm) was recognized by J. Lindenstrauss and A. Pełczyński [**71**] as an appropriate vehicle to reformulate many of the most mysterious of Grothendieck's results from his *Résumé* [**45**]. In addition, Lindenstrauss and Pełczyński recast Grothendieck's "fundamental theorem of the metric theory of tensor products" in a more user-friendly version; they went on to solve a number of long-standing problems in Banach spaces.

The literature related to Grothendieck's inequality (an alias for his fundamental theorem) is enormous. It is probably wisest to recommend to interested readers that they go to the master and study the amazing survey paper of G. Pisier [**102**]. For an exposition of Grothendieck's original argument for his fundamental inequality in its tensorial home, a look at Diestel, Fourie, and Swart [**26**] might help. Alternative views of the inequality, as well as varied applications, can be found in [**24**], [**25**], and [**101**].

The use of compact groups and their Haar measure has been a frequent tool in the local theory of Banach spaces. We recommend the papers of P. Mankiewicz [**75**], P. Mankiewicz and N. Tomczak-Jaegermann [**76**], as well as the book of N. Tomczak-Jaegermann [**127**].

Chapter 7

Haar Measure on Locally Compact Groups

1. Positive Linear Functionals

1.1. Introduction. Let $\mathcal{S}$ be a nonempty locally compact Hausdorff space. Denote by $\mathcal{K}(\mathcal{S})$. the linear space of continuous real valued functions $f : \mathcal{S} \to \mathbb{R}$ such that

$$\overline{\{s \in \mathcal{S} : f(s) \neq 0\}}$$

is compact. $\mathcal{K}^+(\mathcal{S})$ will denote the collection of f's in $\mathcal{K}(\mathcal{S})$ such that $f(s) \geq 0$ for all s and $f(s)$ is not identically zero. Let $\phi : \mathcal{K}(\mathcal{S}) \to \mathbb{R}$ be a linear functional for which

$$\phi(f) \geq 0$$

for all $f \in \mathcal{K}^+(\mathcal{S})$.

Fact 7.1. $|\phi(f)| \leq \phi(|f|)$

Of course, $\phi(f) \geq \phi(g)$ if $f, g \in \mathcal{K}(\mathcal{S})$ and $f \geq g$. It follows that

$$\phi(-|f|) = -\phi(|f|) \leq \phi(f) \leq \phi(|f|).$$

Fact 7.2. If K is a compact subset of $\mathcal{S}$, then there is a $\beta_K > 0$ so that if $f \in \mathcal{K}(\mathcal{S})$ and

$$\{s \in \mathcal{S} : f(s) \neq 0\} \subseteq K,$$

then $|\phi(f)| \leq \beta_K ||f||_\infty$.

Fact 7.3. Let $\mathcal{D}$ be a nonempty subset of $\mathcal{K}^+(\mathcal{S})$, and suppose $\mathcal{D}$ is directed downward (so given $f, g \in \mathcal{D}$ there is an $h \in \mathcal{D}$ so $h \leq f, h \leq g$) and

$$\inf\{f(s) : f \in \mathcal{D}\} = 0$$

for all $s \in \mathcal{S}$. Then given an $\epsilon > 0$, there is an $f_\epsilon \in \mathcal{D}$ so that $f_\epsilon(s) < \epsilon$ for all $s \in S$. Consequently,

$$\inf\{\phi(f) : f \in \mathcal{D}\} = 0.$$

The first assertion of Fact 7.3 can be viewed as a directed version of Dini's theorem.

Proof. Let $\epsilon > 0$ be given. For $f \in \mathcal{D}$ define

$$\mathcal{B}_f = \{s \in S : f(s) \geq \epsilon\}.$$

Each $\mathcal{B}_f$ is compact and, by hypothesis,

$$\bigcap_f \mathcal{B}_f = 0.$$

Hence there is $f_1, f_2, \ldots, f_n \in \mathcal{D}$ so that

$$\mathcal{B}_{f_1} \cap \cdots \cap \mathcal{B}_{f_n} = 0.$$

Let $f_\epsilon \in \mathcal{D}$ be chosen so $f_\epsilon \leq f_i$, for all $1 \leq i \leq n$.

If $g \in \mathcal{D}$, then there is an $h \in \mathcal{D}$ so $h \leq g$ and $h \leq f_\epsilon$. If

$$K = \{s \in \mathcal{S} : g(s) > 0\},$$

then Fact 7.2 says there is $\beta_K > 0$ so that

$$|\phi(f)| \leq \beta_K ||f||_\infty$$

for any $f \in \mathcal{K}^+(G)(S)$ with $\operatorname{supp}(f) \subseteq K$. It follows that

$$|\phi(h)| \leq |\phi(f_\epsilon)| \leq \beta_K ||f_\epsilon||_\infty \leq \epsilon\beta_K. \qquad \square$$

A function $f : S \to [0, \infty]$ is *lower semicontinuous* if for each $s_0 \in \mathcal{S}$ the following holds:

- if $f(s_0) < \infty$, then given $\epsilon > 0$ there is an open set $\mathcal{U}$ containing s_0 so that for any $s \in \mathcal{U}$,

$$f(s) > f(s_0) - \epsilon;$$

- if $f(s_0) = \infty$, then for any $P > 0$ there is an open set $\mathcal{U}$ containing s_0 so if $s \in \mathcal{U}$, then

$$f(s) > P.$$

We denote by $\mathcal{L}^+$ the collection of all lower semicontinuous functions $f : \mathcal{S} \to [0, \infty]$. A basic feature of members of $\mathcal{L}^+$—a motivating factor for their introduction—is the following.

Fact 7.4. If $f \in \mathcal{L}^+$, then $f(s) = \sup\{g(s) : g \in \mathcal{K}^+(s), g \leq f\}$.

The truth of Fact 7.4 is plain if $f(s_0) = 0$. Suppose then that $f(s_0) > 0$. Then for any $P > 0$ with $0 < P < f(s_0)$, there is an open set $\mathcal{U}$ containing s_0 so if $s \in \mathcal{U}$, then $f(s) > P$. We can (and do) assume $\overline{\mathcal{U}}$ is compact. Now Urysohn's lemma is available and on call; it provides us with $g \in \mathcal{K}^+(S)$ so that

$$\overline{\{s : g(s) \neq 0\}} \subseteq \mathcal{U}$$

with $g(s_0) = P$ and $g(s) \leq P$ for all $s \in \mathcal{S}$. Of course, $0 \leq g \leq f$ and P is any number with $0 \leq \mathcal{P} \leq f(s_0)$, so Fact 7.4 is established.

Fact 7.5. Basic facts about the set $\mathcal{L}^+$:

(i) If $f \in \mathcal{L}^+, \alpha > 0$, then $\alpha f \in \mathcal{L}^+$.

(ii) If $f_1, \ldots, f_n \in \mathcal{L}^+$, then $\min\{f_1, \ldots, f_n\} \in \mathcal{L}^+$.

(iii) If $\mathcal{D}$ is a nonempty subset of $\mathcal{L}^+$, then $\sup\{f : f \in \mathcal{D}\} \in \mathcal{L}^+$.

(iv) If $f_1, \ldots, f_n \in \mathcal{L}^+$, then $f_1 + \cdots + f_n \in \mathcal{L}^+$.

Let $f \in \mathcal{L}^+$. Then we know from Fact 7.4 that

$$f(s) = \sup\{g(s) : g \in \mathcal{K}^+(S), g \leq f\}.$$

In light of this, we define $\overline{I}(f)$ by

$$\overline{I}(f) = \sup\{\phi(g) : g \in \mathcal{K}^+(S), g \leq f\}.$$

Keep in mind that $\overline{I}(f) = +\infty$ is a distinct possibility.

Fact 7.6. Basics about $\overline{I}$:

(i) If $f \in \mathcal{K}^+(S)$, then $f \in \mathcal{L}^+$ and $\overline{I}(f) = \phi(f)$.

(ii) If $f_1, f_2 \in \mathcal{L}^+$ and $f_1 \leq f_2$, then $\overline{I}(f_1) \leq \overline{I}(f_2)$.

(iii) If $f \in \mathcal{L}^+$ and $\alpha \geq 0$, then $\overline{I}(\alpha f) = \alpha \overline{I}(f)$.

Now we modify Fact 7.3 to the context of $\overline{I}$.

Theorem 7.7. *Let $\mathcal{D}$ be a nonempty subset of $\mathcal{L}^+$ that is directed upward (so given $f, g \in \mathcal{D}$, there is an $h \in \mathcal{D}$ so $h \geq f$ and $h \geq g$). Then*

$$\overline{I}(\sup\{f : f \in \mathcal{D}\}) = \sup\{\overline{I}(f) : f \in \mathcal{D}\}.$$

Consequently, if $(f_n) \subseteq \mathcal{L}^+$ *is an ascending sequence, then*

$$\overline{I}(\lim_n f_n) = \lim_n \overline{I}(f_n).$$

Keep in mind that $\mathcal{L}^+$ *is permanent under suprema.*

Proof. Maybe $\mathcal{D}$ and $f_0 = \sup\{f : f \in \mathcal{D}\}$ belong inside $\mathcal{K}^+(S)$; in this case, $\{f_0 - f : f \in \mathcal{D}\}$ falls prey to Fact 7.3, and so

$$\begin{aligned} 0 &= \inf\{\phi(f_0 - f) : f \in \mathcal{D}\} \\ &= \inf\{\phi(f_0) - \phi(f) : f \in \mathcal{D}\} \\ &= \phi(f_0) - \sup\{\phi(f) : f \in \mathcal{D}\}, \end{aligned}$$

and all's well.

Generally we call on the fact that if $f \in \mathcal{L}^+$, then

$$f(s) = \sup\{g(s) : g \in \mathcal{K}^+, g \leq f\}.$$

With this in mind, suppose that

$$f_0 = \sup\{f : f \in \mathcal{D}\};$$

that is, suppose that for each $s \in S$,

$$f_0(s) = \sup\{f(s) : f \in \mathcal{D}\}.$$

We want to show that

$$\overline{I}(f_0) \leq \sup\{\overline{I}(f) : f \in \mathcal{D}\}$$

since the reverse inequality follows from Fact 7.6(ii).

To this end, let $\mathcal{C}$ denote the set of all $g \in \mathcal{K}^+(S)$ such that $g \leq f$ for some $f \in \mathcal{D}$. Clearly,

$$\begin{aligned} f_0 &= \sup\{f : f \in \mathcal{D}\} \\ &= \sup\{\sup\{g : g \in \mathcal{K}^+(S), g \leq f\} : f \in \mathcal{D}\} \qquad (30) \\ &= \sup\{g \in \mathcal{K}^+(S) : g \leq f, \text{ for some } f \in \mathcal{D}\}. \end{aligned}$$

If $g_0 \leq f_0, g_0 \in \mathcal{K}^+(S)$, then

$$g_0 = \min\{g_0, f_0\} = \sup\{\min\{g_0, g\} : g \in \mathcal{K}^+(S), g \leq f, \text{ for some } f \in \mathcal{D}\}.$$

By the result of our opening salvo,

$$\phi(g_0) = \sup\{\phi(\min(g_0, g)) : g \in \mathcal{K}^+(S), g \leq f, \text{ for some } f \in \mathcal{D}\}.$$

But now we're in position to call on (30) to conclude that

$$\begin{aligned} \phi(g_0) &= \sup\{\phi(\min(g_0, g)) : g \in \mathcal{K}^+(S), g \leq f, \text{ for some } f \in \mathcal{D}\} \\ &\leq \sup\{\phi(g) : g \in \mathcal{K}^+(S), g \leq f, \text{ for some } f \in \mathcal{D}\} \\ &= \sup\{\sup \phi(g) : g \leq f : f \in \mathcal{D}\} = \sup_D \overline{I}(f). \end{aligned}$$

Since g_0 was any member of $\mathcal{K}^+(S)$ with $g_0 \leq f_0$, it follows that

$$\overline{I}(f_0) = \sup\{\phi(g_0) : g_0 \in \mathcal{K}^+(S), g_0 \leq f_0\} \leq \sup_D \overline{I}(f).$$

This proof is now complete. □

Corollary 7.8. *If $f_1, f_2 \in \mathcal{L}^+$, then $\overline{I}(f_1 + f_2) = \overline{I}(f_1) + \overline{I}(f_2)$.*

This is an easy consequence of ϕ's additivity on $\mathcal{K}^+(\mathcal{S})$, and the fact that

$$f_1 + f_2 = \sup\{g_1 + g_2 : g_1, g_2 \in \mathcal{K}^+, g_1 \leq f_1, g_2 \leq f_2\}.$$

Piggybacking, we get

Corollary 7.9. *For any nonempty $\mathcal{D} \subseteq \mathcal{L}^+$,*

$$\overline{I}\Big(\sum_{f\in\mathcal{D}} f\Big) = \sum_{f\in\mathcal{D}} \overline{I}(f).$$

Next we will extend $\overline{I}$ to the set of all nonnegative extended real-valued functions defined on $\mathcal{S}$.

Let $f : \mathcal{S} \to [0,\infty]$ be any function with nonnegative, extended real values defined on $\mathcal{S}$. Denote by $\mathcal{D}_f$ the set

$$\mathcal{D}_f = \{g : \mathcal{S} \to [0,\infty] : g \in \mathcal{L}^+, f(s) \leq g(s) \text{ for all } s \in \mathcal{S}\}.$$

Trivially $g(s) \equiv +\infty$ belongs to $\mathcal{D}_f$, regardless of f.

Definition 7.10. We define the *upper integral* of f, $\overline{\int} f$, by

$$\overline{\int} f = \inf\{\overline{I}(g) : g \in \mathcal{D}_f\}.$$

Take note that if $f \in \mathcal{L}^+$, then $\overline{\int} f = \overline{I}(f)$.

Fact 7.11. Basic facts about $\overline{\int}$:

(i) If $f_1, f_2 : \mathcal{S} \to [0,\infty]$ and $f_1 \leq f_2$, then $\overline{\int} f_1 \leq \overline{\int} f_2$.

(ii) If $c \geq 0$ and $f : \mathcal{S} \to [0,\infty]$, then $\overline{\int} cf = c\overline{\int} f$.

(iii) If $f_1, f_2 : \mathcal{S} \to [0,\infty]$, then $\overline{\int}(f_1 + f_2) \leq \overline{\int} f_1 + \overline{\int} f_2$.

Theorem 7.12. *If (f_n) is an ascending sequence of functions with $f_n : S \to [0,\infty]$, then*

$$\overline{\int} \sup_n f_n = \sup_n \overline{\int} f_n.$$

Proof. It's plain from Fact 7.11(iii) that

$$\overline{\int} \sup_n f_n \geq \sup_n \overline{\int} f_n,$$

so we will establish the reverse inequality, namely,

$$\sup_n \overline{\int} f_n \geq \overline{\int} \sup_n f_n.$$

If $\sup_n \overline{\int} f_n = +\infty$, then there is nothing to be proved. So we suppose for the sake of argument that $\sup_n \overline{\int} f_n < \infty$. Of course this supposition gives us access to epsilonics.

Let $\epsilon > 0$. For each n, choose $g_n \in \mathcal{L}^+$ so that $f_n \leq g_n$ and

$$\overline{I}(g_n) \leq \overline{\int} f_n + \frac{\epsilon}{2^n}.$$

Set

$$h_1 = g_1, h_2 = \max\{g_1, g_2\}, \ldots, h_n = \max\{g_1, \ldots, g_n\}.$$

Then the sequence (h_n) is ascending, $h_n \geq g_n$ and

$$\begin{aligned} h_{n+1} + \min\{h_n, g_{n+1}\} &= \max\{h_n, g_{n+1}\} + \min\{h_n, g_{n+1}\} \\ &= h_n + g_{n+1}. \end{aligned}$$

So

$$\overline{I}(h_{n+1}) + \overline{I}(\min\{h_n, g_{n+1}\}) = \overline{I}(h_n) + \overline{I}(g_{n+1}).$$

It follows that

$$\begin{aligned} \overline{I}(h_{n+1}) &= \overline{I}(h_n) + \overline{I}(g_{n+1}) - \overline{I}(\min\{h_n, g_{n+1}\}) \\ &\leq \overline{I}(h_n) + \overline{I}(g_{n+1}) - \overline{\int} f_n \\ &\leq \overline{I}(h_n) + \overline{\int} f_{n+1} + \frac{\epsilon}{2^{n+1}} - \overline{\int} f_n; \end{aligned}$$

alternatively, for $n = 1, \ldots, k-1$ (say),

$$\overline{I}(h_{n+1}) - \overline{I}(h_n) \leq \overline{\int} f_{n+1} - \overline{\int} f_n + \frac{\epsilon}{2^{n+1}}.$$

We add these inequalities up from $n = 1$ until $n = k - 1$, and we find that

$$\overline{I}(h_k) - \overline{I}(h_1) \leq \overline{\int} f_k - \overline{\int} f_1 + \frac{\epsilon}{2}.$$

Hence

$$\begin{aligned}\overline{I}(h_k) &\leq \overline{\int} f_k + I(h_1) - \overline{\int} f_1 + \frac{\epsilon}{2}\\ &= \overline{\int} f_k + \left(I(g_1) - \overline{\int} f_1\right) + \frac{\epsilon}{2}\\ &\leq \overline{\int} f_k + \frac{\epsilon}{2} + \frac{\epsilon}{2} = \overline{\int} f_k + \epsilon.\end{aligned}$$

It follows that

$$\begin{aligned}\sup_k \overline{\int} f_k + \epsilon &\geq \sup_k \overline{I}(h_k)\\ &= \overline{I}(\sup_k h_k)\\ &\geq \overline{\int} \sup_k f_k,\end{aligned}$$

with the critical equality of $\sup_k \overline{I}(h_k)$ and $\overline{I}(\sup_k h_k)$, a result of Theorem 7.7. Now $\epsilon > 0$ was arbitrary, so Theorem 7.12 is proved. □

Corollary 7.13. *Given any countable collection* $\{f_j\}$ *of functions* $f_j : \mathcal{S} \to [0, \infty]$, *we have*

$$\overline{\int} \sum_j f_j \leq \sum_j \overline{\int} f_j.$$

Corollary 7.13 is crucial in our generation of outer measures.

1.2. The outer measure associated with a positive linear functional. For any $A \subseteq \mathcal{S}$ the *outer measure* $\mu_\phi^*(A)$ (shortened to $\mu(A)$) is given by

$$\mu(A) = \overline{\int} \chi_A.$$

Fact 7.14. This μ is an outer measure; i.e.,

- $\mu(A) \geq 0$ for any $A \subseteq \mathcal{S}$.
- If $A \subseteq B \subseteq \mathcal{S}$, then $\mu(A) \leq \mu(B)$.
- If (A_n) is a sequence of subsets of $\mathcal{S}$, then

$$\mu\left(\bigcup_n A_n\right) \subseteq \sum_n \mu(A_n).$$

Fact 7.14 follows from the basic facts (announced earlier) about $\overline{\int}$; μ respects the topology of $\mathcal{S}$ and behaves accordingly.

But we know from Theorem 7.15(iv) that

$$\begin{aligned} \mu(A) &= \inf\{\mu(U) : U \text{ is open, } A \subseteq U\} \\ &\geq \mu(U \cap E) + \mu(U \cap E^c) \\ &\geq \mu(A \cap E) + \mu(A \cap E^c). \end{aligned}$$

Theorem 7.17. *All Borel subsets of S are μ-measurable.*

Proof. Suppose V is an open subset of $\mathcal{S}$. Let U be an open subset in $\mathcal{S}$ with $\mu(U) < \infty$, and let $\epsilon > 0$ be given. Use Theorem 7.15 to find an open subset O of $\mathcal{S}$ so that

$$U \cap V^c \subseteq O$$

and

$$\mu(O) < \mu(U \cap V^c) + \frac{\epsilon}{2}. \tag{31}$$

Now appeal to Theorem 7.15 to find an open set W such that $\overline{W}$ is compact, $\overline{W} \subseteq U \cap V$, and

$$\mu(W) + \frac{\epsilon}{2} > \mu(U \cap V). \tag{32}$$

Let

$$W_0 = U \cap O \cap (\overline{W})^c.$$

Then W and W_0 are disjoint open sets. Now

$$U \cup V^c \subseteq W_0 \subseteq O,$$

so (31) tells us that

$$|\mu(W_0) - \mu(U \cap V^c)| < \frac{\epsilon}{2}.$$

In tandem with (32), we see that

$$|\mu(W) + \mu(W_0) - (\mu(U \cap V) - \mu(U \cap V^c))| < \epsilon. \tag{33}$$

But the disjointness of W and W_0 ensures (by Theorem 7.15) that

$$\mu(W \cup W_0) = \mu(W) + \mu(W_0).$$

So since both $W \subseteq U$ and $W_0 \subseteq U$,

$$\begin{aligned} \mu(U) &\geq \mu(W \cup W_0) = \mu(W) + \mu(W_0) \\ &\geq \mu(U \cap V) + \mu(U \cap V^c) - \epsilon \end{aligned}$$

by (33). All in sight is finite and ϵ is arbitrary, so

$$\mu(U) \geq \mu(U \cap V) + \mu(U \cap V^c)$$

holds whenever U is open in $\mathcal{S}$ and $\mu(U) < \infty$. Therefore V is μ-measurable.

Of course, we know that the collection of μ-measurable sets is a σ-algebra, and since it contains every open subset of $\mathcal{S}$, it must contain every Borel set as well. $\square$

Here's a basic regularity feature of this construction.

Fact 7.18. Let A be a μ-measurable set, and suppose $\mu(A) < \infty$. Then

$$\mu(A) = \sup\{\mu(K) : K \text{ is compact}, K \subseteq A\}.$$

Proof. Let $\epsilon > 0$. Choose an open set U so that $A \subset U$ and $\mu(U \cap A^c) < \frac{\epsilon}{4}$. Now appeal to Theorem 7.15 to find a compact set $K \subseteq U$ so $\mu(U \cap K^c) < \frac{\epsilon}{2}$. Now, let W be an open set so that $U \supseteq W \supseteq U \cap A$ and $\mu(W) < \frac{\epsilon}{2}$. Let $F = K \cap W^c$. Then F is compact, and $F \subseteq A$. What's more,

$$\begin{aligned} \mu(A \cap F^c) &= \mu((A \cap K^c) \cup (A \cap W)) \\ &\leq \mu(U \cap K^c) + \mu(W) < \frac{\epsilon}{2} + \frac{\epsilon}{2}, \end{aligned}$$

and so (since $F \subseteq A$ ensures $\mu(A \cap F^c) = \mu(A) - \mu(F)$), it follows that $\mu(F) > \mu(A) - \epsilon$. $\square$

Our next task is to show that the measure μ generated by positive linear functionals ϕ on $\mathcal{K}(S)$ exhibits the comforting duality $L^1(\mu)^* = L^\infty(\mu)$ so useful in uncovering the structure of the space of integrable functions.

To start, we call on some epsilonics to simplify the establishment of a set's measurability.

Lemma 7.19. *A subset A of S is μ-measurable precisely when either of the following conditions is satisfied:*

(i) *$A \cap U$ is μ-measurable for each open set U of finite measure;*

(ii) *$A \cap K$ is μ-measurable for each compact set $K \subseteq S$.*

Naturally, A's μ-measurability entails the measurability of $A \cap U$ and $A \cap K$ for every open U and compact K since all Borel sets are μ-measurable.

Now suppose $A \cap U$ is μ-measurable for each open set U of finite μ-measure. Assume $\mu(B) < \infty$ and $\epsilon > 0$ is given. Pick an open set $U \supseteq B$ with

$$\mu(U) \leq \mu(B) + \epsilon.$$

Since $A \cap U$ *is* μ-measurable,

$$\mu(U) = \mu(U \cap A) + \mu(U \cap A^c),$$

and so

$$\mu(B) + \epsilon \geq \mu(U) \geq \mu(U \cap A) + \mu(U \cap A^c) \geq \mu(B \cap A) + \mu(B \cap A^c),$$

after all, $B \subseteq U$. Thanks to ϵ's arbitrariness,

$$\mu(B) \geq \mu(B \cap A) + \mu(B \cap A^c),$$

and so A is measurable.

To include (ii) we bootstrap on the result of (i). Suppose $A \cap K$ is μ-measurable for each compact set K. Assume U is an open subset of S with $\mu(U) < \infty$, and appealing to Theorem 7.15(v), find an ascending sequence (K_n) of compact subsets of U so that $\mu(U \setminus K_n) \searrow 0$. So $A \cap K_n$ is μ-measurable for each n and $A \cap (U \setminus \bigcup_n K_n)$ satisfies

$$\mu\left(A \cap \left(U \setminus \bigcup_n K_n\right)\right) \leq \mu\left(U \setminus \bigcup_n K_n\right) = 0.$$

So $A \cap (U \setminus \bigcup_n K_n)$ is μ-measurable. But

$$A \cap U = \left(A \cap \left(U \setminus \bigcup_N K_n\right)\right) \cup \left(\bigcup_n (A \cap K_n)\right),$$

and so $A \cap U$ is the countable union of μ-measurable sets and as such is entitled to call itself μ-measurable.

Now we come to the principal technical step we need to make: to overcome the possible non-σ-finiteness of the measure μ. This calls on an exhaustion argument, more or less covering virtually all of S with the well-disposed disjoint compact sets.

Theorem 7.20 (Exhaustion Principle). *There exists a family $\mathcal{C}$ of compact subsets of S such that*

(i) *distinct members of $\mathcal{C}$ are disjoint;*

(ii) *if $K \in \mathcal{C}$, then $\mu(K) > 0$;*

(iii) *if U is open and $K \in \mathcal{C}$ satisfies $K \cap U \neq \emptyset$, then $\mu(K \cap U) > 0$;*

(iv) *if A is measurable with $\mu(A) < \infty$, then $A \cap K \neq \emptyset$ for at most a countable collection of K's in $\mathcal{C}$ and*

$$\mu(A) = \sum_{\mathcal{C}} \mu(A \cap K);$$

(v) *A is measurable if and only if $A \cap K$ is μ-measurable for each $K \in \mathcal{C}$;*

(vi) *$f : S \to \mathbb{R}$ is μ-measurable if and only if $f\chi_K$ is for each $K \in \mathcal{C}$.*

Proof. Let $\mathcal{F}$ be the collection of all families $\mathcal{K}$ of compact subsets of S having the properties:

(a) distinct members of $\mathcal{K}$ are disjoint;

(b) any member of $\mathcal{K}$ has positive μ-measure;

(c) if U is open and $U \cap K \neq \emptyset$ for $K \in \mathcal{K}$, then $U \cap K$ has positive μ-measure.

Partially order $\mathcal{F}$ by inclusion. Note that $\mathcal{F}$ contains the empty set by default, and so $\mathcal{F}$ is itself nonempty(!). Linearly ordered subfamilies of $\mathcal{F}$ have an upper bound (their union) and so Zorn's lemma is available for use. On Zornicating, we find a maximal element $\mathcal{C}$ in $\mathcal{F}$. We will check $\mathcal{C}$ for the properties (i)–(vi).

Properties (i), (ii), and (iii) are all part and parcel of $\mathcal{C}$'s membership in $\mathcal{F}$.

Suppose A is μ-measurable with $\mu(A) < \infty$. By Theorem 7.15(iv) there is an open set U, $A \subseteq U$, with $\mu(U) < \infty$. Look at

$$\mathcal{C}_U = \{K \in \mathcal{C} : \mu(U \cap K) > 0\}.$$

For each $n \in \mathbb{N}$, let

$$\mathcal{C}_{U,n} = \left\{K \in \mathcal{C}_U : \mu(U \cap K) > \frac{1}{n}\right\}.$$

Because $\mu(U) < \infty$, the disjointness of distinct members of $\mathcal{C}$ ensure that each $\mathcal{C}_{U,n}$ is finite and so

$$\mathcal{C}_U = \bigcup_n \mathcal{C}_{U,n}$$

is countable. Now each member of $\mathcal{C}$ meeting A a fortiori meets U so belongs to $\mathcal{C}_U$ by (ii). It follows that there are only countably many members of $\mathcal{C}$ that meet A. List these in a sequence (K_n). Since A is μ-measurable and the K_n's are pairwise disjoint,

$$\begin{aligned} \mu(A) &= \mu\left(A \cap \left(\bigcup_n K_n\right)\right) + \mu\left(A \cap \left(\bigcup_n K_n\right)^c\right) \\ &= \sum_n \mu(A \cap K_n) + \mu\left(A \cap \left(\bigcup_n K_n\right)^c\right). \end{aligned}$$

So we want to show that

$$\mu\left(A \cap \left(\bigcup_n K_n\right)^c\right) = 0.$$

Since A is μ-measurable (as are all Borel sets), the set $A \cap (\bigcup_n K_n)^c)$ is plainly μ-measurable and, sitting as it does inside A, has finite μ-measure. Suppose C is a compact subset of $A \cap (\bigcup_n K_n)^c$ and imagine that $\mu(C) > 0$. Now μ's restriction to the μ-measurable subsets of C is a regular measure on C. The support of this restriction (call it C_μ) is compact, $\mu(C_\mu) = \mu(C) > 0$, and if U is an open subset of S so that $U \cap C_\mu \neq \emptyset$, then $\mu(U \cap C_\mu) > 0$; after all, C_μ is the *support* of μ inside C. But C (and hence C_μ) meets none of the members of $\mathcal{C}$ and so $\mathcal{C} \cup \{C\}$ satisfies (a)–(c), and is therefore a member of $\mathcal{F}$, larger than the maximal member $\mathcal{C}$ of $\mathcal{F}$. Our supposition on $\mathcal{C}$ must be suppressed with $\mu(C) = 0$, the only option left. This establishes (iv).

Let's establish (v). Suppose $A \cap K$ is μ-measurable for each $K \in \mathcal{C}$, and let C be any compact subset of S. By what we just proved in establishing (iv), there is a sequence (K_n) in $\mathcal{C}$ so that

$$\mu(C) = \sum_n \mu(C \cap K_n),$$

and so

$$\mu\left(C \cap \left(\bigcup_n K_n\right)^c\right) = 0.$$

But $A \cap K_n$ is μ-measurable for each n, and so $A \cap C \cap K_n$ is, too. Thus the set $A \cap C \cap (\bigcup_n K_n)^c$ satisfies

$$\mu\left(A \cap C \cap \left(\bigcup_n K_n\right)^c\right) \leq \mu\left(C \cap \left(\bigcup_n K_n\right)^c\right) = 0,$$

and $A \cap C \cap (\bigcup_n K_n)^c$ is therefore μ-measurable. Hence $A \cap C$ is the union of the $A \cap K_n$'s and the μ-null set $A \cap C \cap (\bigcup_n K_n)^c$ is μ-measurable, and this is so for any compact set C in S. An appeal to Lemma 7.19(ii) now tells us that the proof of (v) is complete.

Alas (vi) follows from (v) since for any Borel subset E of $\mathbb{R}$ and any $K \in \mathcal{C}$,

$$f^{\leftarrow}(E) \cap K = (f \cdot \chi_K)^{\leftarrow}(E) \cap K. \qquad \square$$

Before presenting the proof that $L^1(\mu)^*$ is $L^\infty(\mu)$, we make mention of the fact that $L^\infty(\mu)$ *always* sits comfortably inside $L^1(\mu)^*$ via the integration of any $f \in L^1(\mu)$ against an element of $L^\infty(\mu)$. Moreover this inclusion is a linear isometry. The issue is whether or not there is anything else in $L^1(\mu)^*$.

Now we're ready to establish the duality of $L^1(\mu)$ and $L^\infty(\mu)$. It proceeds from the basic fact (a consequence of the Radon–Nikodym theorem for finite measures) that if ν is a finite (positive) measure, then $L^1(\nu)^*$ is identifiable with $L^\infty(\nu)$ via the formula

$$y^*(f) = \int fg \, d\nu, \quad f \in L^1(\nu),$$

relating y^*'s in $L^1(\nu)^*$ and g's in $L^\infty(\nu)$.

What we have in hand is that for any compact set $K \subseteq S$ (with $\mu(K) > 0$), if we restrict μ to the μ-measurable subsets of K, we get $L^1(K, \mu)$ where

(K, μ) serves as a ν to which we can apply the comforting duality of L^1 and L^∞. To be sure we can describe a functional on all of $L^1(\mu)$: we call on our exhaustion principle.

To represent $x^* \in L^1(\mu)^*$ we use the exhaustion principle to "localize" the action of x^*.

Take a typical $f \in L^1(\mu)$. As is always the case for L^1-functions, f has σ-μ-finite support, that is, there is a sequence (A_n) of μ-measurable sets of finite μ-measure such that $f(k) = 0$ if $k \notin \bigcup_n A_n$. An appeal to ((iv)) of the exhaustion principle tells us there is a sequence (K_m) of members of $\mathcal{C}$ so that f vanishes outside $\bigcup_m K_m$. In this way we see that x^*'s value at f depends on a suitably chosen sequence (K_m) from $\mathcal{C}$ and the action of x^* on members of $L^1(K_m, \mu)$.

To be precise, if $K \in \mathcal{C}$, then we can restrict x^* to those f's in $L^1(\mu)$ which vanish outside K, which we will denote (somewhat cavalierly) as $L^1(K, \mu)$. On so doing, this restriction belongs to $L^1(K, \mu)^*$. But μ is a finite measure on K, and so there is a $g_K \in L^\infty(K, \mu) = L^1(K, \mu)^*$ so that

$$x^* f = \int_K f(k) g_K(k) \, d\mu(k),$$

and

$$||g_K||_{L^\infty(K,\mu)} = ||x^*|_{L^1(K,\mu)}|| \leq ||x^*||.$$

We now define g on S by

$$g(s) = \begin{cases} 0 & \text{if } s \notin \bigcup_{\mathcal{C}} K \\ g_K(s) & \text{if } s \in K. \end{cases}$$

By the disjoint nature of distinct members of $\mathcal{C}$, g is well defined. By (iv) of the exhaustion principle, g is μ-measurable, and, thanks to the constraints on $||g_K||_{L^\infty(K,\mu)}$, we have that $g \in L^\infty(\mu)$ with $||g||_\infty \leq ||x^*||$.

Finally if $f \in L^1(\mu)$, then we locate the sequence (K_m) in $\mathcal{C}$ so that f vanishes outside $\bigcup_m K_m$. We see that

$$\begin{aligned} x^*(f) &= x^*\left(f\chi_{\bigcup_m K_m}\right) = \sum_m x^*\left(f\chi_{K_m}\right) \\ &= \sum_m \int_{K_m} f(k) g_{K_m}(k)\, d\mu(k) \\ &= \sum_m \int_{K_m} f(k) g(k)\, d\mu(k) \\ &= \int_{\bigcup_m K_m} f(k) g(k)\, d\mu(k) \\ &= \int f(k) g(k)\, d\mu(k), \end{aligned}$$

and the duality of $L^1(\mu)$ and $L^\infty(\mu)$ is established.

2. Weil's Proof of Existence

In this short section we present Andre Weil's proof of the existence of a left Haar measure for any topological group. The initial stages of the proof are combinatorial averaging properties comparing continuous functions having compact support. Though these properties are uncomplicated and, once formulated, easily established, they are nonetheless basic to establishing the existence of a Haar measure in a general locally compact group.

Throughout G is a locally compact topological group. Let $\mathcal{K}(G)$ denote the linear space of continuous functions $f : G \to \mathbb{R}$ with compact support, and let $\mathcal{K}^+(G)$ denote those nonzero $f \in \mathcal{K}(G)$ with nonnegative values. *Fix, once and for the rest of this discussion,* $\boldsymbol{\omega \in \mathcal{K}^+(G)}$.

Note that if $f, \phi \in \mathcal{K}^+(G)$, then there are numbers $c_1, \ldots, c_n \geq 0$ as well as $g_1, \ldots, g_n \in G$ so that

$$f \leq \sum_{i=1}^n c_{i\,g_i}\phi, \tag{34}$$

where ${}_{g_i}\phi(g) = \phi(g_i^{-1}g)$ denotes a left translation of ϕ. To see why (34) holds, let K be a compact subset of G such that $f(g) = 0$ if $g \notin K$, and let $M > \max\{f(g) : g \in K\}$. Since $\phi \neq 0$ there exists a $g_0 \in G$ for which $\phi(g_0) > 0$; by ϕ's continuity there is an open set U_0 so $g_0 \in U_0$ and $\phi(u) > \frac{\phi(g_0)}{2}$ for all $u \in U_0$. For each $h \in G$, hU_0 is an open set and

$$K \subseteq \bigcup_{h \in G} hU_0.$$

K's compactness leads us to $h_1, \ldots, h_n \in G$ so

$$K = h_1 U_0 \cup \cdots \cup h_n U_0.$$

Letting ${}_h\phi(g) = \phi(h^{-1}g)$, we see that if $g \in K$, then $g \in h_m U_0$ for some $1 \le m \le n$ and $h_m^{-1} g \in U_0$ so

$${}_{h_m}\phi(g) = \phi(h_m^{-1}g) > \frac{\phi(g_0)}{2};$$

hence

$$\frac{2M}{\phi(g_0)}\, {}_{h_m}\phi(g) > M > f(g),$$

for all $g \in G$ and (34) follows.

With this in mind we define

$$(f : \phi) = \inf\left\{ \sum_{i=1}^{n} c_i : f \le \sum_{i=1}^{n} c_i\, {}_{g_i}\phi, \textit{ for some } c_1, \ldots, c_n \ge 0, g_1, \ldots, g_n \in G \right\},$$

and $(f : \phi)$ *is the* functional covering number *of* f *via translates of* ϕ.

Lemma 7.21. *For any* $f, g, \phi, \psi \in \mathcal{K}^+(G)$, *we have*

(i) $0 < (f : \phi) < \infty$;

(ii) *if* $h \in G$, *then* $(f : \phi) = ({}_h f : \phi)$;

(iii) $(f : \phi)$ *is subadditive and positively homogeneous in* f; *i.e.,* $(f + g : \phi) \le (f : \phi) + (g : \phi)$ *and* $(\lambda f : \phi) = \lambda (f : \phi)$ *for any* $\lambda \ge 0$;

(iv) *if* $f \le g$, *then* $(f : \phi) \le (g : \phi)$;

(v) $(f : \phi) \le (f : \psi)(\psi : \phi)$.

Proof. All of the statements follow easily using the definition of $(f : \phi)$. We comment on (v): to show $(f : \phi) \le (f : \psi)(\psi : \phi)$, fix $\epsilon > 0$, and let $c_1, \ldots, c_n \ge 0$ and $g_1, \ldots, g_n \in G$ be chosen so that

$$f \le \sum_{i=1}^{n} c_i\, {}_{g_i}\psi,$$

with $(f : \psi) \le \sum_{i=1}^{n} c_i \le (f : \psi) + \epsilon$ and $d_1, \ldots, d_m \ge 0$ and $h_1, \ldots, h_m \in G$ be chosen so that

$$\psi \le \sum_{j=1}^{m} d_j\, {}_{h_j}\phi$$

with $(\psi : \phi) \le \sum_{j=1}^{m} d_j \le (\psi : \phi) + \epsilon$.

It follows that

$$\begin{aligned} f &\leq \sum_{i=1}^{n} c_{i\,g_i}\psi \\ &\leq \sum_{i=1}^{n} c_i \,{}_{g_i}\!\left(\sum_{j=1}^{m} d_{j\,h_j}\phi\right) \\ &= \sum_{i=1}^{n} c_i \sum_{j=1}^{m} d_{j\,g_i h_j}\phi, \end{aligned}$$

which, since

$${}_{g_i h_j}\phi(g) = {}_{g_i}\phi(h_j^{-1}g) = \phi(g_i^{-1}h_j^{-1}g) = {}_{h_j g_i}\phi(g)$$

says

$$(f:\phi) \leq \sum_{i=1}^{n} c_i \sum_{j=1}^{m} d_j \leq ((f:\psi)+\epsilon)((\psi:\phi)+\epsilon).$$

Let $\epsilon \searrow 0$. Then

$$(f:\phi) \leq (f:\psi)(\psi:\phi). \qquad \square$$

For any $f, \phi \in \mathcal{K}^+(G)$ we define

$$\mu_\phi(f) = \frac{(f:\phi)}{(\omega:\phi)}.$$

We'll see that $\mu_\phi(f)$ is an approximation to our ultimate $\int f \, d\mu$. As we squeeze the support of ϕ to ever smaller sets, we'll get better and better approximations. At least, that is the game plan.

Lemma 7.22. *For any $f, g, \phi \in \mathcal{K}^+(G)$*

(i) $\mu_\phi(f) > 0$;

(ii) $\mu_\phi({}_g f) = \mu_\phi(f)$ *for all* $g \in G$;

(iii) μ_ϕ *is subadditive and positively homogeneous; i.e.,* $\mu_\phi(f+g) \leq \mu_\phi(f) + \mu_\phi(g)$ *and* $\mu_\phi(\lambda f) = \lambda \mu_\phi(f)$ *for all* $\lambda \geq 0$;

(iv) μ_ϕ *is monotone nondecreasing;*

(v) $\dfrac{1}{(\omega:f)} \leq \mu_\phi(f) \leq (f:\omega)$.

Proof. We comment on (v): ω, f and ϕ are all nonzero, so quantities involved in the inequalities from (v) of Lemma 7.21 are positive. Therefore

$$(f:\phi) \leq (f:\omega)(\omega:\phi)$$

and

$$(\omega : \phi) \leq (\omega : f)(f : \phi),$$

and (v) of Lemma 7.22 is immediate from these. □

Recall that any $f \in \mathcal{K}^+(G)$ is both left- and right-uniformly continuous.

Lemma 7.23. *Let $f_1, f_2 \in \mathcal{K}^+(G)$, and fix $\epsilon > 0$. Then there exists a neighborhood V of the identity so that if $\phi \in \mathcal{K}^+(G)$ with the support of ϕ contained in V, then*

$$\mu_\phi(f_1) + \mu_\phi(f_2) \leq \mu_\phi(f_1 + f_2) + \epsilon.$$

In effect (by Lemma 7.22), as we squeeze the support of ϕ toward $\{e\}$, the function μ_ϕ tends to be additive.

Proof. Let K be a compact subset of G so that K contains the support of f_1 and the support of f_2. Choose $f \in \mathcal{K}^+(G)$ so that

$$\chi_K \leq f \leq 1.$$

Fix $\delta > 0$ and consider

$$f_\delta = f_1 + f_2 + \delta f.$$

Define h_1, h_2 by

$$h_1(g) = \begin{cases} \frac{f_1(g)}{f_\delta(g)} & \text{if } g \notin K, \\ 0 & \text{if } g \in K, \end{cases} \qquad h_2(g) = \begin{cases} \frac{f_2(g)}{f_\delta(g)} & \text{if } g \in K, \\ 0 & \text{if } g \notin K. \end{cases}$$

Then $h_1, h_2 \in \mathcal{K}^+(G)$ and each has its support inside that of f_δ. So

$$f_1 = h_1 f_\delta, \quad f_2 = h_2 f_\delta, \quad h_1 + h_2 \leq 1.$$

Because any member of $\mathcal{K}^+(G)$ is (left) uniformly continuous on G given $\epsilon' > 0$, there is an open set V containing the identity so that if $x^{-1}y \in V$, then

$$|h_1(x) - h_1(y)| \leq \epsilon' \text{ and } |h_2(x) - h_2(y)| \leq \epsilon'.$$

Suppose $\phi \in \mathcal{K}^+(G)$ with the support of ϕ contained in V, and suppose

$$f_\delta \leq \sum_{i=1}^{n} c_{i g_i} \phi,$$

for some $c_1, \ldots, c_n \geq 0$ and $g_1, \ldots, g_n \in V$. Then for any g with $g_i^{-1} g \in V$,

$$f_1(g) = f_\delta(g) h_1(g) \leq \sum_{i=1}^{n} c_i \phi(g_i^{-1} g) h_1(g) \leq \sum_{i=1}^{n} c_i \phi(g_i^{-1} g)(h_1(g_i) + \epsilon')$$

and

$$f_2(g) = f_\delta(g)h_2(g) \le \sum_{i=1}^n c_i\phi(g_i^{-1}g)h_2(g) \le \sum_{i=1}^n c_i\phi(g_i^{-1}g)(h_2(g_i)+\epsilon').$$

Since V contains the support of ϕ, it follows that

$$(f_1:\phi) \le \sum_{i=1}^n c_i(h_1(g_i)+\epsilon') \text{ and } (f_2:\phi) \le \sum_{i=1}^n c_i(h_2(g_i)+\epsilon').$$

Adding these together (and remembering that $h_1+h_2 \le 1$) gives

$$(f_1:\phi)+(f_2:\phi) \le \sum_{i=1}^n c_i(1+2\epsilon'),$$

and so by the arbitrariness of $c_1,\ldots,c_n,g_1,\ldots,g_n$, with respect to f_δ we see (by the definition of $(f_\delta:\phi)$) that

$$(f_1:\phi)+(f_2:\phi) \le (1+2\epsilon')(f_\delta:\phi).$$

We divide everything in the inequality by $(\omega:\phi)$, a positive number for sure, and get

$$\begin{aligned} \mu_\phi(f_1)+\mu_\phi(f_2) &\le (1+2\epsilon')\mu_\phi(f_\delta) = (1+2\epsilon')\mu_\phi(f_1+f_2+\delta f) \\ &\le (1+2\epsilon')(\mu_\phi(f_1+f_2)+\delta\mu_\phi(f)) \\ &= \mu_\phi(f_1+f_2)+2\epsilon'\mu_\phi(f_1+f_2)+(1+2\epsilon')\delta\mu_\phi(f). \end{aligned}$$

An elementary application of epsilonics will finish the proof of Lemma 7.23. □

All the ingredients are present to proceed to a proof of the existence of Haar measure on G if we follow A. Weil and call on transfinite methods.

In fact we can use $\mathcal{K}^+(G)$ as an indexing set and keep in mind that for any $\phi, f \in \mathcal{K}^+(G)$,

$$\frac{1}{(\omega:f)} \le \mu_\phi(f) \le (f:\omega). \tag{35}$$

With this in mind, we let $I(f)$ be the closed bounded interval

$$I(f) = \left[\frac{1}{(\omega:f)}, (f:\omega)\right]$$

in $\mathbb{R}$ and call on Tychonoff's theorem (in its original form) to conclude that

$$I = \prod_{\mathcal{K}^+(G)} I(f)$$

is a compact Hausdorff space. For a fixed $\phi \in \mathcal{K}^+(G)$,

$$(\mu_\phi(f))_{f\in\mathcal{K}^+(G)} \in I,$$

thanks to (35). For any open set V in G that contains the identity, let $\mathcal{F}_V$ denote the set of all points $(\mu_\phi(f))_{f\in\mathcal{K}^+(G)}$, where ϕ's support is contained in V. Each $\mathcal{F}_V$ is nonempty.

If V_1, V_2 and V are open sets each of which contains the identity with $V \subseteq V_1 \cap V_2$, then $\mathcal{F}_V \subseteq \mathcal{F}_{V_1} \cap \mathcal{F}_{V_2}$.

By I's compactness there is a point

$$(\mu(f))_{f\in\mathcal{K}^+(G)} \in \bigcap \overline{\mathcal{F}}_V.$$

What does this mean? Well given $f_1, \dots, f_n \in \mathcal{K}^+(G)$, $\epsilon > 0$ and an open set V containing G's identity, there is a $\phi \in \mathcal{K}^+(G)$ with the support of ϕ contained in V so

$$|\mu(f_1) - \mu_\phi(f_1)|, \dots, |\mu(f_n) - \mu_\phi(f_n)| \le \epsilon.$$

An appeal to Lemmas 7.22 and 7.23 shows that

$$\begin{cases} \mu({}_g f) = \mu(f),\ \mu(f_1 + f_2) = \mu(f_1) + \mu(f_2), \\ \mu(\lambda f) = \lambda\mu(f),\ \mu(f) \ge \frac{1}{(\omega : f)} > 0, \end{cases}$$

whenever $f_1, f_2 \in \mathcal{K}^+(G)$ and $\lambda > 0$.

If we define $\mu(0) = 0$, then we have what's needed to define an invariant measure on G.

3. A Remarkable Approximation Theorem of Henri Cartan

Theorem 7.24 (Bochner and Dieudonné, specially tailored to our purposes). *Let G be a locally compact topological group, let K be a nonempty compact subset of G, and suppose $V_1, \dots, V_n$ are open subsets of G each with compact closure so that $K \subseteq V_1 \cup \dots \cup V_n$. Then there are continuous functions $f_1, \dots, f_n : G \to [0, 1]$ such that the support of each f_i is contained in V_i and $\sum_{i=1}^n f_i(k) = 1$ for each $k \in K$.*

The $f_1, \dots, f_n$ are referred to as a *continuous partition of unity subordinate to the cover* $\{V_1, \dots, V_n\}$ *of* K.

Proof. In Chapter 3 the "Introduction to Topological Groups" we proved that G is paracompact; hence, normal. By G's normality we can find another open cover $\{W_1, \dots, W_n\}$ of K with $\overline{W}_i \subseteq V_i, i = 1, \dots, n$. (Induction: $n = 1$ follows by the very definition of normality. If we assume that the statement holds for any compact subset of G and any open cover of the set by m-open sets for any $m < n$, then let's see what we can do if K is a compact set and $K \subseteq V_1 \cup \dots \cup V_n$. Look at the compact set $C = K \cap (V_2 \cap \dots \cap V_n)^c \subseteq V_1$.

Normality gives us a W_1 so that $C \subseteq W_1 \subseteq \overline{W}_1 \subseteq V_1$. Our induction hypothesis applies to

$$K \cap W_1^c \subseteq V_2 \cup \cdots \cup V_n,$$

providing us with the rest of the W_i's, $W_2, \ldots, W_n$.)

Now suppose $K \subseteq V_1 \cup \cdots \cup V_n$; look for $W_1, \ldots, W_n$ so $\overline{W}_i \subseteq V_i$ as in the above. Appeal to Mr. Urysohn, and he generously provides us with continuous functions $g_1, \ldots, g_n : G \to [0,1]$ such that $\chi_{\overline{W}_i} \leq g_i \leq \chi_{V_i}$. Set $f_i = \frac{g_i}{\sum_{i=1}^n g_i}$. □

Lemma 7.25. *Let $f \in \mathcal{K}^+(G)$, let K be a compact set in G, and let $\epsilon > 0$ be given. Then there exist $k_1, k_2, \ldots, k_n \in K$, $h_1, \ldots, h_n \in \mathcal{K}^+(G)$ so that if $g \in G, k \in K$, then*

$$\left| \sum_{i=1}^n h_i(k)_{k_i} f(g) - {}_k f(g) \right| \leq \epsilon.$$

Proof. Let V be a neighborhood of the identity chosen so that if $g_0 \in Vg$, then

$$|f(g) - f(g_0)| \leq \epsilon,$$

remembering that $f \in \mathcal{K}^+(G)$ must be (right) uniformly continuous on G. Now cover K by a finite collection $k_1 V, \ldots, k_n V$ of translates of V where $k_1, \ldots, k_n \in K$. Next, choose a partition of unity $h_1, \ldots, h_n \in \mathcal{K}^+(G)$ subordinate to the covering $\{k_1 V, \ldots, k_n V\}$ of K so that

- each $h_i = 0$ outside of $k_i V$, and
- $\sum_{i=1}^n h_i(k) = 1$ for each and every $k \in K$.

This we can do thanks to the Bochner–Dieudonné partition of unity theorem (Theorem 7.24).

Notice that if $k \in V$ and $k \in k_i V$, then

$$(k_i^{-1} g)(k^{-1} g)^{-1} = k_i^{-1} g g^{-1} k = k_i^{-1} k \in V,$$

and so by choice of V,

$$|f(k_i^{-1} g) - f(k^{-1} g)| \leq \epsilon.$$

If $k \in V$ but $k \notin k_i V$, then $h_i(k) = 0$. In either case

$$|h_i(k)_{k_i} f(g) - h_i(k)_k f(g)| \leq \epsilon h_i(k)$$

and this is so for $i = 1, \ldots, n$. Adding these inequalities up and recalling that $\sum_{i=1}^n h_i(k) = 1$ for each $k \in K$, we get

$$\left| \sum_{i=1}^n h_i(k)_{k_i} f(g) - {}_k f(g) \right| \leq \epsilon$$

for all $k \in K$ and all $g \in G$, as advertised. □

Lemma 7.26. *Let $f_1, f_2, \ldots, f_n \in \mathcal{K}^+(G), 0 \leq \lambda_1, \lambda_2, \ldots, \lambda_n$, and $\epsilon > 0$ be given. Then there is an open set V containing the identity of G so that if $\psi \in \mathcal{K}^+(G)$ with the support of ψ contained in V, then*

$$\sum_{i=1}^n \lambda_i \mu_\psi(f_i) \leq \mu_\psi\left(\sum_{i=1}^n \lambda_i f_i \right) + \epsilon.$$

Proof. Let K be a compact set that contains the support of $f_1, \ldots, f_n$. Choose $f \in \mathcal{K}^+(G)$ so that

$$\chi_K \leq f \leq 1.$$

Let $\delta > 0$. Look at

$$f_\delta = \lambda_1 f_1 + \cdots + \lambda_n f_n + \delta f \in \mathcal{K}^+(G).$$

Define $h_i, 1 \leq i \leq n$ by

$$h_i(g) = \begin{cases} \frac{\lambda_i f_i(g)}{f_\delta(g)} & \text{if } g \in K, \\ 0 & \text{if } g \notin K. \end{cases}$$

Notice $\lambda_i f_i = f_\delta h_i$ and $\sum_{i=1}^n h_i \leq 1$. Moreover $h_i \in \mathcal{K}^+(G)$. The uniform continuity of members of $\mathcal{K}^+(G)$ ensures that if $\epsilon' > 0$ is given, then there is an open set V that contains the identity so that if $y \in xV$, then

$$|h_1(x) - h_1(y)|, \ldots, |h_n(x) - h_n(y)| \leq \epsilon'.$$

Suppose $\psi \in \mathcal{K}^+(G)$ has its support contained in V and suppose $c_1, \ldots, c_m > 0$ and $g_1, \ldots, g_m \in V$ are chosen so that

$$\lambda_1 f_1 + \cdots + \lambda_n f_n + \delta f = f_\delta \leq \sum_{j=1}^n c_{j\, g_j}\psi.$$

If $g \in g_j V$, then for each $i = 1, \ldots, n$

$$\begin{aligned} \lambda_i f_i(g) &= f_\delta h_i(g) \leq \sum_{j=1}^m c_j \psi(g_j^{-1} g) h_i(g) \\ &\leq \sum_{j=1}^m c_j \psi(g_j^{-1} g)(h_i(g_j) + \epsilon'). \end{aligned}$$

It follows that for $i = 1, \ldots, n$,

$$\lambda_i(f_i : \psi) = (\lambda_i f_i : \psi) \leq \sum_{j=1}^{m} c_j(h_i(g_j) + \epsilon'),$$

and so keeping in mind the fact that $\sum_{i=1}^{n} h_i \leq 1$,

$$\sum_{i=1}^{n} \lambda_i(f_i : \psi) = \sum_{i=1}^{n} (\lambda_i f_i : \psi) \leq \sum_{j=1}^{m} c_j(1 + n\epsilon').$$

By the arbitrariness of $c_1, \ldots, c_m > 0$ and $g_1, \ldots, g_m \in V$ *vis-à-vis* f_δ, we see that

$$\sum_{i=1}^{n} \lambda_i(f_i : \psi) \leq (1 + n\epsilon')(f_\delta : \psi).$$

On dividing everything in plain sight by $(\omega : \psi)$, we get

$$\begin{aligned} \sum_{i=1}^{n} \lambda_i \mu_\psi(f_i) &\leq (1 + n\epsilon')(\mu_\psi(f_\delta)) \\ &= (1 + n\epsilon')(\mu_\psi(\lambda_1 f_1 + \cdots + \lambda_n f_n + \delta f)) \\ &\leq (1 + n\epsilon')(\mu_\psi(\lambda_1 f_1 + \cdots + \lambda_n f_n) + \mu_\psi(\delta f)) \\ &= \mu_\psi(\lambda_1 f_1 + \cdots + \lambda_n f_n) + n\epsilon' \mu_\psi(\lambda_1 f_1 + \cdots + \lambda_n f_n) \\ &\quad + \delta(1 + n\epsilon')\mu_\psi(f). \end{aligned}$$

The standard dose of epsilonics finishes the proof. □

A key element in the proof of the existence (and uniqueness) of Haar measure is the following stunning approximation theorem of H. Cartan. Remarkably, there is no measure theory in the statement of the theorem.

Theorem 7.27 (Cartan). *Let $f \in \mathcal{K}^+(G)$, and let $0 < \epsilon$ be given. Then there is an open set V in G that contains the identity and has compact closure such that if $\phi \in \mathcal{K}^+(G)$ with the support of ϕ contained in V, then there exist $g_1, \ldots, g_n \in \overline{\text{supp}}(f), c_1, \ldots, c_n \geq 0$ so that*

$$\left\| f - \sum_{i=1}^{n} c_{i g_i} \phi \right\| \leq \epsilon.$$

Proof. To start, let $0 < \epsilon' < \epsilon$, and choose an open set V with $\overline{V}$ compact, $e \in V$ so that $|f(g) - f(g_0)| \leq \epsilon'$ whenever $g^{-1}g_0 \in V$.

Let K be a compact set containing the support of f. Let $\delta > 0$. Apply Lemma 7.25 to find $g_1, \ldots, g_n \in K$ and $h_1, \ldots, h_n \in \mathcal{K}^+(G)$ so that if

$g \in G, k \in K$, then

$$\left|\sum_{i=1}^{n} h_i(k)_{g_i}\phi(g) - {}_k\phi(g)\right| \leq \delta.$$

Multiplying by $f(k)$, we see that for all $g \in G, k \in K$, we have

$$\left|\sum_{i=1}^{n} {}_{g_i}\phi(g)h_i(k)f(k) - {}_k\phi(g)f(k)\right| \leq \delta f(k); \tag{36}$$

but the support of f is contained in K, so if $k \notin K$, the left hand side of (36) is zero, and so we have (36) holding for all $k, g \in G$.

Now ϕ's support lies in V, so if $k^{-1}g \in V$, we see

$$|f(g) - f(k)| \leq \epsilon',$$

and so

$$|{}_k\phi(g)f(g) - {}_k\phi(g)f(k)| \leq \epsilon' {}_k\phi(g).$$

On the other hand if $k^{-1}g \notin V$, then ${}_k\phi(g) = \phi(k^{-1}g) = 0$. Regardless of where $k^{-1}g$ is located, we have that for every $k \in G$ and $g \in G$,

$$|{}_k\phi(g)f(k) - {}_k\phi(g)f(g)| \leq \epsilon' {}_k\phi(g). \tag{37}$$

In tandem, (36) and (37) tell us that for any $k, g \in G$

$$\begin{aligned} &\left|\sum_{i=1}^{n} {}_{g_i}\phi(g)h_i(k)f(k) - {}_k\phi(g)f(g)\right| \\ &\leq \left|\sum_{i=1}^{n} {}_{g_i}\phi(g)h_i(k)f(k) - {}_k\phi(g)f(k)\right| + |{}_k\phi(g)f(k) - {}_k\phi(g)f(g)| \\ &\leq \delta f(k) + \epsilon' {}_k\phi(g). \end{aligned} \tag{38}$$

Notice that if $\check{\phi}(g) = \phi(g^{-1})$, then

$${}_k\phi(g) = \phi(k^{-1}g) = \check{\phi}(g^{-1}k) = {}_g\check{\phi}(k).$$

If we now replace ${}_k\phi(g)$ by ${}_g\check{\phi}(k)$ in (38), we get that for all $k, g \in G$,

$$\left|\sum_{i=1}^{n} {}_{g_i}\phi(g)h_i(k)f(k) - {}_g\check{\phi}(k)f(g)\right| \leq \delta f(k) + \epsilon' {}_g\check{\phi}(k). \tag{39}$$

Now viewing g as fixed and letting k be free, we get the functional inequality

$$\left|\sum_{i=1}^{n} {}_{g_i}\phi(g)h_i f - {}_g\check{\phi}f(g)\right| \leq \delta f + \epsilon' {}_g\check{\phi}.$$

We now call on the averaging machinery developed earlier.

With the inequality

$$\left|\sum_{i=1}^{n} {}_{g_i}\phi(g)h_i f - {}_g\check{\phi} f(g)\right| \le \delta f + \epsilon' {}_g\check{\phi}$$

in hand, note that for $\psi \in \mathcal{K}^+(G)$, the subadditivity of the functional μ_ψ tells us that

$$\left|\mu_\psi\left(\sum_{i=1}^{n} {}_{g_i}\phi(g)h_i f\right) - f(g)\mu_\psi({}_g\check{\phi})\right| \le \delta\mu_\psi(f) + \epsilon' \mu_\psi({}_g\check{\phi}).$$

Now $\phi \neq 0$ and by Lemma 7.22, $\mu_\psi({}_g\check{\phi}) = \mu_\psi(\check{\phi}) > 0$, so division by $\mu_\psi({}_g\check{\phi})$ is legal (if not ethical). Do it! The result:

$$\left|\mu_\psi\left(\sum_{i=1}^{n} \frac{{}_{g_i}\phi(g)}{\mu_\psi({}_g\check{\phi})} h_i f\right) - f(g)\right| \le \delta\frac{\mu_\psi(f)}{\mu_\psi(\check{\phi})} + \epsilon'. \tag{40}$$

This is so for any $\psi \in \mathcal{K}^+(G)$.

But Lemma 7.25 says that given $\eta > 0$ there is an open set W containing the identity of G so that if $\psi \in \mathcal{K}^+(G)$ and the support of ψ is inside W, then (setting $\lambda_i = \frac{{}_{g_i}\phi(g)}{\mu_\psi(\check{\phi})}$)

$$\left|\mu_\psi\left(\sum_{i=1}^{n} \frac{{}_{g_i}\phi(g)}{\mu_\psi(\check{\phi})} h_i f\right) - \sum_{i=1}^{n} \frac{{}_{g_i}\phi(g)}{\mu_\psi(\check{\phi})}\mu_\psi(h_i f)\right| \le \eta. \tag{41}$$

Note. W acts as a catalyst ensuring near additivity of μ_ψ, but also providing a term

$$\mu_\psi\left(\sum_{i=1}^{n} \frac{{}_{g_i}\phi(g)}{\mu_\psi(\check{\phi})} h_i f\right)$$

to compare with both

$$f(g) \text{ and a sum of the sort } \sum_{i=1}^{n} c_{i g_i}\phi(g).$$

We're in business! Let

$$c_i = \frac{\mu_\psi(h_i f)}{\mu_\psi(\check{\phi})},$$

then each $c_i > 0$ and for any $g \in G$

$$\begin{aligned}
\left|f(g) - \sum_{i=1}^{n} c_{i_{g_i}}\phi(g)\right| &= \left|f(g) - \sum_{i=1}^{n} \frac{\mu_\psi(h_i f)}{\mu_\psi(\check{\phi})}\,{}_{g_i}\phi(g)\right| \\
&\le \left|f(g) - \mu_\psi\left(\sum_{i=1}^{n} \frac{{}_{g_i}\phi(g)}{\mu_\psi(\check{\phi})} h_i f\right)\right| \\
&+ \left|\mu_\psi\left(\sum_{i=1}^{n} \frac{{}_{g_i}\phi(g)}{\mu_\psi(\check{\phi})} h_i f\right) - \sum_{i=1}^{n} \frac{{}_{g_i}\phi(g)}{\mu_\psi(\check{\phi})}\mu_\psi(h_i f)\right| \\
&\le \delta\frac{\mu_\psi(f)}{\mu_\psi(\check{\phi})} + \epsilon' + \eta \text{ (by (40) and (41)).}
\end{aligned}$$

Standard epsilonics takes over to finish the proof. □

4. Cartan's Proof of Existence of a Left Haar Integral

Before presenting Cartan's proof of the existence of a left invariant Haar integral, we'll state a crucial lemma. We delay the proof of the lemma because it's also critical to Cartan's proof of uniqueness, and the pattern of its proof is worthy of close inspection.

Lemma 7.28. *Let $f \in \mathcal{K}^+(G)$ and $\epsilon > 0$ be given. Then there is an open set U containing G's identity for which, given any $h \in \mathcal{K}^+(G)$ with the support of h contained in U, there is $c \ge 0$ and an open set $V = V_U$ containing the identity so that*

$$|\mu_\phi(f) - c\mu_\phi(h)| \le \epsilon$$

whenever $\phi \in \mathcal{K}^+(G)$ with the support of ϕ contained in V.

The existence of a left Haar integral follows from our next result.

Theorem 7.29. *Let $f \in \mathcal{K}^+(G), \epsilon > 0$. Then there is an open set V containing G's identity for which*

$$|\mu_\phi(f) - \mu_\psi(f)| < \epsilon$$

whenever $\phi, \psi \in \mathcal{K}^+(G)$ with the support of ϕ and the support of ψ contained in V.

Proof. Suppose $0 < \delta \le 1/7$. Then according to the dictates of Lemma 7.28 that there is an open set U containing G's identity so that for any $h \in \mathcal{K}^+(G)$ with the support of h contained in U, there is a $c_f \ge 0$ and another open set V_U containing the identity so that

$$|\mu_\phi(f) - c_f\mu_\phi(h)| \le \delta, \tag{42}$$

whenever $\phi \in \mathcal{K}^+(G)$ with the support of ϕ contained in V_U.

If we apply our lemma again (Lemma 7.28) with our beloved ω replacing f and the same δ in place, we find an open set $\widetilde{U}$ containing the identity so that if $h \in \mathcal{K}^+(G)$ with the support of h contained in $\widetilde{U}$, then for some $c_\omega \geq 0$ and some open set $V_{\widetilde{U}}$ containing the identity, we have

$$|1 - c_\omega \mu_\phi(h)| = |\mu_\phi(\omega) - c_\omega \mu_\phi(h)| \leq \delta \tag{43}$$

whenever $\phi \in \mathcal{K}^+(G)$ with the support of ϕ contained in $V_{\widetilde{U}}$.

On replacing $U, \widetilde{U}$ by $U \cap \widetilde{U}$ and $V_U, V_{\widetilde{U}}$ by $V_U \cap V_{\widetilde{U}}$, we can and do suppose that the same U, V work in both (42) and (43).

Notice that $\delta < 1$. So (43) ensures us that $c_\omega > 0$, making division by c_ω a legal operation. Let

$$R = \frac{c_f}{c_\omega}. \tag{44}$$

If $\phi \in \mathcal{K}^+(G)$ with the support of ϕ is contained in V, then both (42) and (43) are in effect. So

$$\begin{aligned}
|\mu_\phi(f) - R| &= |\mu_\phi(f) - c_f\mu_\phi(h) + c_f\mu_\phi(h) - R| \\
&\leq |\mu_\phi(f) - c_f\mu_\phi(h)| + c_f\left|\mu_\phi(h) - \frac{1}{c_\omega}\right| \\
&\leq \delta + \frac{c_f}{c_\omega}|c_\omega\mu_\phi(h) - 1| \\
&\leq \delta + R\delta = (1+R)\delta.
\end{aligned}$$

Appealing again to (42), we see that

$$c_f\mu_\phi(h) \leq \mu_\phi(f) + \delta = \frac{(f:\phi)}{(\omega:\phi)} + \delta \leq (f:\omega) + \delta \tag{45}$$

because $(f:\phi) \leq (f:\omega)(\omega:\phi)$. By (43) we have

$$c_\omega\mu_\phi(h) \geq 1 - \delta. \tag{46}$$

Since $\mu_\phi(h) > 0$ we see (using $\delta \leq 1/2$) that

$$R = \frac{c_f}{c_\omega} = \frac{c_h\mu_\phi(h)}{c_\omega\mu_\phi(h)} \leq \frac{(f:\omega)+\delta}{1-\delta} \leq 2(f:\omega) + 1.$$

Investing this knowledge wisely we find

$$|\mu_\phi(f) - R| \leq (1+R)\delta \leq (1 + 2(f:\omega) + 1)\delta = 2(1 + (f:\omega))\delta$$

so long as $\phi \in \mathcal{K}^+(G)$ with the support of ϕ contained in V.

Cutting to the chase: given $f \in \mathcal{K}^+(G)$ and $\epsilon > 0$ there is an $R \geq 0$ and an open set V containing G's identity so that

$$|\mu_\phi(f) - R| \leq \epsilon$$

whenever $\phi \in \mathcal{K}^+(G)$ with the support of ϕ contained in V. Of course if $\psi \in \mathcal{K}^+(G)$ has its support inside V, then

$$|\mu_\psi(f) - R| \leq \epsilon$$

also.

Conclusion. Given $f \in \mathcal{K}^+(G)$ and $\epsilon > 0$ there is an open set V containing the identity so that if $\phi, \psi \in \mathcal{K}^+(G)$ have their support inside V, then

$$|\mu_\phi(f) - \mu_\psi(f)| \leq 2\epsilon.$$

Among friends, this is good enough to finish the proof. □

Our hard work is done. It's time to reap the rewards.

Existence of a Left Invariant Integral. We now look at all pairs (V, ϕ) where V is an open set in G containing G's identity and $\phi \in \mathcal{K}^+(G)$ with the support of ϕ contained in V. We order the pairs (V, ϕ) by

$$(V_1, \phi_1) \leq (V_2, \phi_2)$$

if $V_2 \subseteq V_1$; so convergence with respect to this direction measures what happens as that V coordinates shrinks to $\{e\}$.

Fixing $f \in \mathcal{K}^+(G)$, we see that our last theorem tells us that

$$\mu(f) \equiv \lim_{(V,\phi)} \mu_\phi(f)$$

exists. By Lemma 7.23, the functional μ is additive on $\mathcal{K}^+(G)$; Lemma 7.22 tells us that μ is positively homogeneous and invariant under left translates. What's more, for any $\phi \in \mathcal{K}^+(G)$,

$$\mu_\phi(\omega) = \frac{(\omega : \phi)}{(\omega : \phi)} = 1,$$

so $\mu(\omega) = 1$. Therefore μ is a left invariant Haar measure on G.

5. Cartan's Proof of Uniqueness

We restate Lemma 7.28 and include its proof.

Lemma 7.30. *Let $f \in \mathcal{K}^+(G)$ and $\epsilon > 0$ be given. Then there is an open set U containing G's identity for which, given any $h \in \mathcal{K}^+(G)$ with the support of h contained in U, there is $c \geq 0$ and an open set $V = V_U$ containing the identity so that*

$$|\mu_\phi(f) - c\mu_\phi(h)| \leq \epsilon$$

whenever $\phi \in \mathcal{K}^+(G)$ with the support of ϕ contained in V.

Proof. Let W be an open set containing the identity with $\overline{W}$ compact, and let $K = \overline{\{x : f(x) \neq 0\}}$ be the support of f. Fix $f' \in \mathcal{K}^+(G)$ so that $f' \geq \chi_{K \cdot W}$.

For any $\eta > 0$, we know (by Cartan's approximation theorem) that there is an open set U containing the identity (we can suppose that $U \subseteq W$) so that if $h \in \mathcal{K}^+(G)$ with the support of h contained in U, then there are $k_1, \ldots, k_n \in K$ and $c_1, \ldots, c_n \geq 0$ so that

$$\left\| f - \sum_{i=1}^{n} c_i \, {}_{k_i}h \right\| \leq \eta.$$

Notice that h vanishes outside of U and so outside of W. It soon follows that each of the functions ${}_{k_i}h$ vanish outside $k_i W$ and so outside $K \cdot W$. But f itself vanishes outside K which lies inside $K \cdot W$. So f and ${}_{k_i}h$ both vanish outside $K \cdot W$. Using this and the behaviour of f' we see that we can upgrade the above inequality to

$$\left| f - \sum_{i=1}^{n} c_i \, {}_{k_i}h \right| \leq \eta f',$$

a functional inequality involving members of $\mathcal{K}^+(G)$. Now if $\phi \in \mathcal{K}^+(G)$, we see that

$$\left| \mu_\phi(f) - \mu_\phi\left(\sum_{i=1}^{n} c_i \, {}_{k_i}h \right) \right| \leq \eta \mu_\phi(f') \leq \eta(f' : \omega) \tag{47}$$

(the last inequality follows by Lemma 7.22(v)).

An appeal to our new additive lemma (Lemma 7.26) tells us that there is an open set $V = V_U$ containing the identity so that if we further stipulate that ϕ have its support contained in V, then we can assure that

$$\left| \mu_\phi\left(\sum_{i=1}^{n} c_i \, {}_{k_i}h \right) - \sum_{i=1}^{n} c_i \mu_\phi({}_{k_i}h) \right| \leq \eta. \tag{48}$$

But $\mu_\phi({}_{k_i}h) = \mu_\phi(h)$, so combining (47) and (48) and letting $c = \sum_{i=1}^n c_i$,

$$\begin{aligned}
&|\mu_\phi(f) - c\mu_\phi(h)| \\
&\quad = \left|\mu_\phi(f) - \mu_\phi\left(\sum_{i=1}^n c_i\, {}_{k_i}h\right) + \mu_\phi\left(\sum_{i=1}^n c_i\, {}_{k_i}h\right) - \sum_{i=1}^n c_i\mu_\phi(h)\right| \\
&\quad \leq \left|\mu_\phi(f) - \mu_\phi\left(\sum_{i=1}^n c_i\, {}_{k_i}h\right)\right| \\
&\quad + \left|\mu_\phi\left(\sum_{i=1}^n c_i\, {}_{k_i}h\right) - \sum_{i=1}^n c_i\mu_\phi(h)\right| \\
&\quad \leq \eta(f^{'} : \omega) + \eta.
\end{aligned}$$

Epsilonics now rules! □

Uniqueness of μ. To prove the uniqueness of left invariant integrals on $\mathcal{K}^+(G)$, we consider a functional ν on $\mathcal{K}^+(G)$ which satisfies $\nu(f) > 0$ for each $f \in \mathcal{K}^+(G)$, ν is additive and positively homogeneous, and ν is left invariant so $\nu(f) = \nu({}_gf)$ for any $g \in G$. We further assume that $\nu(\omega) = 1$, where ω is our chosen one in $\mathcal{K}^+(G)$. What do we want? We want to show that ν is in fact, $\lim_\phi \mu_\phi(f)$.

The path blazed by Cartan leads through the proof of Lemma 7.30. Let us repeat the lemma's proof but with ν in mind instead of μ_ϕ. Let $f \in \mathcal{K}^+(G)$.

We start with an open set W containing the identity and having compact closure, $\overline{W}$. Let $K = \overline{\{x : f(x) \neq 0\}}$ be the support of f. Pick $f^{'} \in \mathcal{K}^+(G)$ so that $f^{'} \geq \chi_{K\cdot W}$.

Exactly as before, we know that given $\eta > 0$ there is an open set U containing the identity (contained, if we wish—and we do—inside W) so that if $h \in \mathcal{K}^+(G)$ with the support of h contained in U, then there are $k_1, \ldots, k_n \in K$ and $c_1, \ldots, c_n \geq 0$ so that

$$\left\|f - \sum_{i=1}^n c_i\, {}_{k_i}h\right\| \leq \eta.$$

As before, analysis of the supports of f and h soon tell us that

$$\left|f - \sum_{i=1}^n c_i\, {}_{k_i}h\right| \leq \eta f^{'}.$$

Now the inequality (47)

$$\left|\mu_\phi(f) - \mu_\phi\left(\sum_{i=1}^n c_i\, {}_{k_i}h\right)\right| \leq \eta\mu_\phi(f^{'}) \leq \eta(f^{'} : \omega)$$

is replaced by

$$\left|\nu(f) - \nu\left(\sum_{i=1}^{n} c_i\, {}_{k_i}h\right)\right| \leq \eta\nu(f').$$

Since ν is additive and positively homogeneous, the inequality (48)

$$\left|\mu_\phi\left(\sum_{i=1}^{n} c_i\, {}_{k_i}h\right) - \sum_{i=1}^{n} c_i\mu_\phi({}_{k_i}h)\right| \leq \eta$$

is replaced by the identity

$$\nu\left(\sum_{i=1}^{n} c_i\, {}_{k_i}h\right) = \sum_{i=1}^{n} c_i\nu({}_{k_i}h).$$

The left invariance of ν tells us that $\nu({}_{k_i}h) = \nu(h)$, so letting $c = \sum_{i=1}^{n} c_i$,

$$\begin{aligned} |\nu(f) - c\nu(h)| &= \left|\nu(f) - \sum_{i=1}^{n} c_i\nu(h)\right| \\ &\leq \left|\nu(f) - \nu\left(\sum_{i=1}^{n} c_i {}_{k_i}h\right)\right| \\ &\leq \eta\nu(f'). \end{aligned}$$

Note. The "c" is the same here as it was in the case of μ_ϕ or at least, it can be chosen to be so.

If we now turn to the proof of the existence of a left invariant integral μ (Theorem 7.29), we can easily see that the constant c_ω found therein also works with ν replacing μ_ϕ, this time without any codicils about ϕ's support. As a result, if $R = \frac{c}{c_\omega}$ as in that proof (see (44)), we can conclude the same truism: there is an open set U containing the identity so that for any $h \in \mathcal{K}^+(G)$ there is a $c_\omega > 0$ so that

$$|1 - c_\omega\nu(h)| = |\nu(\omega) - c_\omega\nu(h)| \leq \delta$$

so long as h's support is contained in U. The point, once again, is that c_ω can be chosen within the realm of epsilonics to be the same for ν as for μ_ϕ; again the roots are in Cartan's approximation theorem. The resulting R for ν is the same as for μ_ϕ and so ultimately, $|\nu(f) - \mu_\phi(f)| \leq 2\epsilon$ just as in the proof of existence.

The upshot: for any $\epsilon > 0$

$$|\nu(f) - \mu_\phi(f)| \leq 2\epsilon.$$

6. Notes and Remarks

Our discussion of positive linear functionals follows M. A Naimark's *Normed Rings* [**91**] with occasional help found in D. Cohn's treasured book [**20**].

The proof in §2 is due to A. Weil [**136**]. Our presentation of Weil's proof (and Cartan's, for that matter) is based on positive linear functionals. Again our functional analytic bias entered our decision. For those who would prefer a more purely measure theoretic approach, we recommend the treatment found in D. Cohn [**20**] or P. Halmos [**50**].

H. Cartan's wonderful proof in [**15**] of the existence and uniqueness (achieved more or less simultaneously) makes no call on the Axiom of Choice. The elegance of his approach far outweighs the added complexity.

There do not appear to be any essentially different proofs for general locally compact groups, other than those of Cartan (with a simplified variation due to Alfsen [**1**]) and Weil; the generalization due to Steinlage is itself a mix of Cartan and Weil with a touch of Banach. We went with Cartan's original proof for historical reasons.

6.1. Existence of a non-Haar measurable set à la Vitali. Let G be a nondiscrete locally compact topological group, and let μ be a left Haar measure defined on G.

Here's how to find a nonmeasurable subset of G.

Start with an open set U that contains G's identity e and has compact closure. Of course

$$0 < \mu(U) < \infty.$$

Let V be an open set containing E such that

$$V = V^{-1} \text{ and } V \cdot V \cdot V \subseteq U.$$

Let D be a countably infinite subset of V. Let H be the subgroup of G generated by D. Then H is also countably infinite.

Look at the collection $\{Hx_\alpha : \alpha \in A\}$ of all *right* cosets of G (modulo H) and denote by A_0 the set

$$A_0 = \{\alpha \in A : Hx_\alpha \cap V \neq \emptyset\}.$$

For each $\alpha \in A_0$, pick exactly one point y_α such that

$$y_\alpha \in Hx_\alpha \cap V = Hy_\alpha \cap V.$$

Our nominee is

$$E = \{y_\alpha : \alpha \in A_0\}.$$

We observe that $H \cap (V \cdot V)$ is countably infinite. It's countable since it's a subset of the countable set H, and it's infinite since $D \subseteq V \subseteq V \cdot V$, and $D \subseteq H$.

Consider $(H \cap V \cdot V) \cdot E$. It's a disjoint union of sets of the form xE, where $x \in H \cap V \cdot V$; each of the countably infinite many sets xE, where $x \in (H \cap V \cdot V) \cdot E$ has the same measure $\mu(E)$.

It follows that the measure of $(H \cap V \cdot V) \cdot E$ has two choices:

$$\mu((H \cap V \cdot V) \cdot E) = 0 \quad \text{if } \mu(E) = 0$$

or

$$\mu((H \cap V \cdot V) \cdot E) = \infty \quad \text{if } \mu(E) > 0.$$

But

$$V \subseteq (H \cap V \cdot V) \cdot E \subseteq U.$$

Indeed, if $v \in V$, then $v \in Hx_\alpha$ for some $\alpha \in A$. Therefore $v \in V \cap Hx_\alpha$, and so $\alpha \in A_0$, which means that $v \in Hy_\alpha$. Thus $v = hy_\alpha$ for some $h \in H$ and $\alpha \in A_0$. Therefore

$$h = vy_\alpha^{-1} \in V \cdot V^{-1} = V \cdot V,$$

and so

$$v \in (H \cap V \cdot V) \cdot E.$$

On the other side of the ledger,

$$(H \cap V \cdot V) \cdot E \subseteq V \cdot V \cdot E \subseteq V \cdot V \cdot V \subseteq U.$$

From this it follows that

$$\mu(V) \leq \mu((H \cap V \cdot V) \cdot E) \leq \mu(U),$$

and this is impossible.

A close look at what has just been done is warranted.

Suppose Σ is a σ-field of subsets of G that contains all the μ-measurable sets, $gE \in \Sigma$ if $g \in G$ and $E \in \Sigma$, and $\tilde{\mu}$ is a countably additive left invariant extension of μ to Σ.

Reasoning as before, we still cannot admit E to Σ without running afoul of the rules of logic, so E just does not gain membership in Σ.

After all, if $E \in \Sigma$, then

$$(H \cap V \cdot V) \cdot E \in \Sigma$$

too. As before

$$\tilde{\mu}((H \cap V \cdot V) \cdot E) = 0 \text{ or } \tilde{\mu}((H \cap V \cdot V) \cdot E) = \infty.$$

But $V \subseteq (H \cap V \cdot V) \cdot E \subseteq U$ makes either possibility untenable. □

P. Erdős and R. Daniel Mauldin [36] showed that if G is an uncountable group, then there does not exist a nontrivial σ-finite countably additive measure on the field of all subsets of G that is left invariant.

Their proof relies on Ulam's proof (using the Axiom of Choice) that the first uncountable cardinal is not a measurable cardinal.

6.2. Von Neumann–Kakutani–Weil Proof of Uniqueness. There is another well-known argument for the uniqueness of Haar measure in the setting of a general locally compact group. This proof revolves around the judicious use of Fubini's theorem. The proof was given independently by J. von Neumann [1936] and S. Kakutani [59]. Each needed an added hypothesis on the group to ensure the availability of the appropriate Fubini theorem. As was his wont, A. Weil [135] realized the availability of a suitable version of the Fubini scheme and gave a general proof of uniqueness. As a result, this might be called the Von Neumann–Kakutani–Weil proof. How's that for pedigree?

We'll be dealing with Haar measures μ and ν on a locally compact topological group G. Here's the utility grade Fubini theorem that we will need.

Theorem 7.31 (Fubini). *Let $h \in \mathcal{K}(G \times G)$. Then*

(i) *for each $x, y \in G$ the functions $f(x,\cdot), f(\cdot,y)$ belong to $\mathcal{K}(G)$;*

(ii) *the functions*

$$x \to \int_G f(x,y)\,d\nu(y)$$

and

$$y \to \int_G f(x,y)\,d\mu(x)$$

belong to $\mathcal{K}(G)$;

(iii) $$\int\int f(x,y)\,d\nu(y)\,d\mu(x) = \int\int f(x,y)\,d\mu(x)\,d\nu(y).$$

For a clear exposition of what's involved, we suggest the beautiful book *Measure Theory* by Donald L. Cohn [20].

Theorem 7.32. *Let μ and ν be left Haar measures on the locally compact group G. Then there is a positive real number c such that $\nu = c\mu$.*

Proof. In fact, let $g \in \mathcal{K}^+(G), g \neq 0$, and let $f \in \mathcal{K}$. Since $\int g d\mu \neq 0$ the ratio

$$\frac{\int f d\mu}{\int g d\mu}$$

is legal. We will show that this ratio depends only on f and g.

Now if $h \in \mathcal{K}(G \times G)$, then Fubini's theorem in its utility grade version tells us that both integrals

$$\int \int h(x,y) d\mu(x) d\nu(y), \int \int h(x,y) d\nu(y) d\mu(x)$$

exist and are equal. We look at the right-hand integral and apply the Fubini theorem, substituting $y^{-1}x$ for x to get

$$\int \int h(x,y) d\nu(y) d\mu(x) = \int \int h(y^{-1}x, y) d\mu(x) d\nu(y),$$

which in reapplying the Fubini theorem and substituting xy for y gives

$$\begin{aligned} \int \int h(x,y) d\nu(y) d\mu(x) &= \int \int h(y^{-1}x, y) d\mu(x) d\nu(y) \\ &= \int \int h(y^{-1}, xy) d\nu(y) d\mu(x). \end{aligned} \tag{49}$$

Now a judicious choice of h comes to the fore: Let

$$h(x,y) = \frac{f(x)g(yx)}{\int g(zx) d\nu(z)}.$$

We have that $h \in \mathcal{K}^+(G)$ and the function $x \to \int g(zx) d\nu(z)$ is continuous and never vanishes. Moreover if $K = \text{support}(f)$ and $\widetilde{K} = \text{support}(g)$, then h's support is contained in the compact set $K \times \widetilde{K} K^{-1}$.

Further we see that

$$h(y^{-1}, xy) = \frac{f(y^{-1})g(x)}{\int g(zy^{-1}) d\nu(z)},$$

and so looking to (49) we get

$$\int \int \frac{f(x)g(yx)}{\int g(zx) d\nu(z)} d\nu(y) d\mu(x) = \int \int \frac{f(y^{-1})g(x)}{\int g(zy^{-1}) d\nu(z)} d\nu(y) d\mu(x).$$

Since

$$\int \int \frac{f(x)g(yx)}{\int g(zx) d\nu(z)} d\nu(y) d\mu(x) = \int f(x) \int \frac{g(yx)}{\int g(zx) d\nu(z)} d\mu(x) = \int f(x) dx,$$

we have

$$\begin{aligned}\int f(x)d\mu(x) &= \int\int \frac{f(y^{-1})g(x)}{\int g(zy^{-1})d\nu(z)}d\nu(y)d\mu(x)\\ &= \int g(x)\int \frac{f(y^{-1})}{\int g(zy^{-1})d\nu(z)}d\nu(y)d\mu(x)\\ &= \int g(x)d\mu(x)\int \frac{f(y^{-1})}{\int g(zy^{-1})d\nu(z)}d\nu(y).\end{aligned}$$

In other words,

$$\frac{\int f(x)d\mu(x)}{\int g(x)d\mu(x)} = \int \frac{f(y^{-1})}{\int g(zy^{-1})d\nu(z)}d\nu(y),$$

regardless of ν and regardless of μ. □

6.3. Form of Invariant Measures on Some Classical Groups. Following the directions of Hewitt and Ross [**55**] we "compute" the Haar measure in case G is a topological group with easily checked qualifications:

(i) G is an open set in $\mathbb{R}^n$ for some n, so if $x = (x_1, \ldots, x_n)$, $y = (y_1, \ldots, y_n) \in G$, then

$$x \cdot y = F(x_1, \ldots, x_{n,1}, \ldots, y_n) \in G$$

for $F : G \times G \to G$.

(ii) $\frac{dF_j}{dx_k}, \frac{dF_j}{dy_k}$ exist and are continuous throughout G.

Denote by l_a, r_a the left and right translations by $a \in G$,

$$l_a(x) = ax, r_a(x) = xa.$$

(iii) The Jacobians of l_a and r_a are constant, that is, depend only on a.

If $J(\tau)$ denotes the Jacobian of the transformation τ, let

$$L(a) = |J(l_a)|, \text{ and } R(a) = |J(r_a)|.$$

The nature of Jacobians ensures that

$$J(l_a \circ l_b) = J(l_a) \cdot J(l_b)$$

(since $l_a \circ l_b = l_{ab}$), and so

$$L(a)L(b) = L(ab).$$

Similarly,

$$R(a)R(b) = R(ab).$$

Moreover $l_e = \mathrm{id}_G = r_e$ so

$$L(e) = 1 = R(e).$$

Each of the maps $L, R : G \to (0, \infty)$ is a continuous homomorphism of G onto the multiplicative group $(0, \infty)$.

Left Haar integration? Take $\phi : G \to \mathbb{K}$ continuous with compact support, and denote by

$$\int_G \phi(x)\, dx$$

the Riemann integral of ϕ; this makes sense because ϕ has compact support. Set

$$I(f) = \int_G \frac{f(x)}{L(x)}\, dx$$

so $I(f)$ is a left Haar integral. Indeed using the classical change of variables formulas, we see that for any $a \in G$

$$\begin{aligned} I({}_{a^{-1}}f) &= \int_G ({}_{a^{-1}}f)(x) \frac{dx}{L(x)} \\ &= \int_G ({}_{a^{-1}}f \circ l_a)(y) \frac{1}{(L \circ l_a)(y)}\, dy \\ &= \int_G f(y) \frac{dy}{L(y)} = I(f). \end{aligned}$$

So I is a left invariant functional on the collection of continuous functions with compact support.

Similarly

$$J(f) = \int_G f(x)\, \frac{dx}{R(x)}$$

is a right invariant integral.

Let's see this Haar measure machine in action. We start with $\mathcal{GL}(2, \mathbb{R})$, the general linear group on $\mathbb{R}^2$. Here

$$\begin{aligned} l_a(x) &= l_{\begin{pmatrix} a_{11} & a_{12} \\ a_{21} & a_{22} \end{pmatrix}} \begin{pmatrix} x_{11} & x_{12} \\ x_{21} & x_{22} \end{pmatrix} \\ &= \begin{pmatrix} a_{11}x_{11} + a_{12}x_{21} & a_{11}x_{12} + a_{12}x_{22} \\ a_{21}x_{11} + a_{21}x_{21} & a_{21}x_{12} + a_{22}x_{22} \end{pmatrix}. \end{aligned}$$

So

$$J(l_a) = \begin{pmatrix} a_{11} & 0 & a_{12} & 0 \\ 0 & a_{11} & 0 & a_{12} \\ a_{21} & 0 & a_{22} & 0 \\ 0 & a_{21} & 0 & a_{22} \end{pmatrix}$$

and

$$\begin{aligned}
\det J(l_a) &= a_{11}\det\begin{pmatrix} a_{11} & 0 & a_{12} \\ 0 & a_{22} & 0 \\ a_{21} & 0 & a_{22} \end{pmatrix} + a_{12}\det\begin{pmatrix} 0 & a_{11} & a_{12} \\ a_{21} & 0 & 0 \\ 0 & a_{21} & a_{22} \end{pmatrix} \\
&= a_{11}(a_{11}a_{22}a_{22} - a_{12}a_{21}a_{22}) + a_{12}(a_{12}a_{21}a_{21} - a_{11}a_{21}a_{22}) \\
&= a_{11}^2a_{22}^2 - 2a_{11}a_{12}a_{21}a_{22} + a_{12}^2a_{21}^2 \\
&= (a_{11}a_{22} - a_{12}a_{21})^2 = (\det x)^2,
\end{aligned}$$

and thus the left invariant acts like

$$\phi \to \int_{\mathcal{GL}(2,\mathbb{R})} \phi(x)\frac{dx}{L(x)} = \int \frac{\phi(x)}{(\det x)^2}\, dx.$$

What about the right integral?

$$\begin{aligned}
r_a(x) &= r_{\begin{pmatrix} a_{11} & a_{12} \\ a_{21} & a_{22} \end{pmatrix}} \begin{pmatrix} x_{11} & x_{12} \\ x_{21} & x_{22} \end{pmatrix} \\
&= \begin{pmatrix} a_{11}x_{11} + a_{21}x_{12} & a_{12}x_{11} + a_{22}x_{12} \\ a_{11}x_{21} + a_{21}x_{22} & a_{12}x_{21} + a_{22}x_{22} \end{pmatrix}.
\end{aligned}$$

So

$$J(r_a) = \begin{pmatrix} a_{11} & a_{21} & 0 & 0 \\ a_{12} & a_{22} & 0 & 0 \\ 0 & 0 & a_{11} & a_{21} \\ 0 & 0 & a_{12} & a_{22} \end{pmatrix},$$

and

$$\begin{aligned}
\det J(r_a) &= a_{11}\det\begin{pmatrix} a_{22} & 0 & 0 \\ 0 & a_{11} & a_{21} \\ 0 & a_{12} & a_{22} \end{pmatrix} - a_{21}\det\begin{pmatrix} a_{12} & 0 & 0 \\ 0 & a_{11} & a_{21} \\ 0 & a_{12} & a_{22} \end{pmatrix} \\
&= a_{11}(a_{22}a_{11}a_{22} - a_{12}a_{21}a_{22}) - a_{21}(a_{12}a_{11}a_{22} - a_{21}a_{12}a_{12}) \\
&= (\det x)^2.
\end{aligned}$$

Since J and I are the same and each, evaluated at suitable continuous f's is

$$\int_{\mathcal{GL}(2,\mathbb{R})} \frac{f(x)}{(\det x)^2}\, dx,$$

$\mathcal{GL}(2,\mathbb{R})$ is unimodular.

Similar computations will result in $\mathcal{GL}(n,\mathbb{R})$ also being unimodular with Haar integral of

$$f : \mathcal{GL}(n,\mathbb{R}) \to K$$

that is continuous and has compact support being given by

$$\text{Riemann integral} \int \frac{f(x)}{|\det(x)|^n}\, dx.$$

6.4. Existence of a Complete System of Irreducible Representations. Let G be a locally compact topological group.

Definition 7.33. A *unitary representation* of G is a homomorphism from G into the group of unitary representations, $g \to U_g$, on the same Hilbert space H such that if $g_\alpha \to g_0$ in G, then for each $x \in H$,

$$U_{g_\alpha} x \to U_{g_0} x.$$

The unitary representation $g \to U_g$ is *irreducible* if the only closed linear subspaces S of H (the underlying Hilbert space of representations) that are left invariant (i.e., $U_g S \subseteq S$) by each U_g are $S = \{0\}$ and $S = H$.

Definition 7.34. We say that G *admits a complete system of irreducible unitary representations* if given $g_0 \in G$ such that $g_0 \neq e$ (G's identity), there is an irreducible unitary representation $g \to U_g$ of G over some Hilbert space H such that $U_{g_0} \neq \mathrm{id}_H$.

The Peter–Weyl theorem asserts that every compact topological group admits a complete system of finite-dimensional irreducible unitary representations (i.e., unitary representations over a finite-dimensional Hilbert space). The original proof of the Peter–Weyl theorem, valid for compact Lie groups, relied on the existence of left-invariant differential forms. As soon as Haar discovered his measure, the Peter–Weyl theorem was understood to extend to the proof that every compact metrizable topological group admits a complete system of finite-dimensional irreducible unitary representations. When von Neumann's proof of the existence of a Haar measure over any compact group made its appearance, the extension of the Peter–Weyl theorem for general compact groups was an (almost) immediate consequence.

In a parallel development, von Neumann showed that every *abelian* locally compact group satisfying the second axiom of countability admits a complete system of one-dimensional irreducible unitary representations—such representations are usually called *characters*. Prophetically, I. M. Gelfand and D. A. Raikov soon presented a proof of this result based on developing machinery of commutative Banach algebras.

Soon after Weil's proof in [**136**] that every locally compact topological group admits a (left-) Haar measure, Gelfand and Raikov showed in [**42**] that every locally compact topological group admits a complete system of irreducible unitary representations. Their approach is a delightful mix of group theoretic ideas, basic notions in Banach algebras, Hilbert space spectral theory, and measure theory; their paper is a model of clarity. We indicate an outline of the Gelfand–Raikov paper and enthusiastically recommend it for those who wish to see truly beautiful mathematics done beautifully.

The Gelfand–Raikov proof proceeds in stages.

Stage one establishes the connection between unitary representations of G and positive definite functions $\phi : G \to \mathbb{C}$. Recall that $\phi : G \to \mathbb{C}$ is *positive definite* if

$$\sum_{k=1}^{n} \sum_{l=1}^{n} \phi(g_l^{-1} g_k) \lambda_k \overline{\lambda_l} \geq 0$$

for any finite collection $g_1, \dots, g_n \in G$ and $\lambda_1, \dots, \lambda_n \in \mathbb{C}$. Suppose $\phi : G \to \mathbb{C}$ is a continuous positive definite function. Assume $a, b : G \to \mathbb{C}$ are finitely nonzero functions, and define the inner product $\langle a, b \rangle$ of a and b by

$$\langle a, b \rangle = \sum_{h \in G} \sum_{k \in G} \phi(k^{-1} h) a(h) \overline{b(k)}.$$

If we identify a and b whenever $\langle a - b, a - b \rangle = 0$, and complete the resulting inner product space, we get a Hilbert space which we denote by $L^2(\phi)$. For any $g \in G$, let U_g be the mapping that takes the finitely nonzero $a : G \to \mathbb{C}$ to the function $U_g a$

$$(U_g a)(h) = a(g^{-1} h). \tag{50}$$

It is easy to see that U_g extends to a unitary operator from $L^2(\phi)$ to $L^2(\phi)$ and $g \to U_g$ is a unitary representation of G over the Hilbert space $L^2(\phi)$.

On the other hand if $g \to U_g$ is a unitary representation of G over the Hilbert space H and

$$\phi(g) = \langle U_g x, x \rangle,$$

then ϕ is a continuous positive definite function on G which satisfies $\phi(g) \neq 0$ if $x \neq 0$. If x_0 is a cyclic vector for the representation $g \to U_g$ (meaning $U_G(x_0)$ is dense in H), then H is isomorphic to $L^2(\phi_0)$ where

$$\phi_0(g) = \langle U_g x_0, x_0 \rangle,$$

and U_g corresponds to the translation operator in (50) acting on $L^2(\phi_0)$.

Definition 7.35. If ϕ, χ are positive definite functions on G, we say that $\chi \ll \phi$ if $\phi - \chi$ is positive definite. We say that ϕ is an *elementary positive definite function* if $\chi \ll \phi$ (χ also positive definite) implies

$$\chi(g) = c\phi(g)$$

for some constant c.

Theorem 7.36. *If the unitary representation $g \to U_g$ arises from an elementary continuous positive definite function ϕ, then this representation is irreducible. If ϕ is a continuous positive definite function and ϕ gives rise to an irreducible unitary representation, then ϕ is elementary.*

Briefly if P is a projection on $L^2(\phi)$ that commutes with each U_g $(g \in G)$ (i.e., if P's range is invariant under all the U_g's), then

$$\chi(g) \equiv \langle U_g P x_0, x_0 \rangle$$

is positive definite and $\chi \ll \phi$. So if ϕ is elementary,

$$\chi(g) = c\phi(g)$$

and

$$\langle Pa, a \rangle = c\langle a, a \rangle$$

for any finitely nonzero $a : G \to \mathbb{C}$. Such a's are dense in $L^2(\phi)$, and so

$$\langle Px, x \rangle = c\langle x, x \rangle$$

for all $x \in L^2(\phi)$, that is, $P = c\,\mathrm{id}_{L^2(\phi)}$. Thus P is either 0 or $\mathrm{id}_{L^2(\phi)}$.

On the other hand, if the continuous positive definite function ϕ gives rise to an irreducible representation $g \to U_g$ over the Hilbert space H, then every nonzero vector $x \in H$ is cyclic. Take any $x_0 \in H, x_0 \neq 0$, and suppose $\chi \ll \phi$ where $\phi(g) = \langle U_g x_0, x_0 \rangle$. If $a, b : G \to \mathbb{C}$ are finitely nonzero, we can define B by

$$\langle Ba, b \rangle = \sum_h \sum_k \chi(k^{-1}h) a(h) \overline{b(k)}$$

and extend B to a bounded self-adjoint operator on $L^2(\phi)$ which commutes with each U_g. By $g \to U_g$'s irreducibility, B must be $c\,\mathrm{id}_{L^2(\phi)}$, and so $\chi(g) = c\phi(g)$.

It is immediate from this that we have

Corollary 7.37. *In order that G admit a compete system of irreducible unitary transformations, it is both necessary and sufficient of given $g_0 \in G$, $g_0 \neq e$ that there is an elementary continuous positive definite function $\phi_0 : G \to \mathbb{C}$ so that $\phi_0(g_0) \neq \phi_0(e)$.*

Stage two discusses the group algebra $L^1(G)$ and relates positive linear functionals thereupon to positive linear functionals.

Let μ be the left invariant Haar measure on G. For each μ-measurable set $E \subseteq G$, Eh is also μ-measurable for any $h \in E$, and

$$\mu(Eh) = l_h \mu(E)$$

for some real number l_h that is independent of E. The function $h \to l_h$ is a continuous function from G into $\mathbb{R}$ that satisfies

$$l_{hk} = l_h l_k.$$

Playing a central role in the Gelfand–Raikov story is the *group algebra* $L^1(G)$, the space of (equivalence classes of) of μ-integrable complex-valued functions defined on G. In addition to its Banach space structure (the norm is $||x||_1 = \int_G |x(g)|\, d\mu(g) = \int |x|\, d\mu$), $L^1(G)$ comes equipped with a multiplication (convolution) and a $*$-operation: if $x, y \in L^1(G)$, then $x \cdot y$ is given by

$$(x \cdot y)(g) = \int_G x(h)y(h^{-1}g)\, d\mu(h),$$

and the $*$-operation is

$$x^*(g) = l_{g^{-1}}\overline{x(g^{-1})}.$$

Here's what's so about the operations and how they treat members of $L^1(G)$:

Fact 7.38. For any $x, y, z \in L^1(G), \alpha \in \mathbb{C}$, we have

- $x \cdot y \in L^1(G)$ and $||x \cdot y||_1 \leq ||x||_1||y||_1$;
- $\alpha(x \cdot y) = (\alpha x) \cdot y = x \cdot (\alpha y)$, $(x \cdot y) \cdot z = x \cdot (y \cdot z)$, $x \cdot (y + z) = x \cdot y + x \cdot z$, and $(x + y) \cdot z = xz + yz$.

Fact 7.39. For any $x \in L^1(G), x^* \in L^1(G)$, and $||x^*||_1 = ||x||_1$. Moreover for any $x, y \in L^1(G)$, $\alpha \in \mathbb{C}$, we have

- $(\alpha x)^* = \bar{\alpha} x^*$,
- $(x + y)^* = x^* + y^*$,
- $(x \cdot y)^* = y^* \cdot x^*$,
- $(x^*)^* = x$.

The operator T_g defined by

$$T_g x(h) = x(g^{-1}h),$$

acts as a bounded linear operator from $L^1(G)$ to $L^1(G)$, and it also acts on $L^2(G)$ to $L^2(G)$. It is easy to see that T_g is an isometry of $L^1(G)$ onto itself, and on $L^2(G)$, T_g is a unitary operator. In fact $g \to T_g$ is a unitary representation of G over $L^2(G)$.

Definition 7.40. A linear functional L defined on $L^1(G)$ is called *positive* if for any $x \in L^1(G)$,

$$L(x^* \cdot x) \geq 0.$$

Now the existence of the left invariant measure μ makes its presence felt.

Definition 7.41. A function $\phi : G \to \mathbb{C}$ is *integrally positive definite* if ϕ is essentially bounded (with respect to μ) and satisfies

$$\int\int \phi(h^{-1}g)x(g)\overline{x(h)}\, d\mu(g)\, d\mu(h) \geq 0$$

for all $x \in L^1(G)$.

With this definition in hand it is easy to see that the functional L on $L^1(G)$ given by

$$L(x) = \int \phi(g)x(g)\,d\mu(g)$$

is a positive linear functional in the sense described above.

We can now state the following

Theorem 7.42 (Key Theorem). *Every positive linear functional L arises from a continuous positive definite function ϕ via the formula*

$$L(x) = L_\phi(x) = \int \phi(g)x(g)\,d\mu(g).$$

It follows from this that any integrally positive definite function is equal μ-almost everywhere to a continuous positive definite function.

We say that the continuous positive definite function ϕ is *normalized* if $\phi(e) = 1$.

Let $\mathcal{P}$ denote the collection of all positive definite functions with $\phi(e) \leq 1$. Then $\mathcal{P}$ is plainly a convex set and the identically zero function 0 is an extreme point of $\mathcal{P}$. Critical to the Gelfand–Raikov scheme is the following:

Theorem 7.43. *Any nonzero extreme point of $\mathcal{P}$ is a normalized elementary positive definite function, and conversely, any normalized elementary positive definite function on G is an extreme point of $\mathcal{P}$.*

Stage three utilizes the correspondences derived in stage two to set up an application of the Krein–Milman theorem.

To get started, we define the weak topology via telling what a weak neighborhood of $\phi_0 \in \mathcal{P}$ looks like: given $\epsilon > 0$ and $x_1, \ldots, x_n \in L^1(G)$, we define the weak neighborhood $W(\phi_0; x_1, \ldots, x_n, \epsilon)$ by

$$W = \left\{\phi \in \mathcal{P} : \left|\int (\phi_0(g) - \phi(g))x_k(g)\,d\mu(g)\right| < \epsilon, \text{ for } 1 \leq k \leq n\right\}.$$

By Real$L^1(G) = \mathcal{R}L^1$, we mean the real linear space which is the closed real linear space spanned by those $x \in L^1(G)$ for which $x = x^*$. If L is a positive linear functional on $L^1(G)$, then $L(x^*) = \overline{L(x)}$; using this, such an L can be interpreted as a bounded real linear functional on $\mathcal{R}L^1$, and since

$$L(x) = L\left(\frac{x + x^*}{2}\right) + iL\left(\frac{x - x^*}{2i}\right),$$

the restriction of L to $\mathcal{R}L^1$ is a one-to-one operation between positive linear functionals on the Banach $\star$-algebra $L^1(G)$ and positive real linear functionals on $\mathcal{R}L^1$.

If $L = L_\phi$, $\phi \in \mathcal{P}$, then

$$||L_\phi|| = ||\phi||_{L^\infty}(\mu)\phi(e) \leq ||\phi||_{L^\infty}(\mu),$$

and so, thanks to the Key Theorem, we can identify $\mathcal{P}$ in its weak topology with the weak*-closed convex set of all positive linear functionals on $\mathcal{R}L^1$ having norm less than or equal to one, a weak* convex set. The Krein–Milman theorem announces its presence! If we interpret this in terms of the weak topology on $\mathcal{P}$ as described above, then here is what we get:

Theorem 7.44. *The set $\mathcal{P}$ of continuous positive definite functions ϕ satisfying $\phi(e) \leq 1$ is the smallest weakly closed convex set containing all the normalized elementary continuous positive definite functions and the zero function $O(g) := 0$.*

Because G is locally compact, we have a well-tested manner of constructing nontrivial positive definite functions: $L^2(G)$ is a Hilbert space with the usual inner product (x, y) of $x, y \in L^2(G)$

$$(x, y) = \int x(g)\overline{y(g)}\, d\mu(g);$$

the operator $U_g : L^2(G)\to L^2(G)$ given by

$$(U_g x)(h) = x(g^{-1}h)$$

is a unitary operator and $g\to U_g$ is a unitary representation of G over $L^2(G)$.

If we take a $g_0 \in G$ such that $g_0 \neq e$, let W be an open set containing the identity such that $gW \cap W = \emptyset$, and choose $x_W \in \mathcal{K}^+(G)$ so that $(x_W, x_W) = 1$ yet $x_W(h) = 0$ outside W, then the function

$$\phi(g) = (U_g x_W, x_W)$$

is a continuous positive definite function such that

$$\phi(e) = (x_W, x_W) = 1,$$

so ϕ is normalized. What's more—$g_0 W \cap W = \emptyset$, so

$$\phi(g_0) = (U_{g_0} x_W, x_W) = \int x_W(g_0^{-1}h) x_W(h)\, d\mu(h) = 0,$$

since $g_0^{-1}h \in W$ precisely when $h \in g_0 W$ which for $h \in W$ does not happen.

There are plenty of normalized continuous positive definite functions on G!

To finish the argument, Gelfand and Raikov now argue by contradiction that if some $g_0 \in G$ is not e but there is no irreducible unitary representation $g\to U_g$ so that $U_g \neq \mathrm{id}$, it is because every normalized continuous positive definite ϕ has $\phi(g) = \phi(e)$ and from this so, too, does each normalized continuous positive definite function ϕ.

6.5. The Pontrjagin–van Kampen Duality Theorem. We close this Notes and Remarks section on a high note: a brief description of the still-wonderful Pontrjagin–van Kampen duality theorem.

Let G be a locally compact abelian topological group. We denote by Γ the collection of all continuous homomorphisms of G into the circle group

$$\Pi = \{z \in \mathbb{C} : |z| = 1\}.$$

Then Γ is an abelian group under pointwise multiplication; we can equip Γ with the Pontrjagin topology: if $\gamma_0 \in \Gamma$, then a basic open set about γ_0 is generated by a compact subset K of G and a $\delta > 0$, and it looks like

$$P(\gamma_0; K, \delta) = \{\gamma \in \Gamma : |\gamma(g) - \gamma_0(g)| < \delta \text{ for all } g \in K\}.$$

With this topology Γ is a locally compact abelian topological group called the *dual group* of G.

There is a natural pairing of G and Γ, namely $(g, \gamma) = \gamma(g)$. Here $\gamma \in \Gamma$ is viewed as a function of $g \in G$ but g can also be viewed as a function of γ. It's here that magic happens. Since Γ is a locally compact abelian topological group we can apply to Γ the procedure just established to G and and the result will be ostensibly another locally compact abelian topological group, the dual group of Γ. It is a fundamental phenomenon of mathematics that the result is a topological group that is isomorphic as a topological group to G! Every locally compact abelian topological group is its own second dual group.

This was discovered in case the locally compact topological group also satisfies the second axiom of countability by L. S. Pontrjagin in 1934 [**103**] and in general by E. R. van Kampen in 1935 [**131**]. The original proof proceeded through the evolving structure of locally compact abelian groups. An excellent modern source for this approach is Sid Morris's *Pontrjagin Duality and the Structure of Locally Compact Abelian Groups* [**86**].

Naturally the Pontrjagin–van Kampen duality theorem suggests that each isomorphic invariant of a locally compact abelian topological group has a dual version and the classification and delineation of dual properties has been an active pursuit for many years. We mention a few of the more famous such results:

- G is compact if and only if Γ is discrete (this is due to Pontrjagin and van Kampen).
- The weight of G and that of Γ are always the same, where the weight of the topological space is the least cardinality of a base for its topology.

- If G is a compact abelian topological group, then G is metrizable if and only if Γ is countable.
- G is connected if and only if Γ contains no nontrivial compact subgroups.

A list of interesting/important duality statements is long and interesting of and by itself. Hewitt and Ross's *Abstract Harmonic Analysis* [**55**] serves as an excellent *start* toward a complete list. The mention of the Hewitt–Ross volume hints at one of the most enduring consequences of the Pontrjagin–van Kampen duality theorem: abstract harmonic analysis.

A few words about this direction are worth making. For one thing, the duality theorem allows one to define the Fourier transform. Indeed, if $x \in L^1(G)$ and $\gamma \in \Gamma$, then we can define $\hat{x}(\gamma)$ as

$$\hat{x}(g) = \int_G x(g)\overline{\gamma(g)}\, dg,$$

where dg indicates integrating against the Haar measure. Many old friends from classical Fourier analysis extend to this abstract setting. To wit:

- If $x, y \in L^1(G)$, then $x \cdot y \in L^1(G)$ with $||x \cdot y||_1 \leq ||x||_1 ||y||_1$, where
$$(x \cdot y)(g) = \int_G x(g-h)y(h)\, dh$$
(we now write the operation in G in an additive manner because of the abelian nature of G); of course this is just an extension of what happens in general.
- If $x \in L^1(G)$, then the Fourier transform $\hat{x}$ defined by
$$\hat{x}(\gamma) = \int_G x(g)\overline{\gamma(g)}\, dg$$
is continuous and vanishes at ∞ with $||\hat{x}||_\infty \leq ||x||_1$.
- The map $x \to \hat{x}(\gamma)$ is a complex-valued multiplicative linear functional on $L^1(G)$ and, in fact, any such functional on $L^1(G)$ comes about in this fashion.
- The Fourier transform of $x \cdot y, \widehat{x \cdot y}$ is $\hat{x} \cdot \hat{y}$; that is, for any $\gamma \in \Gamma$ and any $x, y \in L^1(G)$,
$$x \hat{\cdot} y(\gamma) = \hat{x}(\gamma)\hat{y}(\gamma).$$
- If $x \in L^1(G)$ and $\hat{x}(\gamma) \equiv 0$, then $x = 0$.

The basic principles of classical harmonic analysis often have interpretations in this general setting, including Plancherel's theorem: the Fourier transform restricted to $L^1(G) \cap L^2(G)$ is an isometry (with respect to the L^2-norm) onto a dense linear subspace of $L^2(\Gamma)$ and may be uniquely extended to an isometry of $L^2(G)$ onto $L^2(\Gamma)$.

In fact, developing the Fourier transform as above allows one to derive the Pontrjagin–van Kampen duality theorem via the Fourier transform. A key feature of this approach is an alternative definition of the topology of the dual group Γ, namely if $\gamma_0 \in \Gamma, \epsilon > 0$ and $x_1, \ldots, x_n \in L^1(G)$, then a basic neighborhood of γ_0 is given by

$$N(\gamma_0; x_1, \ldots, x_n, \epsilon) = \{\gamma \in \Gamma : |\hat{x}_i(\gamma) - \hat{x}_i(\gamma_0) < \epsilon, i = 1, \ldots, n\}.$$

This topology is the same as the Pontrjagin topology introduced earlier, and this fact uses the notions of the Fourier analysis in essential ways.

This approach to the duality theorem via the Fourier transform is due to H. Cartan and R. Godement [**16**]. To see some of the places this takes you, Walter Rudin's *Fourier Analysis in Groups* [**108**] is highly recommended.

It is natural to wonder how tight is the hypothesis of local compactness to constructing a Haar measure. As we will see in Chapter 10, lack of local compactness, even in very well-behaved groups (such as Polish groups) that are *not* locally compact leads to pathological conclusions. Nevertheless, local compactness is always nearby in case the underlying measure-theoretics behave, as shown by A. Weil in the appendices of [**136**].

A *measurable group* is a group G with a σ-ring $\mathcal{M}$ and a σ finite, not identically zero, nonnegative extended real-valued measure m defined on $\mathcal{M}$ where both $\mathcal{M}$ and m are left invariant and the map that takes $(x, y) \in G \times G$ to $(x, xy) \in G \times G$ is measurable. With Weil we suppose that G is separated, so given any $x \in G$ other than the identity there is a measurable set A in G such that the m-measure of the symmetric difference of A and xA is positive.

Weil showed how to introduce a topology (called the *Weil topology*) in G making G into a locally bounded topological group with the added features that $m(U) > 0$ for each open U in G and $m(E)$ is finite for each set E that is bounded in G. (Local boundedness just means that there is a locally compact topological group of which G is a dense subgroup.) The point is that if G is a measurable group, then G is a dense subgroup of a locally compact topological group G' and m is the trace of a left Haar measure m' defined on G' in the sense that $m'(E \cap S) = m(E)$.

Chapter 8

Metric Invariance and Haar Measure

Suppose G is a locally compact metrizable group. Then G admits a metric ρ which is left invariant and generates G's topology; G also admits a left invariant regular Borel measure (a left Haar measure). Is there any connection between these left invariant objects? In this chapter we show the answer is an emphatic "yes."

The highlight of the chapter is the wedding of Haar measure and Hausdorff measure; presiding at the ceremony is C. Bandt. The result is a beautiful offspring: *fractional Hausdorff measure.* A consequence is that if λ is a left Haar measure on G and ρ is a left invariant metric generating G's topology, then ρ-isometric subsets of G have the same λ-outer measure.

Now the assumption that G has a left invariant metric that generates its group topology is, of course, not an assumption at all. The Birkhoff–Kakutani theorem assures us that G's metrizability is enough to ensure such a metric exists. However, since we are looking for relationships between left invariant measures on G and left invariant metrics on G, it is instructive (and fun!) to show how to generate a left invariant metric that generates G's topology directly from its left Haar measure. This proof is due to R. A. Struble [**124**] and leads off the chapter.

Theorem 8.1 (R. A. Struble). *Let G be a locally compact group with left Haar measure λ. Let (V_n) be a decreasing sequence of open sets that form a neighborhood basis of the identity e in G where $\overline{V}_n$ compact for each n.*

Then

$$\rho(x, y) = \sup_n \lambda(xV_n \triangle yV_n)$$

defines a left invariant metric on G *which is compatible with the topology of* G.

Proof. It's clear that $\rho(x, y)$ is well defined and that $\rho(x, y) = \rho(y, x)$. Moreover $\rho(x, y) \geq 0$ and $\rho(x, y) < \infty$ regardless of $x, y \in G$ since each V_n is a Borel set with compact closure. Further, $\rho(zx, zy)$ and $\rho(x, y)$ coincide because λ is left invariant.

If $x \neq y$, then there must be an m so that $xV_m \cap yV_m = \emptyset$ since G is Hausdorff; but now

$$\rho(x, y) \geq \lambda(xV_m \triangle yV_m) = 2\lambda(V_m) > 0.$$

On noticing that for any n and any $x, y, z \in G$,

$$xV_n \triangle yV_n \subseteq (xV_n \triangle zV_n) \cup (zV_n \triangle yV_n),$$

we see that for any $x, y, z \in G$,

$$\begin{aligned} \lambda(xV_n \triangle yV_n) &\leq \lambda((xV_n \triangle zV_n) \cup (zV_n \triangle yV_n)) \\ &\leq \lambda(xV_n \triangle zV_n) + \lambda(zV_n \triangle yV_n) \\ &\leq \rho(x, z) + \rho(z, y), \end{aligned}$$

and with this

$$\rho(x, y) \leq \rho(x, z) + \rho(z, y).$$

In sum, ρ is a left invariant metric on G.

If G's topology is discrete, then $\lambda(\{e\}) > 0$ so $V_m = \{e\}$ for some m; hence if $x \neq y$,

$$\rho(x, y) \geq \lambda(xV_m \triangle yV_m) = \lambda(\{x, y\}) = 2\lambda(\{e\}) > 0,$$

and the topology induced by ρ is discrete.

If G's topology is *not* discrete, then $\lambda(V_n) \searrow \lambda(\bigcap_n V_n) = \lambda(\{e\}) = 0$. If V is any open set containing e, then there is an $m \in \mathbb{N}$ so that $V_m V_m^{-1} \subseteq V$.

Claim 1. $x \in V$ *whenever* $\rho(x, e) < \lambda(V_m)$. To see this, let $\rho(x, e) < \lambda(V_m)$. Then

$$\lambda(xV_m \triangle V_m) \leq \rho(x, e) < \lambda(V_m),$$

a positive number. Were $xV_m \cap V_m = \emptyset$, then

$$\lambda(xV_m \triangle V_m) = 2\lambda(V_m) < \lambda(V_m),$$

oops! So $xV_m \cap V_m \neq \emptyset$ and thus there are $v_1, v_2 \in V_m$ so $xv_1 = v_2 \in xV_m \cap V_m$ and

$$x = v_2 v_2^{-1} \in V_m V_m^{-1} \subseteq V.$$

This is so whenever $\rho(x, e) < \lambda(V_m)$, and our claim is justified.

Let's look at all of the points x such that $\rho(x, e) < r$, where $r \in \mathbb{Q}, r > 0$. There must be an $m \in \mathbb{N}$ so that $\lambda(V_n) < \frac{r}{4}$, whenever $n \geq m$. Each of the functions

$$f_k(x) = \lambda(xV_k \triangle V_k)$$

is continuous and satisfies $f_k(e) = \lambda(V_k \triangle V_k) = \lambda(\emptyset) = 0$. But now we know there is an $l \in \mathbb{N}$ so that if $x \in V_l$, then $f_1(x), \ldots, f_{m-1}(x) < r$.

Claim 2. If $x \in V_l$, then $\rho(x, e) < r$. Why is this so? Well if $x \in V_l$, then by choice of $l \in \mathbb{N}$, we have

$$\lambda(xV_1 \triangle V_1), \ldots, \lambda(xV_{m-1} \triangle V_{m-1}) < r.$$

What about $\lambda(xV_k \triangle V_k)$ for $k \geq m$? In this case,

$$\lambda(xV_k \triangle V_k) \leq 2\lambda(V_k) < 2 \cdot \frac{r}{4} < r.$$

It follows that $\rho(x, e) = \sup_n \lambda(xV_m \triangle V_n) < r$.

Our two claims taken in tandem show that ρ generates G's topology about e. Since ρ is left invariant and since G's topology is too, this is enough to say that ρ generates G's topology everywhere. □

Theorem 8.2 (C. Bandt [**8**]). *If ρ is a left invariant metric on the locally compact metrizable group G defining the topology of G, then any two subsets of G that are ρ-isometric have the same left Haar measure.*

To prove Theorem 8.2, we need to develop a fractional Hausdorff measure. Let A be a fixed compact set with nonempty interior.

We'd like to construct a Hausdorff gauge function h so that the associated Hausdorff measure μ^h on G satisfies

$$0 < \mu^h(A) < \infty.$$

Sadly, finding such a gauge function is elusive. Happily Bandt found a way around this—the fractional Hausdorff measure. Given a Hausdorff gauge function h, we define the *fractional Hausdorff measure* ν^h by

$$\nu^h(E) = \liminf_{t \searrow 0} \Bigg\{ \sum_j c_j h(\operatorname{diam}(B_j)) : c_j > 0, \\ \chi_E \leq \sum_j c_j \chi_{B_j}, \operatorname{diam}(B_j) \leq t \Bigg\}.$$

Mimicking the proofs encountered in Hausdorff measures, we see that ν^h is a metric outer measure (ensuring us that Borel sets are ν^h-measurable) and

ν^h is left invariant reflecting ρ's left invariance. The issue is to judiciously choose h so that

$$0 < \nu^h(A) < \infty.$$

Once this has been achieved, we see that for any compact set

$$K \subseteq G, \nu^h(K) < \infty$$

(K can be covered by finitely many of A's left translates, each of which has the same ν^h-measure), and for any nonempty open subset U of G, $\nu^h(U) > 0$ (we can cover A by finitely many left translates of U, each having the same ν^h-measure).

Before proceeding, we take note that if $0 < s < t$, then for any $E \subseteq G$

$$\text{(51)} \qquad \inf\left\{\sum_j c_j h(\operatorname{diam}(B_j)) : c_j > 0, \chi_E \leq \sum_j c_j \chi_{B_j}, \operatorname{diam}(B_j) \leq t\right\}$$

is at least as large as

$$\inf\left\{\sum_j c_j h(\operatorname{diam}(B_j)) : c_j > 0, \chi_E \leq \sum_j c_j \chi_{B_j}, \operatorname{diam}(B_j) \leq s\right\};$$

after all, there are at least as many fractional coverings of E when the sets B_j have diameter less than or equal to s as there are with sets B_j having diameter less than or equal to t. So, as a function of t, (51) ascends as t descends to zero. It follows that $\nu^h(E)$ *is determined by what happens when the fractional coverings involve sets of small diameter.*

The gauge function that works is

$$h(t) = \sup\{\lambda(B) : \operatorname{diam}(B) \leq t\},$$

a function that takes finite values on some open interval $(0, T_0)$. Indeed, G is locally compact, and so G has a basis for its open sets consisting of sets with compact closure. It follows that at any point of G, we have a basis of open balls with compact closure all with diameter less than T_0, for some $T_0 > 0$. The left invariance of ρ ensures that the same T_0 works throughout G.

Note 8.3. ν^h is left invariant.

As a matter of fact, the metric ρ of Theorem 8.1 that generates G's topology is left invariant, and so for any subset B of G, the diameter of B and gB are the same, since they're ρ-isometric.

By the same token, sets in G that are isometric with respect to ρ are assigned the same ν^h-values.

The real issue with ν^h is to show that it's nontrivial, that is, $0 < \nu^h(A) < \infty$. Once we know this to be so, then ν^h is a left Haar measure on G and so is but a multiple of λ. Hence sets in G that are isometric (with respect to ρ) have the same λ-measure as well, and we will have proved Theorem 8.2. We proceed with several lemmas to get that ν^h is nontrivial.

Lemma 8.4. *For any Borel set $B \subseteq G$,*

$$\lambda(B) \leq \nu^h(B).$$

Proof. Suppose $\chi_B \leq \sum_j c_j \chi_{B_j}$. Then $\chi_B \leq \sum_j c_j \chi_{\overline{B}_j}$ and so

$$\begin{aligned}
\lambda(B) &= \int_B d\lambda \leq \int \sum c_j \chi_{\overline{B}_j}\, d\lambda \\
&= \sum c_j \lambda(\overline{B}_j) \leq \sum c_j h(\operatorname{diam}(\overline{B}_j)) \\
&= \sum c_j h(\operatorname{diam}(B_j)).
\end{aligned}$$

It follows that

$$\lambda(B) \leq \nu^h(B).$$

□

Of course a particular consequence of this lemma is

Corollary 8.5. $0 < \lambda(A) < \nu^h(A)$.

Lemma 8.6. *Let $E(t)$ be defined by*

$$E(t) = \inf\left\{ \sum_{j=1}^{n} c_j : n \in \mathbb{N}, \chi_A \leq \sum_{j=1}^{n} c_j \chi_{B_j}, c_j \geq 0, \mathit{diam}(B_j) \leq t \right\}.$$

If

$$\liminf_t h(t)E(t) < \infty,$$

then $\nu^h(A) < \infty$.

Proof. There is a $c > 0$ and a sequence (t_k), $t_k > 0$ with $t_k \searrow 0$ so that

$$h(t_k)E(t_k) < c$$

for all k. In other words,

$$E(t_k) < \frac{c}{h(t_k)}$$

for all k. For each k, choose a fractional covering of A,

$$\chi_A \leq \sum_{j=1}^{n(k)} c_j^{(k)} \chi_{B_j^{(k)}}, \quad c_j^{(k)} \geq 0$$

with

$$\sum_{j=1}^{n(k)} c_j^{(k)} \leq \frac{c}{h(t_k)}.$$

Of course, the definition of the fractional Hausdorff measure $\nu^h(A)$ ensures that

$$\nu^h(A) \leq \inf\left\{\sum_{j=1}^{n} c_j h(\mathrm{diam}(B_j)) : c_j \geq 0, \chi_A \leq \sum c_j \chi_{B_j}, \mathrm{diam}(B_j) \leq t_k\right\},$$

so

$$\nu^h(A) \leq \sum_{j=1}^{n(k)} c_j^{(k)} h(t_j) \leq c < \infty. \qquad \square$$

Lemmas 8.4 and 8.6 show us the way to the end, namely the proof Bandt's theorem, which will follow from the following.

Lemma 8.7 (Principal Lemma). *For each $\epsilon > 0$ there is a $t_0 > 0$ so that if U is an open subset of G with diam$(U) \leq t_0$, then for some $s_1, \ldots, s_n \in G$ and $\alpha_1, \ldots, \alpha_n > 0$ we have*

$$\chi_A \leq \sum_{i=1}^{n} \alpha_i \chi_{s_i \cdot U}$$

and

$$\lambda(A) \leq \sum_{i=1}^{n} \alpha_i \lambda(s_i \cdot U) \leq (1+\epsilon)\lambda(A).$$

This proof depends on Cartan's approximation theorem (Theorem 7.27) and is rather delicate. We postpone the proof until after seeing what it buys us—the completion of the proof of Bandt's theorem. So with the principal lemma in hand, let's show how to put Lemmas 8.4 and 8.6 into play.

Corollary 8.8. *The fractional Hausdorff measure is nontrivial; in fact,*

$$0 < \nu^h(A) < \infty.$$

Proof. If $\epsilon > 0$, then we can choose $t > 0$ so that $t < \min\{t_0, \epsilon\}$ and h is continuous at t. We can do this since h is monotone and so is continuous at all but countably many points of $(0, \min\{t_0, \epsilon\})$.

We can find an open set B in G with diam$(B) \leq t$ so that

$$h(t) \leq (1+\epsilon)^2 \lambda(B).$$

How can we do this? Well if we pick $t' < t$ so that $h(t') \leq h(t)$ and since h is continuous at t,

$$h(t) < (1+\epsilon)h(t'),$$

then we choose C so that $\operatorname{diam}(C) \leq t'$ and $h(t') \leq (1+\epsilon)\lambda(C)$. Then

$$B = \left\{x \in G : \rho(x, C) < \frac{t - t'}{2}\right\}$$

will do. By our Principal Lemma we have a functional covering of A: there is $g_1, \ldots, g_n \in G$ and $\alpha_1, \ldots, \alpha_n > 0$ so that

$$\chi_A \leq \sum_{i=1}^{n} \alpha_i \chi_{g_i \cdot B}$$

with

$$\lambda(B)\left(\sum_{i=1}^{n} \alpha_i\right) \leq \sum_{i=1}^{n} \alpha_i \lambda(\alpha_i \cdot B) \leq (1+\epsilon)\lambda(A).$$

It follows that

$$\begin{aligned} h(t)E(t) &= h(t)E(t) \leq h(t)\sum_{i=1}^{n} \alpha_i \\ &\leq (1+\epsilon)^2 \lambda(U) \sum_{i=1}^{n} \alpha_i \\ &\leq (1+\epsilon)^3 \lambda(A). \end{aligned}$$

Since $\epsilon > 0$ was arbitrary,

$$\liminf_{t \searrow 0} h(t)E(t) \leq \lambda(A) < \infty,$$

and so by Lemma 8.6, $\nu^h(A) < \infty$. We already know that $0 < \nu^h(A)$ and so by Note 8.3, ν^h is ρ-invariant with $0 < \nu^h(A) < \infty$. □

Proof of the Principal Lemma—Lemma 8.7. Since λ is regular and A is compact, we can find an open set V such that

$$V \subseteq \{x \in G : \rho(x, A) < b\}$$

that contains A, has $\overline{V}$ compact, and satisfies

$$\lambda(V) \leq (1+\epsilon)^{1/3} \lambda(A).$$

To see this, for each $a \in A$, let U_a be an open set containing a such that

$$U_a \subseteq \overline{U}_a \subseteq \{x : \rho(x, A) < b\}.$$

Let $a_1, \ldots, a_n \in A$ so that

$$A \subseteq U_{a_1} \cup \cdots \cup U_{a_n}.$$

Then

$$V = U_{a_1} \cap \cdots \cap U_{a_n}$$

is open, contains A, $\overline{V}$ is compact, and $\overline{V} \subseteq \{x : \rho(x, A) < b\}$.

Let

$$W = \{x \in U : \rho(x, A) < b/2\}.$$

Let $f : G \to [0, 1]$ be a continuous function that is one on A and vanishes outside of W. Choose $\alpha > 0$ so that $\alpha[1 + (1+\epsilon)^{1/3}] < (1+\epsilon)^{1/3} - 1$, that is, $\alpha < 1 - (1+\epsilon)^{-1/3}$. Then

$$\frac{1+\alpha}{1-\alpha} < (1+\epsilon)^{1/3}.$$

We appeal to Cartan's approximation scheme to get an open set U_0 that contains G's identity e for which if $\phi \in \mathcal{K}^+(G)$ and U_0 contains the support of ϕ, then for some $g_1, \ldots, g_n \in \operatorname{supp}(f)$ and some $c_1, \ldots, c_n \geq \alpha$ we have

$$\left| f(g) - \sum_{i=1}^{n} c_i \, {}_{g_i}\phi(g) \right| \leq \alpha$$

for all $g \in G$. Let $t_0 < \min\{b/2, d(e, U_0^c)\}$ be a positive number. We'll show that this is the t_0 claimed in the Principal Lemma.

Let B be an open set with $\operatorname{diam}(B) \leq t_0$. Notice that if $g \in B$, then $g^{-1}B$ is an open set that contains e and $\operatorname{diam}(g^{-1}B) = \operatorname{diam}(B) \leq t_0$, since d is left invariant. Each point of B is within t_0 of g and so each point of $g^{-1}B$ is within t_0 of e. Can any $x \in g^{-1}B$ also be in U_0^c? If we try to imagine such an x, then

$$d(e, x) \leq t_0 < d(e, U_0^c),$$

an impossibility. So (replacing B with $g^{-1}B$ if necessary) we can assume our B in the opening line of this paragraph contains e and is open with $\operatorname{diam}(B) \leq t_0$ and $B \subseteq U_0$.

Now λ is inner regular so we can choose a compact $C \subseteq B$ so $\lambda(C)$ is almost $\lambda(B)$, that is,

$$\lambda(B) \leq (1+\epsilon)^{1/3}\lambda(C).$$

Suppose $\phi : G \to [0, 1]$ is a continuous function for which

$$\chi_C \leq \phi \leq \chi_B \leq \chi_{U_0}.$$

We know that ϕ is a member of $\mathcal{K}^+(G)$ with $\operatorname{supp}(\phi) \subseteq U_0$ and so Cartan's scheme tells us we can find $g_1, \ldots, g_n \in \operatorname{supp}(f)$ and $c_1, \ldots, c_n \geq 0$ so that for any $g \in G$

$$\left| f(g) - \sum_{i=1}^{n} c_i \, {}_{g_i}\phi(g) \right| \leq \alpha.$$

Alternatively,

$$f(g) - \alpha \le \sum_{i=1}^{n} c_i \, {}_{g_i}\phi(g) \le f(g) + \alpha$$

for any $g \in G$. Since $\chi_A \le f$,

$$\chi_A(g) - \alpha \le \sum_{i=1}^{n} c_i \, {}_{g_i}\phi(g)$$

for any $g \in G$. Let $\phi_i = {}_{g_i}\phi$ and $d_i = \frac{c_i}{1-\alpha}$. Plainly $\phi_i \le \chi_{g_i B}$ (since $\phi \le \chi_B$) and

$$\chi_A \le \sum_{i=1}^{n} d_i \phi_i \le \sum_{i=1}^{n} d_i \chi_{g_i B}.$$

Now $\sum_{i=1}^{n} d_i \chi_{g_i B}$ is zero in V^c; after all, $g_1, \ldots, g_n$ are in the support of f and

$$d(W, V^c) > b/2 \ge \operatorname{diam}(B),$$

so

$$\sum_{i=1}^{n} d_i \phi_i \le (1+\epsilon)^{1/3} \chi_V.$$

It follows that

$$\begin{aligned}
\sum_{i=1}^{n} d_i \int \phi \, d\lambda &= \sum_{i=1}^{n} d_i \int \phi_i \, d\lambda = \int \sum_{i=1}^{n} d_i \phi_i \, d\lambda \\
&\le (1+\epsilon)^{1/3} \int \chi_V \, d\lambda = (1+\epsilon)^{1/3} \lambda(V).
\end{aligned}$$

Consequently,

$$\begin{aligned}
\sum_{i=1}^{n} d_i \lambda(g_i B) &= \lambda(B) \sum_{i=1}^{n} d_i \le (1+\epsilon)^{1/3} \lambda(C) \sum_{i=1}^{n} d_i \\
&\le (1+\epsilon)^{1/3} \int \phi \, d\lambda \sum_{i=1}^{n} d_i \le (1+\epsilon)^{2/3} \lambda(V) \le (1+\epsilon) \lambda(A),
\end{aligned}$$

and that's that. □

Proof of Bandt's Theorem—Theorem 8.2. By Corollary 8.8, ν^h is a nontrivial left invariant Haar measure on G, so it is a multiple of λ. Since sets in G which are ρ-isometric have the same ν^h values, it follows that these sets have the same left Haar measure. □

1. Notes and Remarks

Whenever G is a locally compact metrizable topological group, G has a base for its topology consisting of open sets with compact closure; the collection of open balls with respect to the metric generates G's topology also forms a base for its topology. When can one find a left-invariant metric generating G's topology all of whose balls have compact closure? Of course, for such a thing to be so, G must be separable. After all, G is the union of the n-balls centered at the identity; if each of these have compact closure then it's easy to see that G has a countable dense subset—the union of the countable dense subsets of each n-ball will do.

Our next result, also due to R. A. Struble, tells us that this is the whole story. Though the topic of this result of Struble is not germane to the study of Haar measure, it is simply too satisfying a result not to be included.

Here's the theorem of R. A. Struble that led him to consider Theorem 8.1.

Theorem 8.9. *A locally compact group metrizable topological group has a left invariant metric that generates its topology in which all its open balls have compact closure if and only if G satisfies the second axiom of countability.*

Lemma 8.10. *Let G be a locally compact, second countable (hence metrizable, separable) group. Then there exists a family $\{U_r : r > 0\}$ such that*

(i) for each r, each U_r is open and $\overline{U}_r$ is compact;

(ii) $U_r = U_r^{-1}$;

(iii) $U_rU_s \subseteq U_{r+s}$ (so if $r < s$, then $U_r \subseteq U_rU_{r-s} \subseteq U_s$),;

(iv) $\{U_r : r > 0\}$ is a base for the open sets about e;

(v) $\bigcup_{r>0} U_r = G$.

Once Lemma 8.10 is established, we're ready for business. Indeed, let $\{U_r : r > 0\}$ be the family of open sets about e generated from Lemma 8.10. For $x, y \in G$, set

$$d(x,y) = \inf\{r : y^{-1}x \in U_r\}.$$

- Since $G = \bigcup_{r>0} U_r$, for a pair $x, y \in G$, we have $y^{-1}x \in U_r$ for some $r > 0$. It follows that $d(x,y) \geq 0$.
- $e \in U_r$ for each $r > 0$ so $d(x,x) = 0$.

- If $y^{-1}x \neq e$, then there is an $r_0 > 0$ so that $y^{-1}x \notin U_{r_0}$ (Lemma 8.10(iv) tells us this). But whenever $0 < r_0 < r$, we have (by Lemma 8.10(iii))
$$U_{r_0} \subseteq U_{r_0}U_{r-r_0} \subseteq U_r,$$
so $d(x,y) \geq r_0 > 0$.
- $U_r = U_r^{-1}$ so $y^{-1}x \in U_r$ precisely when $x^{-1}y \in U_r$; consequently, $d(x,y) = d(y,x)$.
- Suppose $x, y, z \in G$ with $y^{-1}x \in U_r, z^{-1}y \in U_s$. Then
$$z^{-1}x = z^{-1}yy^{-1}x \in U_sU_r \subseteq U_{r+s},$$
so $d(x,z) \leq r+s$. This is so whenever $y^{-1}x \in U_r$ so $d(x,z) \leq d(x,y)+s$; again this is so whenever $z^{-1}y \in U_s$ so $d(x,z) \leq d(x,y)+d(y,z)$.
- Finally, if $x, y, z \in G$, then
$$d(zx,zy) = \inf\{r : (zy)^{-1}zx \in U_r\} = \inf\{r : y^{-1}x \in U_r\} = d(x,y).$$

To summarize: d is a left invariant metric on G.

Since $d(x,e) < r$ means $x = e^{-1}x \in U_r$, the open d-ball of radius r centered at e is contained in U_r. Also this same d-ball contains $U_{r'}$ for any $0 < r' < r$ since if $x \in U_{r'}$, then
$$e^{-1}x = x \in U_{r'} \subseteq U_{r'}U_{\frac{r-r'}{2}} \subseteq U_{\frac{r+r'}{2}},$$
and so $d(x,e) \leq \frac{r+r'}{2} < r$. Therefore if $0 < r' < r$, then
$$U_{r'} \subseteq \{x : d(x,e) < r\} \subseteq U_r.$$
It follows that the open d-balls of radius r about e are cofinal with the collection $\{U_r : r > 0\}$ so the closure of each open d-ball is compact and d generates G's topology.

Proof of Lemma 8.10. Let ρ be the left invariant metric resulting from Theorem 8.1. We can assume that each of the open balls
$$B_r = \{x \in G : \rho(x,e), r\}$$
has compact closure for $0 < r \leq 2$; after all, there is an r_0 so that for $r < r_0$, $\overline{B_{r_0}}$ is compact by G's locally compact nature, so recalibrate ρ to make $r_0 = 2$ if necessary.

For $0 < r < 2$ we let $U_r = B_r$. This assures us clearly of (iv) and since we'll keep these U_r's, (iv) is assumed henceforth. Also (i), (ii), and (iii) hold when $r + s < 2$ by ρ's left invariant metric nature.

G is locally compact and satisfies the second countability axiom so G admits a countable open base

$$\{W_{2^n} : n \in \mathbb{N}\}$$

for its topology, where we can (and do) assume that $\overline{W}_{2^n}$ is compact for each n. We define

$$U_2 = B_2 \cap W_2.$$

It's easy to verify that (i) and (ii) hold for $0 < r < 2$ and if $r + s < 2$, then (iii) holds as well.

We'll now inch our way from from (i), (ii), and (iii), $(r + s \leq 2)$ holding for $0 < r \leq 2$ to $0 \leq r \leq 4$.

First we have to define U_r for $2 < r < 2^2$. Let $0 < r < 2^2$. Set

$$U_r = \bigcup U_{t_1} \cdots U_{t_m},$$

where the union extends over all $t_1, \ldots, t_m$ so that each t_i satisfies $0 < t_i \leq 2$ and $t_1 + \cdots + t_m = r$.

If $2 < r < 2^2$ and $t_1 + \cdots + t_m = r$ where each $t_i > 0$, then there must be $k, l \in \mathbb{N}$ so that $1 \leq k < l < m$ and $t_1 + \cdots + t_k \leq 2$, $t_{k+1} + \cdots + t_l \leq 2$, and $t_{l+1} + \cdots + t_m \leq 2$. Why is this so? Well let k be the least j_1 so that $t_1 + \cdots + t_{j_1} \leq 2$, and let l be the least j_2 so that $t_{j_1+1} + \cdots + t_{j_2} \leq 2$. Then $\sum_{j_2+1}^{m} t_j \leq 2$ because otherwise, $t_{j_k+1} + \cdots + t_m > 2$ and $t_1 + \cdots + t_{j_1+1} \geq 2$ too, where $j_1 + 1 < j_2 + 1$.

It follows that

$$\begin{aligned} U_{t_1} \cdot U_{t_2} \cdots U_{t_m} &\subseteq (U_{t_1} \cdots U_{t_k})(U_{t_{k+1}} \cdots U_{t_l})(U_{t_{l+1}} \cdots U_{t_m}) \\ &\subseteq U_{t_1+\cdots+t_k} U_{t_{k+1}+\cdots+t_l} U_{t_{l+1}+\cdots+t_m} \quad \text{(by (iv))} \\ &\subseteq U_2 \cdot U_2 \cdot U_2, \end{aligned}$$

so $U_r \subseteq U_2 \cdot U_2 \cdot U_2$ whenever $0 < r < 2^2$.

Now $\overline{U}_2$ is compact so $\overline{U_2 \cdot U_2 \cdot U_2} \subseteq \overline{U}_2 \cdot \overline{U}_2 \cdot \overline{U}_2$ is also, and $\overline{U}_r$ is compact for $0 < r < 2^2$.

Since

$$(U_{t_1} \cdots U_{t_m})^{-1} = U_{t_m}^{-1} \cdots U_{t_1}^{-1} = U_{t_m} \cdots U_{t_1},$$

we see that

$$\begin{aligned} U_r^{-1} &= \left(\bigcup U_1 \cdots U_{t_m}\right)^{-1} \\ &= \bigcup (U_1 \cdots U_{t_m})^{-1} \\ &= \bigcup U_{t_m}^{-1} \cdots U_1^{-1} \\ &= \bigcup U_{t_m} \cdots U_1 = U_r \end{aligned}$$

for $2 < r < 2^2$.

Finally if $r > 0, s > 0$ and $r + s < 2^2$, then upon supposing $t_1, \ldots, t_m$, $\tau_1, \ldots, \tau_j$ are positive and satisfying $t_1 + \cdots + t_m = r, \tau_1 + \cdots + \tau_j = s$, then

$$(U_{t_1} \cdots U_{t_m})(U_{\tau_1} \cdots U_{\tau_j}) = U_{t_1} \cdots U_{t_m} U_{\tau_1} \cdots U_{\tau_j}$$

so

$$U_r \cdot U_s \subseteq U_{r+s}.$$

What if $r, s > 0$ and $r + s = 2^2$? Suppose $r = s = 2$. Then $U_r U_s = U_2 U_2$. If $2 < r$, then $s < 2$. But $r < 2^2$ so

$$U_r \subseteq U_2 U_2 U_2$$

and so

$$U_r U_s \subseteq (U_2 U_2 U_2) U_2.$$

Either way define

$$U_{2^2} = (U_2 U_2 \cdot U_2 U_2) \cup W_{2^2}.$$

Here's what is so for $0 < r < 2^2$:

- each U_r is open and $\overline{U}_2$ is compact;
- $U_r^{-1} = U_r$;
- if $r, s > 0$ and $r + s \leq 2^2$, then $U_r U_s \subseteq U_{r+s}$.

We still have $\{U_r : 0 \leq r \leq 2^2\}$ as a basis for the topology of G about e and of course, $W_{2^2} \subseteq U_{2^2}$.

We continue from U_{2^2} to U_{2^3} in a similar fashion and inch our way forward in a straightforward modification of the above procedure.

The fact that at each stage we ensure $W_{2^n} \subseteq U_{2^n}$ allows us to conclude that

$$G = \bigcup_{n \in \mathbb{N}} W_{2^n} \subseteq \bigcup_{n \in \mathbb{N}} U_{2^n} \subseteq G,$$

which is Lemma 8.10(v). □

Theorem 8.11 (Braconnier [**12**]). *If G is a locally compact topological group and G admits a bi-invariant metric d that determines its topology, then G is unimodular.*

Proof. Let λ be left Haar measure on G and assume that G is not unimodular. Let U be an open set containing e such that $\lambda(U) < \infty$. Let B be an open ball centered at e (of radius ρ) so that $\overline{B}$ is compact, and $\overline{B} \subseteq U$. Then for any $x \in G$, the set xBx^{-1} is just B, thanks to d's bi-invariance, so

$$xBx^{-1} \subseteq U.$$

Let $x_0 \in G$ satisfy

$$\Delta(x_0^{-1})\lambda(B) > \lambda(U),$$

where Δ is the modular function of G. Then $x_0 B x_0^{-1} \subseteq U$ and

$$\lambda(x_0 B x_0^{-1}) = \Delta(x_0^{-1})\lambda(x_0 B) = \Delta(x_0^{-1})\lambda(B) > \lambda(U).$$

Oops! □

Example 8.12. The general linear group $\mathcal{GL}(2;\mathbb{R})$ is unimodular (see §6.3). For each m, let

$$X_m = \begin{pmatrix} \frac{1}{m} & \frac{1}{m} \\ 0 & m \end{pmatrix} \text{ and } Y_m = \begin{pmatrix} m & \frac{1}{m} \\ 0 & \frac{1}{m} \end{pmatrix} \in \mathcal{GL}(2;\mathbb{R}).$$

Then

$$X_m Y_m = \begin{pmatrix} \frac{1}{m} & \frac{1}{m} \\ 0 & m \end{pmatrix}\begin{pmatrix} m & \frac{1}{m} \\ 0 & \frac{1}{m} \end{pmatrix} = \begin{pmatrix} 1 & \frac{2}{m^2} \\ 0 & 1 \end{pmatrix} \to \begin{pmatrix} 1 & 0 \\ 0 & 1 \end{pmatrix}.$$

But

$$Y_m X_m = \begin{pmatrix} m & \frac{1}{m} \\ 0 & \frac{1}{m} \end{pmatrix}\begin{pmatrix} \frac{1}{m} & \frac{1}{m} \\ 0 & m \end{pmatrix} = \begin{pmatrix} 1 & 2 \\ 0 & 1 \end{pmatrix}.$$

Therefore no bi-invariant metric can be found so that it generates the topology of $\mathcal{GL}(2;\mathbb{R})$. An easy modification of this example shows the same for $\mathcal{GL}(n;\mathbb{R})$, when $n \geq 2$ as well.

Theorem 8.13. *If G is a locally compact metrizable topological group and ρ is a left invariant metric that generates G's topology, then (G,ρ) is a complete metric.*

Indeed if U is an open set with compact closure and if $e \in U$, then there is an open ball B centered at e with $\overline{B}$ both compact and contained in U. Suppose R is the radius of B, and let (g_n) be a ρ-Cauchy sequence in (G,ρ). Then there is an $N \in \mathbb{N}$ so for $m, n \leq N$,

$$\rho(g_n, g_m) < \frac{R}{3}.$$

It soon follows that for $n \geq N$,

$$\rho(g_n, g_N) < \frac{R}{3}.$$

Therefore for $n \geq N$, g_n lies in the compact, closed ball of radius $\frac{R}{3}$ and so (g_n) must converge.

Observation 8.14. Let U be an open set in the topological group G with $e \in G$, and suppose that K is a compact subset of G. Then there is an open set V in G with $e \in V$ so that

$$xVx^{-1} \subseteq U$$

for every $x \in K$.

Proof. To see this, let $\mathcal{W}$ denote the collection of all open sets W in G such that $W = W^{-1}$. We claim that for any $y \in G$ there is a $V \in \mathcal{W}$ so that if $x \in Vy$, then

$$xVx^{-1} \subseteq U.$$

In fact, we can pick $V_1 \in \mathcal{W}$ so that $V_1 \cdot V_1 \cdot V_1 \subseteq U$, and we can pick $V_2 \in \mathcal{W}$ so that

$$yV_2y^{-1} \subseteq V_1.$$

(This is thanks to the continuity of $x \to ax \to axa^{-1}$ for any $a \in G$.) Let

$$V = V_1 \cap V_2.$$

Then if $x \in Vy$,

$$xy^{-1} \in V \subseteq V_1$$

and

$$yx^{-1} = (x^{-1}y)^{-1} \in V^{-1} = V \subseteq V_1,$$

and so

$$xVx^{-1} \subseteq xV_2x^1 = (xy^{-1})\Big(y(V_2y^{-1})\Big)(yx^{-1}) \subseteq V_1 \cdot V_1 \cdot V_1 \subseteq U.$$

So for each $y \in K$ there is a $V_y \in \mathcal{W}$ so that $x \in V_x y$ implies $xV_yx^{-1} \subseteq U$. But

$$K \subseteq \bigcup_{y \in K} V_y Y,$$

and each $V_y y$ is open; hence we can find $y_1, \ldots, y_n \in K$ so that

$$K \subseteq \left(V_{y_1} y_1\right) \cup \cdots \cup \left(V_{y_n} y_n\right).$$

Let

$$V = V_{y_1} \cap \cdots \cap V_{y_n}.$$

Then if $x \in K$, it must be that $x \in V_{y_k} y_k$ for some $k, 1 \leq k \leq n$, and so

$$xVx^{-1} \subseteq xV_{y_k} y_k \subseteq U.$$ □

Theorem 8.15. *A compact metrizable topological group G admits a bi-invariant metric that generates its topology.*

Proof. Let ρ be a left invariant metric on G that generates G's topology. For $x, y \in G$ define

$$d(x, y) = \sup\{\rho(xz, yz) : z \in G\}.$$

Then d is finite for all $x, y \in G$ and is easily seen to be bi-invariant.

Suppose $\epsilon > 0$. By our observation above, there is a $\delta > 0$ so that

$$z^{-1} \cdot \{x \in G : \rho(x, e) < \delta\} \cdot z \subseteq \{x \in G : \rho(x, e) < \epsilon\},$$

for all z. It follows that $\rho(x, e) < \delta$ ensures that $d(z^{-1}xz, e) = d(xz, z) < \epsilon$ for all z and so $d(x, e) \leq \epsilon$. The open ρ-ball of radius δ centered at e is contained in the closed d-ball of radius ϵ centered at e.

It is plain that the open d-ball of radius ϵ centered at e is contained in the open ρ-ball of radius ϵ centered at e. Therefore ρ and d generate the same topology. □

Moreover, we state the following:

- If G is a topological group of the second category and H is a subgroup of G, then $G\backslash H$ is either empty or of the second category in G.

  Let $y \in G\backslash H$. Then $yH \in G\backslash H$ (distinct cosets are disjoint). Therefore, should $G\backslash H$ be of the first category, then so is yH, and from this we conclude that $G = yH \cup (G\backslash H)$ is of the first category.

- If G is a topological group of the second category and H is a dense $\mathcal{G}_\delta$ subgroup of G, then $H = G$.

  After all, $H = \bigcap_n H_n$ where each H_n is a dense open subset of G and so $G\backslash H_n$ is closed and nowhere dense for each n. It follows that $G\backslash H = \bigcup_n (G\backslash H_n)$ is of the first category, and so by the previous remark, $G\backslash H = \emptyset$.

Theorem 8.16 (V. Klee [**67**]). *Let G be a topological group with a bi-invariant metric ρ which generates G's topology. Suppose (G, ρ) admits a complete metric d that generates G's topology. Then G is actually complete under ρ.*

Proof. Let (G^*, ρ^*) be the completion of (G, ρ). Then (G^*, ρ^*) is a topological group into which (G, ρ) is naturally isomorphically and isometrically embedded as a dense subgroup. But topological complete metric spaces are always $\mathcal{G}_\delta$-sets in any super metric space, thanks to an oldie but goodie of Sierpinski. Hence $G = G^*$ and so G is ρ-complete. □

Chapter 9

Steinlage on Haar Measure

1. Uniform Spaces: The Basics

Definition 9.1. A *uniformity* for a set X is a nonvoid family $\mathcal{U}$ of subsets of $X \times X$ such that the following hold.

(i) If $U \in \mathcal{U}$, then
$$\Delta := \{(x, x) : x \in X\} \subseteq U.$$

(ii) If $U \in \mathcal{U}$, then
$$U^{-1} := \{(y, x) : (x, y) \in U\} \in \mathcal{U}.$$

(iii) If $U \in \mathcal{U}$, then there exists $V \in \mathcal{U}$ so that
$$V \circ V := \{(x, z) : (x, y), (y, z) \in V, \text{ for some } y \in X\} \subseteq U.$$

(iv) If $U, V \in \mathcal{U}$, then
$$U \cap V \in \mathcal{U}.$$

(v) If $U \in \mathcal{U}$ and $U \subseteq V \subseteq X \times X$, then
$$V \in \mathcal{U}.$$

We then say that $(X, \mathcal{U})$ is a *uniform space.* A subfamily $\mathcal{B}$ of a uniformity $\mathcal{U}$ is a *base* for $\mathcal{U}$ if each member of $\mathcal{U}$ contains a member of $\mathcal{B}$. A base $\mathcal{B}$ for $\mathcal{U}$ determines membership in $\mathcal{U}$: $U \in \mathcal{U}$ precisely when there is a $B \in \mathcal{B}$ so that $B \subseteq U$.

Theorem 9.2. *A nonvoid family $\mathcal{B}$ of subsets of $X \times X$ is a base for some uniformity for X if and only if $\mathcal{B}$ satisfies the following:*

(i) *Each member of $\mathcal{B}$ contains Δ.*

(ii) *If $U \in \mathcal{B}$, then U^{-1} contains a member of $\mathcal{B}$.*

(iii) *If $U \in \mathcal{B}$, then $V \circ V \subseteq U$ for some $V \in \mathcal{B}$.*

(iv) *The intersection of two members of $\mathcal{B}$ contains a member of $\mathcal{B}$.*

Example 9.3. Let (X, d) be a metric space, and let $\mathcal{B}$ consist of all subsets U_ϵ of $X \times X$ where $\epsilon > 0$ and

$$U_\epsilon := \{(x, y) \in X \times X : d(x, y) < \epsilon\}.$$

$\mathcal{B}$ is a base for some uniformity for X, namely the *metric uniformity.*

Example 9.4. Let G be a topological group, and let $\mathcal{B}_L$ consist of all sets of the form

$$\{(x, y) \in G \times G : y \in xU\},$$

where U is an arbitrary neighborhood of the identity of G. B_L is called the *left uniformity* of G. We can naturally consider the *right uniformity* and the *two-sided uniformity* of G.

These examples are, in fact, the *raison d'etre* for uniform spaces; indeed, André Weil created theory of uniform spaces with an eye to providing a common generalization of metric spaces and topological groups. In so doing, he provided a viewpoint from which to discuss Cauchy sequences/nets, completeness, total boundedness, and uniform continuity.

Definition 9.5. A family $\mathcal{S}$ is a *subbase* for the uniformity $\mathcal{U}$ when the family $\mathcal{B}$ of all finite intersections of members of $\mathcal{S}$ is a base for $\mathcal{U}$.

The next result provides us with enough firepower to recognize subbases when we see them.

Theorem 9.6. *Suppose $\mathcal{S}$ is a family of subsets of $X \times X$ that satisfies the following conditions:*

(i) *each member of $\mathcal{S}$ contains Δ;*

(ii) *if $U \in \mathcal{S}$, then U^{-1} contains a member of $\mathcal{S}$;*

(iii) *if $U \in \mathcal{S}$, then there is a $V \in \mathcal{S}$ so that $V \circ V \subseteq U$.*

Then $\mathcal{S}$ is a subbase for some uniformity for $\mathcal{S}$.

Corollary 9.7. *The union of any nonvoid collection of uniformities for X is a subbase for a uniformity for X.*

Suppose $(X,\mathcal{U})$ is a uniform space. The *uniform topology* is the collection of all $T \subseteq X$ such that for each $x \in T$ there is a $U \in \mathcal{U}$ so that

$$U[x] = \{y \in X : (x,y) \in U\} \subseteq T.$$

The uniform topology *is* a topology.

Theorem 9.8. *Let $A \subseteq X$. Then the interior of A relative to the uniform topology is the set*

$$\{x \in X : U[x] \subseteq A \text{ for some } U \in \mathcal{U}\}.$$

Proof. If O is an open set that is contained in A, it's because for each $x \in O$ there is a $U \in \mathcal{U}$ so that $U[x] \subseteq O \subseteq A$. It follows that the set

$$B = \{x : U[x] \subseteq A \text{ for some } U \in \mathcal{U}\}$$

contains each and every open set O for which $O \subseteq A$.

To see that the set B is open, note that if $x \in B$, then by definition there is a $U \in \mathcal{U}$ so that $U[x] \subseteq A$. With U we an find a $V \in \mathcal{U}$ so that $V \circ V \subseteq U$. Notice that if $y \in V[x]$, then $(x,y) \in V$, and it follows that

$$\begin{aligned} V[y] &= \{z : (y,z) \in V\} \\ &\subseteq \{z : (x,y),(y,z) \in V, \text{ for some } y \in X\} \\ &= (V \circ V)[x] \subseteq U[x] \subseteq A. \end{aligned}$$

Therefore if $y \in V[x]$, then $V[y] \subseteq A$, and so $V[x] \subseteq B$. In other words, if $x \in B$, then there is a $V \in \mathcal{U}$ so that $V[x] \subseteq B$. Thus B is open in the uniform topology.

Since B is open and contains every open set that is contained in A, B is A's interior. □

So given a uniform space $(X,\mathcal{U})$, $U[x]$ is a neighborhood of x for each $U \in \mathcal{U}$, and the family $\{U[x] : U \in \mathcal{U}\}$ is a base for the neighborhood system of x. It's an easy consequence of this and the definition of a base for $\mathcal{U}$ to see that if $\mathcal{B}$ is a base for $\mathcal{U}$, then for any $x \in X$,

$$\{U[x] : U \in \mathcal{B}\}$$

is a basis for the neighborhood system of x in the uniform topology.

Theorem 9.9. *If $U \in \mathcal{U}$, then the interior of U (in the product topology) is also in $\mathcal{U}$. Consequently, the family of all open symmetric members of $\mathcal{U}$ is a base for $\mathcal{U}$.*

We will rely on the following simple features of symmetric sets.

Lemma 9.10. *If $V \subseteq X \times X$ is symmetric (i.e., $V = V^{-1}$), then for any $U \subseteq X \times X$,*

$$V \circ U \circ V = \bigcup_{(x,y)\in U} V[x] \times V[y].$$

Proof.

$$\begin{aligned}
V \circ U \circ V &= \{(u,v) : (u,x) \in V, (x,y) \in U, (y,v) \in V, \text{for some } x, y \in X\} \\
&= \bigcup_{(x,y)\in U} \{(u,v) : (u,x), (y,v) \in V\} \\
&= \bigcup_{(x,y)\in U} \{(u,v) : (x,u), (y,v) \in V\} \ \ \text{(by } V\text{'s symmetry)} \\
&= \bigcup_{(x,y)\in U} V[x] \times V[y],
\end{aligned}$$

and we have proved the lemma. □

Now to prove the theorem, note that the interior of a set $M \subseteq X \times X$ is the set

$$\Big\{(x,y) : U[x] \times V[y] \subseteq M \ \text{ for some } \ U, V \in \mathcal{U}\Big\}.$$

But once $U, V \in \mathcal{U}$, we have $W = U \cap V \in \mathcal{U}$ also, so M's interior in $X \times X$ is

$$\{(x,y) : W[x] \times W[y] \subseteq M \ \text{ for some } \ W \in \mathcal{U}\}.$$

But $W \in \mathcal{U}$ so there is a $\widetilde{W} \in \mathcal{U}$ so

$$\widetilde{W} \circ \widetilde{W} \circ \widetilde{W} \subseteq W,$$

where $\widetilde{W}$ is symmetric (this takes a moment or two of reflection on the consequences of the axioms of uniformities). Now our lemma steps in to tell us that

$$\widetilde{W} \circ \widetilde{W} \circ \widetilde{W} = \bigcup_{(x,y)\in \widetilde{W}} \widetilde{W}[x] \times \widetilde{W}[y].$$

Each point of $\widetilde{W}$ is an interior point of W, so W's interior contains $\widetilde{W}$, which belongs to $\mathcal{U}$. It follows that the interior of W is in $\mathcal{U}$ as well.

To see that symmetric members of $\mathcal{U}$ form a base for $\mathcal{U}$ is an easy consequence of the above. After all, we showed that if $U \in \mathcal{U}$, then U's interior is the set

$$\{(x,y) : W[x] \times W[y] \subseteq U \ \text{ for some symmetric } \ W \in \mathcal{U}\}.$$

But if U is itself symmetric then this set is itself symmetric and open *and* in $\mathcal{U}$. Of course being U's interior put this set squarely inside of U, too. $\square$

Theorem 9.11. *Let $A \subseteq X$. Then the uniform closure of A is precisely $\bigcap_{U \in \mathcal{U}} U[A]$, where*

$$U[A] = \{y \in X : (x, y) \in U \text{ for some } x \in A\}.$$

Proof. Note that $x \in \overline{A}$ if and only if $U[x] \cap A \neq \emptyset$ for each $U \in U$. But $U[x] \cap A \neq \emptyset$ if and only if $x \in U^{-1}[A]$. Since each member of $\mathcal{U}$ contains a symmetric member of $\mathcal{U}$, $x \in \overline{A}$ if and only if $x \in U[A]$ for each member $U \in \mathcal{U}$. $\square$

Definition 9.12. A function $f : (X,\mathcal{U}) \to (Y, \mathcal{V})$ is *uniformly continuous* if for each $V \in \mathcal{V}$,

$$\{(x, y) : (f(x), f(y)) \in V\} \in \mathcal{U}.$$

Note that in testing for uniform continuity, one need only test for V's in a base for $\mathcal{V}$. So $f : (X,\mathcal{U}) \to \mathbb{R}$ is uniformly continuous if and only if for each $\epsilon > 0$ there is a $U \in \mathcal{U}$ so $|f(x) - f(y)| < \epsilon$ if $(x, y) \in U$.

The composition of two uniformly continuous functions is uniformly continuous as usual.

Theorem 9.13. *Each uniformly continuous function is continuous with respect to the uniform topologies.*

Proof. Suppose $f : (X,\mathcal{U}) \to (Y, \mathcal{V})$ is uniformly continuous, and let O be a neighborhood of $f(x)$. Then there is a $V \in \mathcal{V}$ so that $V[f(x)] \subseteq O$ and

$$f^{\leftarrow}(V[f(x)]) = \{y : f(y) \in V[f(x)]\} = \{y : (f(x), f(y)) \in V\}.$$

If we consider the induced map

$$F : X \times X \to Y \times Y, \quad F(x_1, x_2) = (f(x_1), f(x_2)),$$

then it's plain that f's uniform continuity says that for each $V \in \mathcal{V}$ there is a $U \in \mathcal{U}$ so $F(U) \subseteq V$. But

$$\{y : (f(x), f(y)) \in V\}$$

is naught else but $F^{\leftarrow}(V)[x]$, and this is a neighborhood of x. So $f^{\leftarrow}(O)$ is a neighborhood of x for any neighborhood O of $f(x)$ and f is continuous. $\square$

If $Y \subseteq X$ and $(X,\mathcal{U})$ is a uniform space, then the inclusion of Y into X induces a smallest uniformity making this map uniformly continuous; this is the *relative uniformity*. It consists of subsets of $Y \times Y$ of the form

$$(Y \times Y) \cap U,$$

where $U \in \mathcal{U}$. Naturally the relative uniform topology on Y is the uniform relative topology as you would want it to be.

If $\{(X_\alpha, \mathcal{U}_\alpha) : \alpha \in A\}$ is a family of uniform spaces, then we introduce a uniformity on ΠX_α by looking at the smallest uniformity such that $p_{\alpha_0} : \Pi X_\alpha \to X_{\alpha_0}$ is uniformly continuous for each $\alpha_0 \in A$; it has as a subbase the sets of the form $p_\alpha^{\leftarrow}(U_\alpha)$ where $U_\alpha \in \mathcal{U}_\alpha$. The resulting uniformity if the *product uniformity*. It's plain and easy-to-see that the uniform product topology is the product uniform topology—again, things behave. More is so, and we record this thusly:

Theorem 9.14. *The topology of the product uniformity is the product topology. A function f on a uniform space (into a product of uniform spaces) is uniformly continuous if and only if the composition of f with each coordinate projection is uniformly continuous.*

Theorem 9.15. *Let $(X, \mathcal{U})$ be a uniform space, and let d be a pseudo-metric for X. Then d is uniformly continuous on $X \times X$ relative to the product uniformity if and only if the set*

$$\{(x, y) : d(x, y) < r\}$$

is a member of $\mathcal{U}$ for each $r > 0$.

Proof. We need to show that d's uniform continuity is equivalent to

$$V_{d,r} = \{(x, y) : d(x, y) < r\} \in \mathcal{U}$$

for each $r > 0$.

Preface. Let $U \in \mathcal{U}$. Then the sets

$$\{(x, y), (u, v) : (x, u) \in U\}, \{(x, y), (u, v) : (y, v) \in U\}$$

are preimages under the coordinate projections, and so each belongs to the product uniformity. In fact, the family of sets of the form

$$\{((x, y), (u, v)) : (x, u) \in U, (y, v) \in U\}_{U \in \mathcal{U}}$$

is a base for the product uniformity.

Since d is uniformly continuous, for each $r > 0$ there is a $U \in \mathcal{U}$ so that if $(x, u), (y, v) \in U$, then $|d(x, y) - d(u, v)| < r$. But $(u, v) = (y, y)$ is always in U so this ensures us that if $(x, y) \in U$, then $d(x, y) = |d(x, y) - d(y, y)| < r$, and $U \subseteq V_{d,r}$, and so $V_{d,r} \in \mathcal{U}$.

Conversely, notice that if $(x, u), (y, v) \in V_{d,r}$, then

$$|d(x, y) - d(u, v)| < 2r.$$

To see this, note that

$$d(x,y) \le d(x,u) + d(u,v) + d(v,y)$$

and

$$d(u,v) \le d(u,x) + d(x,y) + d(y,v),$$

which after an inspection tells us

$$d(x,y) - d(u,v) \le d(x,u) + d(v,y) \text{ and } d(u,v) - d(x,y) \le d(u,x) + d(y,v),$$

which thanks to the symmetry of d says

$$|d(x,y) - d(u,v)| \le d(x,u) + d(v,y) < r + r = 2r.$$

So if $V_{d,r} \in \mathcal{U}$ for each $r > 0$, then d is uniformly continuous. □

Lemma 9.16 (Metrization Lemma). *Let (U_n) be a sequence of subsets of $X \times X$ such that $U_0 = X \times X$, each U_n contains the diagonal Δ, and $U_{n+1} \circ U_{n+1} \circ U_{n+1} \subseteq U_n$ for each n. Then there is a nonnegative real-valued function d on $X \times X$ such that*

(i) *$d(x,z) \le d(x,y) + d(y,z)$ for all x, y, z, and*

(ii) *$U_n \subseteq \{(x,y) : d(x,y) < 2^{-n}\} \subseteq U_{n-1}$ for all $n \ge 1$.*

If each U_n is symmetric, then d can be chosen to be a pseudo-metric satisfying (ii).

Proof (Outline). Define $f : X \times X \to \mathbb{R}$ by

$$f(x,y) = \begin{cases} 2^{-n} & \text{precisely when } (x,y) \in U_{n-1} \setminus U_n \\ 0 & \text{when } (x,y) \in \bigcap_n U_n \end{cases}$$

and

$$d(x,y) = \inf\left\{ \sum_{i=0}^{n} |f(x_i, x_{i+1})| : x = x_0, y = x_{n+1} \right\}.$$

It is easy to see that d satisfies the triangle inequality so we have (i). Since $d(x,y) \le f(x,y)$, we see that

$$U_n \subseteq \{(x,y) : d(x,y) < 2^{-n}\}.$$

Note here that if each U_n is symmetric (i.e., $U_n = U_{n-1}$), then $f(x,y) = f(y,x)$ and so $d(x,y) = d(y,x)$ so d is a pseudometric. What remains? Well, we'll show that

$$f(x_0, x_{n+1}) \le 2 \sum_{i=0}^{n} f(x_i, x_{i+1}). \tag{52}$$

From this, it follows that if $d(x,y) < 2^{-n}$, then $f(x,y) < 2 \cdot 2^{-n}$, and so

$(x, y) \in U_{n-1}$ thereby ensuring that

$$\{(x, y) : d(x, y) < 2^{-n}\} \subseteq U_{n-1},$$

and (ii) holds.

If $n = 0$, then (52) is plain. Let $n > 0$, and let k be the largest integer such that

$$\sum_{i=0}^{k-1} f(x_i, x_{i+1}) \leq \frac{1}{2} \sum_{i=0}^{n} f(x_i, x_{i+1}).$$

Of course,

$$\sum_{i=0}^{k} f(x_i, x_{i+1}) > \frac{1}{2} \sum_{i=0}^{n} f(x_i, x_{i+1}),$$

and so

$$\sum_{i=k+1}^{n} f(x_i, x_{i+1}) \leq \frac{1}{2} \sum_{i=0}^{n} f(x_i, x_{i+1}).$$

Since

$$f(x_k, x_{k+1}) \leq \sum_{i=0}^{n} f(x_i, x_{i+1}),$$

we see that

$$\begin{aligned} f(x_0, x_{n+1}) &\leq \sum_{i=0}^{k-1} f(x_i, x_{i+1}) + f(x_k, x_{k+1}) + \sum_{i=k+1}^{n} f(x_i, x_{i+1}) \\ &\leq \frac{1}{2} \sum_{i=0}^{n} f(x_i, x_{i+1}) + \sum_{i=0}^{n} f(x_i, x_{i+1}) + \frac{1}{2} \sum_{i=0}^{n} f(x_i, x_{i+1}) \\ &= 2 \sum_{i=0}^{n} f(x_i, x_{i+1}), \end{aligned}$$

and (52) holds. □

Of course, if a uniformity $\mathcal{U}$ for X has a countable base $(V_n)_{n \geq 1}$, then the axioms of a uniformity allow us to construct symmetric members $(U_n)_{n \geq 0}$ of $\mathcal{U}$ such that $U_n \subseteq V_n$ and $(U_n)_{n \geq 0}$ satisfy the hypotheses of Lemma 9.16, the metrization lemma.

So we get that *a uniformity with a countable base is a pseudo-metrizable uniformity.* Of course the converse is also so: if d is a pseudo-metric, then $\{(x, y) : d(x, y) < 1/n\}$ is a countable base for the pseudo-metric uniformity. Thus we have

Theorem 9.17. *A uniform space is pseudo-metrizable if and only if its uniformity has a countable base.*

Keep in mind our characterization of those pseudo-metrics that are uniformly continuous with respect to the product uniformity: if $(X,\mathcal{U})$ is a uniform space and d is a pseudo-metric on $X \times X$, then d is uniformly continuous of $X \times X$ relative to the product uniformity if and only if for each $r > 0$ the set $\{(x,y) : d(x,y) < r\} \in \mathcal{U}$.

So suppose $(X,\mathcal{U})$ is a uniform space, and let $\mathcal{P}$ be the family of all pseudo-metrics on X that are uniformly continuous on $X \times X$ (with respect to the product uniformity, of course). The uniformity generated by $\mathcal{P}$ is no bigger than $\mathcal{U}$ by our characterization of uniformly continuous pseudo-metrics.

On the other hand, our metrization lemma assures us that given a member $U \in \mathcal{U}$ there is a member $d \in \mathcal{P}$ such that

$$\{(x,y) : d(x,y) < 1/4\} \subseteq U,$$

and so $\mathcal{U}$ is no larger than the uniformity generated by $\mathcal{P}$. In other words, we have the following.

Theorem 9.18. *Each uniformity for X is generated by the family of all pseudo-metrics that are uniformly continuous on $X \times X$.*

A bit of standard argumentation soon reveals the following.

Corollary 9.19 (A. Weil). *Each uniform space is uniformly isomorphic to a subspace of the product of pseudo-metric spaces. Consequently, a topology τ for a set X is the uniform topology for* some *uniformity for X if and only if τ is a completely regular topology.* (Warning: "Completely regular" is sans Hausdorff!)

2. Some Miscellaneous Facts and Features about Uniform Spaces

We don't propose to study uniform spaces in detail, but since we mentioned "Cauchy sequences/nets" and "completeness" we feel obligated to at least define these terms and state a few of the results of immediate interest to analysts.

Definition 9.20. A net $(x_\alpha)_D$ in a uniform space $(X,\mathcal{U})$ is a *Cauchy net* if given $U \in \mathcal{U}$ there is a $d_U \in D$ so that if $d, d' \geq d_U$, then $(x_d, x_{d'}) \in U$.

Convergent nets relative to the uniform topology are Cauchy nets. $(X,\mathcal{U})$ is *complete* if Cauchy nets are convergent.

It is important to understand that completeness entails *nets* and not just sequences. This distinction is alive and dangerous to the ill informed, especially in analysis. For instance, consider an infinite-dimensional Banach space X in its weak topology. The associated uniformity is generated by the collection of sets of the form

$$U_{x_1^*,\dots,x_n^*;\epsilon} = \{(x,y) : |x_1^*(x-y)|,\dots,|x_n^*(x-y)| < \epsilon\},$$

where $x_1^*,\dots,x_n^* \in X^*, \epsilon > 0$.

It's a fact that an infinite-dimensional Banach space in its weak topology is *never* complete. All is not lost though because it is a phenomenon often (but not always) encountered that a Banach space in its weak uniformity be sequentially complete; indeed, in addition to the reflexive spaces, $L^1(\mu), L^1(\mathbb{T})/H^1, H^1(\mathbb{D})^*, C(S)^*$, Orlicz spaces generated by Δ_2-formation, preduals of von Neumann algebras, find themselves enjoying the weak completeness of weakly Cauchy sequences.

Again if X is a Banach space and X^* is its dual then the *weak*-uniformity* is generated by sets of the form

$$U_{x_1,\dots,x_n;\epsilon} = \{(x^*,y^*) : |x^*(x_1) - y^*(x_1)|,\dots,|x^*(x_n) - y^*(x_n))| < \epsilon\},$$

where $x_1,\dots,x_n \in X, \epsilon > 0$. Remarkably enough, we again have that if X is an infinite-dimensional Banach space, then its dual space X^* with the weak*-uniformity is *never* complete but *always* sequentially complete.

3. Compactness in Uniform Spaces

If $(X,\mathcal{U})$ is a compact uniform space, then every neighborhood of the diagonal Δ in $X \times X$ is a member of $\mathcal{U}$.

Let $\mathcal{B}$ be the family of closed members of $\mathcal{U}$, and let $\mathcal{V}$ be an open set in $X \times X$ that contains Δ.

Suppose $(x,y) \in \bigcap_{U\in\mathcal{B}} U$. Since $\mathcal{B}$ is a base for $\mathcal{U}$, $y \in U[x]$, we see that y belongs to every open set in X that contains x. Hence (x,y) lies in any open set in $X \times X$ that contains Δ. Therefore

$$\bigcap_{U\in\mathcal{B}} U \subseteq V.$$

Claim. There exists $U_1,\dots,U_n \in \mathcal{B}$ so $U_1 \cap \dots \cap U_n \subseteq V$.

Suppose not. Then for any $U_1,\dots,U_n \in \mathcal{B}$, we can find a point

$$x_{U_1,\dots,U_n} \in (U_1 \cap \dots \cap U_n) \setminus V.$$

With this natural ordering

$$(U_1, \ldots, U_n) \geq (U_1', \ldots, U_m') \text{ when } \bigcap_{I=1}^{n} U_i \subseteq \bigcap_{j=1}^{m} U_j',$$

we get a net in X: $(x_{U_1}, \ldots, x_{U_n})$ which necessarily has a limit point

$$x \in \bigcap_{U \in \mathcal{B}} U \subseteq V.$$

But $x \in V$ says for $(U_1, \ldots, U_n)$ big enough,

$$(x_{U_1}, \ldots, x_{U_n}) \subseteq V,$$

and so there are

$$U_1, \ldots, U_n \in \mathcal{B},$$

so $U_1 \cap \cdots \cap U_n \subseteq V$, and as a result $V \in \mathcal{U}$.

It follows that

$$\bigcap_{U \in \mathcal{B}} U \subseteq V.$$

But each member U of $\mathcal{B}$ is closed in the compact space X, and so each is compact and V is open. Thus the intersection of some finite subfamily of $\mathcal{B}$ is also a subset of V, and from this we see that V itself belongs to $\mathcal{U}$.

If d is a pseudo-metric on X that is continuous on $X \times X$, then for any $\epsilon > 0$

$$\{(x, y) : d(x, y) < \epsilon\}$$

is an open set that contains the diagonal. This is tantamount to d being uniformly continuous.

Therefore if $(X,\mathcal{U})$ is a compact uniform space, then every continuous $f : (X,\mathcal{U}) \to \mathbb{R}$ is uniformly continuous.

Look at (f, f) where $(f, f) : X \times X \to \mathbb{R} \times \mathbb{R}$; if we look at d,

$$d(x, y) = |f(x) - f(y)|,$$

then d is a continuous pseudo-metric on $X \times X$, and so is uniformly continuous. A quick check shows that this is just f's uniform continuity.

Corollary 9.21. *If $(X,\mathcal{U})$ is a compact uniform space and $f : X \to \mathbb{R}$ is continuous, then f is uniformly continuous. Indeed $d(x, y) = |f(x) - f(y)|$ is a continuous pseudo-metric on $X \times X$, and hence uniformly continuous.*

4. From Contents to Outer Measures

Let S be a locally compact Hausdorff space. Denote by $\mathcal{C}(S)$ the collection of all compact subsets of S.

Definition 9.22. A *content* is a set function λ on $\mathcal{C}(S)$ which satisfies

(i) $0 \leq \lambda(C) < \infty$;

(ii) if $C_1, C_2 \in \mathcal{C}(S)$ with $C_1 \subseteq C_2$, then $\lambda(C_1) \leq \lambda(C_2)$;

(iii) if $C_1, C_2 \in \mathcal{C}(S)$, then $\lambda(C_1 \cup C_2) \leq \lambda(C_1) + \lambda(C_2)$;

(iv) if C_1, C_2 are disjoint members of $\mathcal{C}(S)$, then $\lambda(C_1 \cup C_2) = \lambda(C_1) + \lambda(C_2)$.

Since $\lambda(\emptyset) + \lambda(\emptyset) = \lambda(\emptyset \cup \emptyset) = \lambda(\emptyset) < \infty$, $\lambda(\emptyset) = 0$.

A content λ induces λ_* on the topology of S:

$$\lambda_*(U) = \sup\{\lambda(C) : U \supset C \in \mathcal{C}(S)\}.$$

Theorem 9.23. *The function λ_* vanishes at $\emptyset$, is monotone nondecreasing, countably subadditive, and countably additive (on disjoint open sets, of course).*

Proof. The fact that $\lambda_*(\emptyset) = 0$ and the monotone nature of λ_* are both plain.

Suppose U, V are open and C is a compact subset of $U \cup V$. Then $C \setminus U$ and $C \setminus V$ are disjoint compact sets in S, and so there exist disjoint open sets $\widetilde{U}$ and $\widetilde{V}$ such that $C \setminus U \subseteq \widetilde{U}$ and $C \setminus V \subseteq \widetilde{V}$. Let $D = C \setminus \widetilde{U}$ and $E = C \setminus \widetilde{V}$. Then $D \subseteq U, E \subseteq V$, and both D and E are compact. Since $\widetilde{U} \cap \widetilde{V} = \emptyset$,

$$D \cup E = (C \setminus \widetilde{U}) \cup (C \setminus \widetilde{V}) = C \setminus (\widetilde{U} \cap \widetilde{V}) = C.$$

Now

$$\lambda(C) \leq \lambda(D) + \lambda(E) \leq \lambda_*(U) + \lambda_*(V),$$

and so

$$\lambda_*(U \cup V) = \sup\{\lambda(C) : C \subseteq U \cup V\} \leq \lambda_*(U) + \lambda_*(V),$$

and λ_* is (finitely) subadditive.

If (U_n) is a sequence of open sets in S and C is a compact set such that $C \subseteq \bigcup_n U_n$, then there is an N so that

$$C \subseteq \bigcup_{n=1}^{N} U_n.$$

It follows that

$$\lambda(C) \leq \lambda_*\Big(\bigcup_{n=1}^{N} U_n\Big) \leq \sum_{n=1}^{N} \lambda_*(U_n) \leq \sum_n \lambda_*(U_n).$$

Therefore, since $C \in \mathcal{C}$ is an arbitrary compact set contained in $\bigcup_n U_n$,

$$\lambda_*\Big(\bigcup_n U_n\Big) \leq \sum_n \lambda_*(U_n),$$

and λ_* is countably subadditive.

Next suppose that U and V are disjoint open sets, and let C, D be compact sets such that $C \subseteq U$ and $D \subseteq V$. Then $C \cup D \subseteq U \cup V$, and so

$$\lambda(C) + \lambda(D) = \lambda(C \cup D) \leq \lambda_*(U \cup V).$$

It follows that

$$\lambda_*(U) + \lambda_*(V) \leq \lambda_*(U \cup V)$$

and λ_* is additive on disjoint open sets. Of course if $U_1, \ldots, U_n, \ldots$ is a sequence of pairwise disjoint open sets, then for any N

$$\lambda_*\Big(\bigcup_n U_n\Big) \geq \lambda_*\Big(\bigcup_{n=1}^{N} U_n\Big) = \sum_{n=1}^{N} \lambda_*(U_n).$$

It follows that

$$\lambda_*\Big(\bigcup_n U_n\Big) \geq \sum_n \lambda_*(U_n),$$

and so the proof is complete. □

From λ we pass to λ_* and now we define μ^*

$$\mu^*(E) = \inf\{\lambda_*(U) : E \subseteq U, U \text{ is open}\}.$$

Theorem 9.24. *μ^* is an outer measure.*

Proof. Since $\emptyset$ is open and $\lambda_*(\emptyset) = 0$, $\mu^*(\emptyset) = 0$. Clearly μ^* is monotone nondecreasing.

If (E_n) is a sequence of subsets of S with each having $\mu^*(E_n) < \infty$, then given an $\epsilon > 0$ there is an open set U_n containing E_n and such that

$$\mu^*(E_n) \leq \lambda_*(U_n) \leq \mu^*(E_n) + \frac{\epsilon}{2^n}.$$

It follows that

$$\begin{aligned}\mu^*\Big(\bigcup_n E_n\Big) &\le \lambda_*\Big(\bigcup_n U_n\Big) \le \sum_n \lambda_*(U_n)\\ &\le \sum_n \Big(\mu^*(E_n) + \frac{\epsilon}{2^n}\Big)\\ &= \sum_n \mu^*(E_n) + \epsilon,\end{aligned}$$

and μ^* is countably subadditive. □

Fact 9.25. A set E is μ^**-measurable* if and only if

$$\mu^*(U) \ge \mu^*(U \cap E) + \mu^*(U \cap E^c) \tag{53}$$

for each open set $U \subseteq S$.

In fact, take any $A \subseteq S$, and suppose $A \subseteq U$, U open. Then assuming (53),

$$\begin{aligned}\lambda_*(U) = \mu^*(U) &\ge \mu^*(U \cap E) + \mu^*(U \cap E^c)\\ &\ge \mu^*(A \cap E) + \mu^*(A \cap E^c).\end{aligned}$$

It follows that

$$\begin{aligned}\mu^*(A) &= \inf\{\lambda_*(U) : A \subseteq U, U \text{ is open}\}\\ &\ge \mu^*(A \cap E) + \mu^*(A \cap E^c).\end{aligned}$$

Naturally the reverse inequality follows from μ^* subadditivity and the very definition of μ^*-measurability. □

Fact 9.26. If $C \in \mathcal{C}(S)$, then

$$\mu^*(\text{int}(C)) \le \lambda(C) \le \mu^*(C).$$

Proof. We first show that

$$\mu^*(U) = \lambda_*(U).$$

Since $U \subseteq U$ it follows that $\mu^*(U) \le \lambda_*(U)$.

On the other hand, if V is open and $U \subseteq V$, then $\lambda_*(U) \le \lambda_*(V)$, and so

$$\lambda_*(U) \le \inf\{\lambda_*(V) : U \supseteq V, V \text{ is open}\} = \mu^*(U).$$

There is only one conclusion:

$$\mu^*(U) = \lambda_*(U).$$

If $C \in \mathcal{C}(S)$ and U is open with $C \subseteq U$, then $\lambda(C) \le \lambda_*(U)$ so

$$\lambda(C) \le \inf\{\lambda_*(U) : C \subseteq U, U \text{ is open}\} = \mu^*(C).$$

If $C, D \in \mathcal{C}(S)$ and $D \subseteq \text{int}(C)$, then $\lambda(D) \leq \lambda(C)$, and so

$$\mu^*(\text{int}(C)) = \lambda_*(\text{int}(C)) = \sup\{\lambda(D) : D \subseteq \text{int}(C)\} \leq \lambda(C).$$

In summary,

$$\mu^*(\text{int}(C)) \leq \lambda(C) \leq \mu^*(C).$$

□

Fact 9.27. Compact sets are μ^*-measurable.

Proof. Let $C \in \mathcal{C}(S)$. We need to show that for any open set U,

$$\mu^*(U) \geq \mu^*(U \cap C) + \mu^*(U \cap C^c).$$

Let D be a compact subset of $U \cap C^c$, and let E be a compact subset of $U \cap D^c$. Then $U \cap C^c$ and $U \cap D^c$ are open and $D \cap E = \emptyset$ with $D \cup E \subseteq U$. so

$$\mu^*(U) = \lambda_*(U) \geq \lambda(D \cup E) = \lambda(D) + \lambda(E).$$

It follows that

$$\begin{aligned}
\mu^*(U) &\geq \lambda(D) + \sup\{\lambda(E) : E \subseteq U \cap D^c\} \\
&= \lambda(D) + \lambda_*(U \cap D^c) \\
&= \lambda(D) + \mu^*(U \cap D^c) \\
&\geq \lambda(D) + \mu^*(U \cap C).
\end{aligned}$$

In turn, this says

$$\begin{aligned}
\mu^*(U) &\geq \mu^*(U \cap C) + \sup\{\lambda(D) : D \subseteq U \cap C^c\} \\
&= \mu^*(U \cap C) + \lambda_*(U \cap C^c) \\
&= \mu^*(U \cap C) + \mu^*(U \cap C^c),
\end{aligned}$$

and D is μ^*-measurable. □

Here's an important feature of μ^*: $\mu^*(C) < \infty$ for each $C \in \mathcal{C}(S)$. In fact, we know that there is a compact set $F \subseteq S$ so that $C \subseteq \text{int}(F) \subseteq F$. It follows that

$$\mu^*(C) \leq \mu^*(\text{int}(F)) \leq \lambda(F) < \infty.$$

Finally we take note of the following.

Theorem 9.28. *If $\phi : S \to S$ is a homeomorphism of S onto S and λ is a content, then*

$$\lambda_\phi(C) = \lambda(\phi(C))$$

is a content too. If μ^*_ϕ is the outer measure generated by λ_ϕ, then

$$\mu^*_\phi = \mu^*(\phi(E)).$$

So if λ is invariant under ϕ, then so is μ^*.

5. Existence of G-invariant Contents

Definition 9.29. Let X be a locally compact Hausdorff space, and let G be a group of homeomorphisms of X onto itself. We say that G is *transitive* if given $x, y \in X$ there is a $g \in G$ so that $gx = y$. We say that G is *weakly transitive* if for any nonempty open set $U \subseteq X$,

$$\{gU : g \in G\}$$

covers X. We say that G *separates compact sets* if given a pair A, B of disjoint compact sets in X and $g \in G$, there is a nonempty open set $U \subseteq X$ such that gU never meets both A and B.

Example 9.30. Let $X = \mathbb{R}$, and let G be translation by rational numbers. Then G is a weakly transitive group of homeomorphisms from X onto itself that is not transitive.

Here's the first central result of Steinlage; it pertains to the existence of a G-invariant measure. The attentive reader will soon realize that this proof is modeled on Banach's proof.

Lemma 9.31. *Let G be a weakly transitive group of homeomorphisms of X onto X, and suppose that G separates compact sets. Then there is a content λ on X such that λ is G-invariant and $\lambda(C) > 0$ whenever C is a compact subset of X with nonempty interior.*

Recall that a *content* is a nonnegative real-valued map defined on $\boldsymbol{\mathcal{C}(X)}$, the compact subsets of X, which is monotone nondecreasing, subadditive, and additive on disjoint sets.

Proof. To build λ we proceed by stages.

Stage 1. Fix $x_0 \in X$, and let

$$\Omega(x_0) = \{U \subseteq X : x_0 \in U, U \text{ open}\}.$$

Let A and B be disjoint compact sets in X. There exists a nonempty set V such that for any $g \in G$, gV can intersect A or B—*but not both.* Since G is weakly transitive, there is a $g \in G$ so that gV contains x_0. Therefore $gV \in \Omega(x_0)$ and gV cannot intersect both A and B. In other words, *we can separate compact sets by members of $\Omega(x_0)$.*

Let E be a compact subset of X and B a subset of X with nonempty interior. Define

$$h(E, B) = \text{the least number of images } gB \text{ needed to cover } E \text{ with } g \in G.$$

Plainly, $h(E, B)$ is well defined and $0 \leq h(E, B) < \infty$. If $E \neq \emptyset$, then $h(E, B) \geq 1$.

Fact 9.32. More is so: if E, E_1, E_2 are compact subsets of X, then

(i) $h(E_1, B) \leq h(E_2, B)$ whenever $E_1 \subseteq E_2$;

(ii) $h(gE, B) = h(E, B)$ for every $g \in G$;

(iii) $h(E_1 \cup E_2, B) \leq h(E_1, B) + h(E_2, B)$;

(iv) if D is compact with nonempty interior, then $h(E, B) \leq h(E, D) \cdot h(D, B)$;

(v) if $E_1 \cap E_2 = \emptyset$, then there is a $B_0 \in \Omega(x_0)$ such that for any $B \in \Omega(x_0)$ with $B \subseteq B_0$,

$$h(E_1 \cup E_2, B) = h(E_1, B) + h(E_2, B).$$

As one might expect, all but Fact 9.32(v) are self-explanatory and easily established; Fact 9.32(v) needs some work. We know by the fact that G separates compact sets that there is a $B_0 \in \Omega(x_0)$ so that gB_0 never meets both E_1 and E_2. Let

$$n = h(E_1 \cup E_2, B_0).$$

Then there are points $g_1, \ldots, g_n \in G$ so that

$$\begin{aligned} E_1 \cup E_2 &= \bigcup_{i=1}^{n} g_i B_0 \\ &= \left(\bigcup_i \{g_i B_0 : g_i B_0 \cap E_1 \neq \emptyset\}\right) \cup \left(\bigcup_i \{g_i B_0 : g_i B_0 \cap E_2 \neq \emptyset\}\right). \end{aligned}$$

If $1 \leq i \leq n$, then either $g_i B_0 \cap E_1 \neq \emptyset$ or $g_i B_0 \cap E_2 \neq \emptyset$, but not both. Now $\{g_i B_0 : g_i B_0 \cap E_1 \neq \emptyset\}$ must cover E_1, and $\{g_i B_0 : g_i B_0 \cap E_2 \neq \emptyset\}$ must cover E_2. So

$$n \geq h(E_1, B_0) + h(E_2, B_0),$$

and Fact 9.32(v) is established.

Stage 2. Fix a compact set $A_0 \subseteq X$ with a nonempty interior. For each compact E and for each $U \in \Omega(x_0)$ define

$$l_U(E) = \frac{h(E, U)}{h(A_0, U)}.$$

The attentive reader will realize that a similar normalization occurred in Banach's presentation and in the Cartan–Weil construction. It goes back, in fact, to Haar's original ground-breaking work.

Here's the truth:

Fact 9.33. If E, E_1 and E_2 are compact sets in X, then

(i) if $E_1 \subseteq E_2$, then $l_U(E_1) \leq l_U(E_2)$;

(ii) $l_U(gE) = l_U(E)$ for $g \in G$;

(iii) $l_U(E_1 \cup E_2) \leq l_U(E_1) + l_U(E_2)$;

(iv) $0 \leq l_U(E) \leq h(E, A_0)$ and if E has nonempty interior, then

$$\frac{1}{h(A_0, E)} \leq l_U(E) \leq h(E, A_0);$$

(v) if E_1 and E_2 are disjoint compact sets, then there is a $U_0 \in \Omega(x_0)$ such that

$$l_U(E_1 \cup E_2) = l_U(E_1) + l_U(E_2)$$

for any $U \in \Omega(x_0)$ with $U \subseteq U_0$;

(vi) $l_U(\emptyset) = 0$ for all $U \in \Omega(x_0)$.

Stage 3. Let E be a compact set in X. Set

$$\delta_E = \begin{cases} 0 & \text{if } E \text{ has empty interior} \\ \frac{1}{h(A_0,E)} & \text{if } E \text{ has nonempty interior.} \end{cases}$$

Then

$$K = \prod_{E \in \mathcal{C}(X)} [\delta_E, h(E, A_0)]$$

is a compact Hausdorff space. For $U \in \Omega(x_0)$, we let l_U be the point

$$l_U := (l_U(E) : E \in \mathcal{C}(X)) \in K.$$

For each $V \in \Omega(x_0)$, let

$$\Lambda(V) = \{l_U : U \in \Omega(x_0), U \subseteq V\} \subseteq K.$$

The family

$$\{\Lambda(V) : V \in \Omega(x_0)\}$$

has the finite intersection property since the finite intersection of members of $\Omega(x_0)$ is also in $\Omega(x_0)$. So by K's compactness there is a λ

$$\lambda \in \bigcap_{V \in \Omega(x_0)} \overline{\Lambda(V)}.$$

This λ does the trick. $\square$

What does λ do?

To start, $\lambda \in K$ so for any $E \in \mathcal{C}(X)$,

$$\delta_E \leq \lambda(E) \leq h(E, A_0).$$

It follows that $0 \leq \lambda(E) < \infty$ for any $E \in \mathcal{C}(X)$. Since λ lies in each $\overline{\Lambda(V)}$, $V \in \Omega(x_0)$, and $\Lambda(V)$ consists of all the points l_U where $U \in \Omega(x_0)$ satisfies

$U \subseteq V$. Hence the facts that each l_U is G-invariant and monotone nondecreasing soon reveal that the same is so for λ. Again each l_U is subadditive, and so the same holds for all λ. Finally, if E_1 and E_2 are disjoint compact sets, then there is a $U_0 \in \Omega(x_0)$ such that for any $U \in \Omega(x_0)$ with $U \subseteq U_0$,

$$\lambda_U(E_1 \cup E_2) = \lambda_U(E_1) + \lambda_U(E_2).$$

Since

$$\lambda \in \bigcap_{V \in \Omega(x_0)} \overline{\Lambda(V)},$$

this additivity passes to λ.

An aside:

Proposition 9.34. *Let G be a group of homeomorphisms from the topological space X onto itself. Then G is weakly transitive if and only if for each $x \in X$, the orbit*

$$Gx := \{gx : g \in G\}$$

of x is dense in X.

Proof. Suppose G is weakly transitive but there is an $x_0 \in X$ such that $\overline{Gx_0}$ is not all of X. Look at the open (nonempty) set $X \setminus \overline{Gx_0}$. By weak transitivity there is a $g \in G$ so that

$$x_0 \in g\left(X \setminus \overline{Gx_0}\right).$$

It follows that

$$g^{-1}x_0 \in X \setminus \overline{Gx_0}.$$

Oops!

Now suppose that G is not weakly transitive. Then there is a nonempty open set U so

$$GU := \bigcup_{g \in G} gU \neq X.$$

Plainly, $gGU = GU$ for each $g \in G$ so

$$g(X \setminus GU) = X \setminus GU$$

for each $g \in G$. If $x \in X \setminus GU$, then

$$\overline{Gx} \subseteq X \setminus GU,$$

and Gx is not dense. □

Recall that if $(X, \mathcal{U})$ is a uniform space, then the *uniform topology* is the collection of all $T \subseteq X$ such that for each $x \in T$ there is a $U \in \mathcal{U}$ such that

$$U[x] := \{y \in X : (x, y) \in U\} \subseteq T.$$

Definition 9.35. Let $\mathcal{F}$ be a family of functions from a topological space X into a uniform space $(Y, \mathcal{V})$. Suppose $x \in X$. We say that $\mathcal{F}$ is *equicontinuous at $\boldsymbol{x}$* $\in X$ provided, given $V \in \mathcal{V}$, there is an open set $O \subseteq X$ containing x such that

$$f(O) \subseteq V[f(x)]$$

for each and every $f \in \mathcal{F}$. We say that $\mathcal{F}$ is *equicontinuous* if $\mathcal{F}$ is equicontinuous at each point $x \in X$; i.e., regardless of $x \in X$, given $V \in \mathcal{V}$ there is an open set $O \subseteq X$ that contains x and satisfies

$$f(O) \subseteq V[f(x)]$$

for each $f \in \mathcal{F}$.

Equicontinuity provides new insights into weak transitivity.

Lemma 9.36. *Suppose $(X, \mathcal{U})$ is a uniform space and G is an equicontinuous group of homeomorphisms from X (in its uniform topology) onto itself. If Gx_0 is dense for some $x_0 \in X$, then G is weakly transitive.*

Proof. Suppose that G is not weakly transitive. Then there is an $x \in X$ and a nonempty open set $U \subseteq X$ so that

$$x \notin \bigcup_{g \in G} gU.$$

Since Gx_0 is dense in X, there is a $g_1 \in G$ so that

$$Gx_0 \cap g_1 U \neq \emptyset.$$

Hence (again by Gx_0's by density) there is a $g_2 \in G$ so that

$$g_2 x_0 \in g_1 U,$$

and so

$$x_0 \in g_2^{-1} g_1 U.$$

Set $g_0 = g_2^{-1} g_1$, and so

$$x_0 \in g_0 U.$$

There is a symmetric $V \in \mathcal{U}$ so that

$$V[x_0] \subseteq g_0 U.$$

By equicontinuity there is an open set $O \subseteq X$ that contains x such that

$$gO \subseteq V[gx]$$

for all $g \in G$. But since

$$x \notin \bigcup_{g \in G} gU,$$

it follows that

$$Gx \cap g_0 U = \emptyset.$$

Since $V[x_0] \subseteq g_0 U$,

$$Gx \cap V[x_0] = \emptyset.$$

It follows that for any $g \in G$, $x_0 \notin V[gx]$; otherwise, if $x_0 \in V[gx]$, then $(gx, x_0) \in V$, which by symmetry tells us $(x_0, gx) \in V$ or $gx \in V[x_0]$ yielding, $Gx \cap V[x_0] \neq \emptyset$, which is impossible.

By a similar argument it follows that since $x_0 \notin V[gx]$ for all $g \in G$,

$$Gx_0 \cap V[gx] = \emptyset$$

for each $g \in G$, and since $gO \subseteq V[gx]$ for all $g \in G$,

$$Gx_0 \bigcap \left(\bigcup_{g \in G} gO \right) = \emptyset,$$

contradicting the density of the orbit Gx_0 of x_0.

Our conclusion is clear: G must be weakly transitive. □

We can lighten the load of equicontinuity on G's shoulders. Here's how.

Lemma 9.37. *Let $(X, \mathcal{U})$ be a uniform space, and let G be a group of homeomorphisms of X (in its uniform topology) onto itself. Suppose that there is an $x_0 \in X$ such that G is equicontinuous at x_0 and for any open set U containing x_0,*

$$GU = X.$$

Then G is equicontinuous.

Proof. Fix $U \in \mathcal{U}$. There is a symmetric $V \in \mathcal{U}$ such that

$$V \circ V \subseteq U.$$

Since G is equicontinuous at x_0 there open set $O \subseteq X$, containing x_0 such that

$$gO \subseteq V[gx_0]$$

for all $g \in G$.

Fix $x \in X$. We'll show that G is equicontinuous at x. Since $GO = X$, there is a $g_o \in G$ so that $x \in g_0 O$. Fix $g \in G$. Therefore

$$gg_0 O \subseteq V[gg_0 x_0],$$

so that

$$gx \in gg_0 O \subseteq V[gg_0 x_0].$$

Since

(54) $$(gg_0 x_0, gx) \in V,$$

by symmetry

$$(gx, gg_0x_0) \in V.$$

If $y \in g_0O$, then

$$gy \in gg_oO \subseteq V[gg_0x_0];$$

as above (by (54))

$$(gg_0x_0, gy) \in V.$$

Since

$$(gx, gg_0x_0), (gg_0x_0, gy) \in V,$$

we have

$$(gx, gy) \in V \circ V \subseteq U.$$

All of this is true for any arbitrary $g \in G$.

Therefore, for any $g \in G$, if $y \in g_0O$, then

$$(gx, gy) \in U;$$

i.e., if $y \in g_0O$, then

$$gy \in U[gx]$$

for all $g \in G$. Therefore for any $g \in G$,

$$gg_0U \subseteq U[gx],$$

and G is equicontinuous at x. □

From uniform spaces, we borrow the following Urysohn-like result.

Lemma 9.38 (Uniform Space Lemma). *Let $(X,\mathcal{U})$ be a uniform space, and let A, B be disjoint compact subsets of X (in its uniform topology). Then there is a symmetric $U \in \mathcal{U}$ such that for any $x \in X$ either $U[x] \cap A = \emptyset$ or $U[x] \cap B = \emptyset$.*

Of course the key use of this lemma is to lighten the load of establishing that a group of homeomorphisms on X separates the compact subsets of X.

Theorem 9.39. *Suppose $(X,\mathcal{U})$ is a uniform space whose uniform topology is locally compact and Hausdorff. Let G be an equicontinuous group of homeomorphisms of X onto itself. Then there is a nonzero G-invariant content on X.*

Proof. Let $x_0 \in X$, and set

$$X_0 = \overline{Gx_0}.$$

Then X_0 is a locally compact Hausdorff space in its relative topology which is the uniform topology generated by the relative uniformity $\mathcal{U}_0$ on $X_0 \times X_0$ inherited from $\mathcal{U}$, and G is an equicontinuous group of homeomorphisms taking X_0 into X_0. Of course the orbit of x_0 is dense in X_0, and so by Lemma 9.36, G is weakly transitive in X_0. Thus in order to use Lemma 9.31, we just need to show that G separates compact sets of X_0.

If C and D are disjoint compact subsets of X_0, then by the Uniform Space Lemma, we can find a symmetric $V \in \mathcal{U}_{X_0}$ such that for any $x \in X_0$, either

$$V[x] \cap C = \emptyset \text{ or } V[x] \cap D = \emptyset.$$

By equicontinuity, there is an open set O containing x_0 such that

$$gO \subseteq V[gx_0]$$

for any $g \in G$. It follows that either $gO \cap C = \emptyset$ or $gO \cap D = \emptyset$ for each $g \in G$. Thus G separates compact sets in X_0.

We can appeal to Lemma 9.31 to conclude that there is a nonzero G-invariant content λ on X_0. We can extend λ to a G-invariant content $\tilde{\lambda}$ on X simply by defining

$$\tilde{\lambda}(K) = \lambda(K \cap X_0)$$

for any compact set $K \subseteq X$. □

6. Steinlage: Uniqueness and Weak Transitivity

Theorem 9.40 (Steinlage). *Let G be an equicontinuous group of homeomorphisms of the nonempty locally compact space X onto itself, where X's topology is generated by the uniformity $\mathcal{U}$. Then the G-invariant Haar measure on X is unique if and only if G is weakly transitive.*

On the other hand we have

Note 9.41. Let $(X, \mathcal{U})$ be a uniform space whose uniform topology is locally compact and Hausdorff, and assume that G is an equicontinuous group of homeomorphisms of X onto itself. If G does not act in a weakly transitive manner on X, then the G-invariant content of Theorem 9.39 is not unique.

Proof. If G is not weakly transitive on X, then by Lemma 9.36 *no* point of X has dense orbit. Let $x_0 \in X$, and set

$$X_0 = \overline{Gx_0}.$$

As in the proof of Theorem 9.39, X_0 is a locally compact Hausdorff space in its relative topology which is the uniform topology generated by the relative uniformity $\mathcal{U}_0$ on $X_0 \times X_0$ inherited from $\mathcal{U}$ and G acts as a weakly transitive, equicontinuous group of homeomorphisms of X_0 onto itself (again G's weak transitivity holds on X_0 through the good graces of Lemma 9.36). Hence by

Lemma 9.31 there is a nonzero G-invariant content λ_0 on X_0 which extends to all of X to a content $\tilde{\lambda}_0$ where values at a $C \in C(X)$ are given by

$$\tilde{\lambda}_0(C) = \lambda(C \cap X_0).$$

Now

$$\overline{Gx_0} \neq X,$$

so pick a point $x_1 \in X \setminus \overline{Gx_0}$, and look at

$$X_1 = \overline{Gx_1}.$$

Again as in the proof of Theorem 9.39, we get a locally compact Hausdorff space in the relative topology which is the uniform topology of X_1 generated by the relative uniformity $\mathcal{U}_1$ on $X_1 \times X_1$ inherited from $\mathcal{U}$, and G acts as a weakly transitive, equicontinuous group of homeomorphisms of X_1 onto itself (thanks again to Lemma 9.36 for this). As before, apply Lemma 9.31 to get a nonzero G-invariant content λ_1 on X_1 which extends to a content $\tilde{\lambda}_1$ on all of X by letting

$$\tilde{\lambda}_1(C) = \lambda_1(C \cap X_1)$$

for any $C \in C(X)$.

Neither of these contents it a scalar multiple of the other; indeed $\tilde{\lambda}_0$ is supported on X_0 and $\tilde{\lambda}_1$ is supported on X_1 with $X_1 \subseteq X \setminus X_0$.

The lack of weak transitivity costs the uniqueness of nontrivial G-invariant contents on X. □

Now we're ready to prove the uniqueness aspect of Steinlage's work:

Theorem 9.42. *If $(X, \mathcal{U})$ is a uniform space whose topology is locally compact and Hausdorff and if G is an equicontinuous weakly transitive group of homeomorphisms of the space X (in its uniform topology) onto itself, then there is but one G-invariant integral (up to multiplicative constants) on X.*

The fact is that uniqueness is evidenced, if at all, on the collection of continuous nonnegative real-valued, nonzero functions defined on X having compact support.

We'll refer to members of this collection as *test functions*.

We start with a few basic facts, features regarding the notions of equicontinuity and weak transitivity.

Lemma 9.43. *Let $(X, \mathcal{U})$ be a uniform space, and let G be an equicontinuous group of homeomorphisms of X (in its uniform topology) onto itself. For each $U \in \mathcal{U}$ and $x_0 \in X$, there is a nonnegative real-valued function f defined on X such that*

(i) $f(x_0) \neq 0$;

(ii) $f(x) = 0$ *for each* $x \in X \setminus U[x_0]$;

(iii) $f \circ g(x_0) = f \circ g^{-1}(x_0)$ *for each* $g \in G$.

Proof. Fix $U \in \mathcal{U}$, and $x_0 \in X$. There is a uniformly continuous pseudo-metric d on X such that for each $r > 0$,

$$U_{d,r} = \{(x, y) : d(x, y) < r\} \in \mathcal{U},$$

and there is an $r_0 > 0$ so that $U_{d,r_0} \subseteq U$.

For $x, y \in X$, define

$$h^*(x, y) = \sup\{d(g(x), g(y)) : g \in G\}$$

and

$$h(x, y) = \min\{h^*(x, y), r_0\}.$$

Clearly, h is a pseudo-metric on $X \times X$ and h satisfies

$$h(x, y) = h(g(x), g(y))$$

for each $g \in G$.

Moreover, $h(x, x_0)$ is actually a continuous function of x. Indeed, if we fix x and let $0 < \epsilon < r_0$, then the equicontinuity of G at x assures us that there is an open set O containing x such that for all $g \in G$,

$$gO \subseteq U_{d,\epsilon}[g(x)].$$

It follows that for any $y \in O$,

$$|h(y, x_0) - h(x, x_0)| \leq h(x, y) \leq \epsilon.$$

Define $f : X \to [0, \infty)$ by

$$f(x) = r_0 - h(x, x_0),$$

so f is continuous. Also

$$f(x_0) = r_0 - h(x_0, x_0) = r_0.$$

Further, if $x \notin U[x_0]$, then $x \notin U_{d,r_0}[x_0]$; hence, $d(x, x_0) \geq r_0$ and $h(x, x_0) = r_0$, forcing $f(x) = 0$. Finally, regardless of $g \in G$, we have

$$\begin{aligned} f(g(x_0)) &= r_0 - h(g(x_0), x_0) \\ &= r_0 - h(g^{-1}g(x_0), g^{-1}(x_0)) \\ &= r_0 - h(x_0, g^{-1}(x_0)) = f(g^{-1}(x_0)), \end{aligned}$$

and we have proved (iii). □

Lemma 9.44. *Suppose G is a weakly transitive group of homeomorphisms of the locally compact space X onto itself. Suppose f, h are test functions defined on X. Suppose that h is not identically zero. Then there exist $g_1, \ldots, g_n \in G$ and $c > 0$ so that*

$$f(x) \leq c \sum_{i=1}^{n} h(g_i(x))$$

for each $x \in X$.

Proof. Let $K = \operatorname{supp} f$. There is an $\eta > 0$ and a nonempty open set $O \subseteq X$ so that $h(x) > \eta$ for all $x \in O$. Since G is weakly transitive,

$$K \subseteq \bigcup_{g \in G} gO$$

so there are $g_1, \ldots, g_n \in G$ such that

$$K \subseteq g_1 O \cup \cdots \cup g_n O.$$

Let

$$M = \sup\{f(x) : x \in K\},$$

and set $c = M/\eta$.

If $x \in K$, then $x \in g_i O$ for some i, $1 \leq i \leq n$. It follows that $g_i^{-1}(x) \in O$ and so $h(g_i^{-1}(x)) \geq \eta$. Hence, $g_i O$

$$\frac{M}{\eta} h(g_i^{-1}(x)) \geq M \geq f(x).$$

The constant $c = M/\eta$ works.

Now we're ready to prove Theorem 9.42.

Fix $x_0 \in X$ and $U \in \mathcal{U}$. By equicontinuity of G at x_0, there is a $V \in \mathcal{U}$ such that for each $g \in G$,

$$gV[x_0] \subseteq U[g(x_0)].$$

We may assume that $\overline{V[x_0]}$ is compact.

Let f^* be a test function that vanishes off $V[x_0]$, is positive at x_0, and satisfies

$$f^* g(x_0) = f^* g^{-1}(x_0)$$

for each $G \in G$. Such an f^* exists by Lemma 9.43.

Let I be any G-invariant Haar integral on X, and let f be a test function. For each $g \in G$, let

$$h(g) = I(f \cdot (f^* \circ g^{-1})),$$

and let K_U denote the oscillation of f on U; i.e.,

$$K_U = K = \sup\{|f(x) - f(y)| : (x, y) \in U\} < \infty.$$

It's plain that for any $x \in X$, we have

$$f(g(x_0)) - K \leq f(x),$$

and so

$$\begin{aligned} h(g) &= I(f \cdot (f^* \circ g^{-1})) \\ &\geq I([f(g(x_0)) - K] \cdot (f^* \circ g^{-1})) \\ &= [f(g(x_0)) - K] I(f^* \circ g^{-1}) \\ &= [f(g(x_0)) - K] I(f^*) \end{aligned} \tag{55}$$

by I's invariance.

Let C be any compact set containing f's support, and let ϕ be a $[0,1]$-valued test function that's identically one on C.

Let $\epsilon > 0$. Since f^* is uniformly continuous, there is a $U^* \in \mathcal{U}$ so that if $(x, y) \in U^*$, then

$$|f^*(x) - f^*(y)| < \epsilon.$$

Equicontinuity steps up to provide a $V^* \in \mathcal{U}$ such that for each $g \in G$,

$$gV^*[x_0] \subseteq U^*[gx_0].$$

We can partition unity over C via $[0,1]$-valued test functions $f_1, \ldots, f_n$, where $f_k(x)$ vanishes off $g_k(V^*[x_0])$. If $x \in g_k V^*[x_0]$, then for every $g \in G$

$$g^{-1}(x) \in g^{-1} g_k(V^*[x_0])$$

and

$$g^{-1}(x) \in U^*[g^{-1} g_k(x_0)].$$

Since this is true for each and every $g \in G$, we conclude that

$$|f^*(g^{-1}(x)) - f^*(g^{-1} g_k(x_0))| < \epsilon$$

for all $g \in G$. So for any $g \in G$ and any $x \in X$, we have

$$f_k(x) \cdot f(x) \cdot f^*(g^{-1}(x)) \leq f_k(x) \cdot f(x) \cdot (f^*(g^{-1} g_k(x_0)) + \epsilon).$$

With (55) in mind, we see that

$$\begin{aligned} h(g) &= I\left(\left(\sum_{k \leq n} f_k\right) \cdot f \cdot \left(f^* \circ g^{-1}\right)\right) \\ &= \sum_{k \leq n} I(f_k \cdot f \cdot (f^* \circ g^{-1})) \\ &\leq \sum_{k \leq n} [f^*(g^{-1} g_k(x_0)) + \epsilon] I(f_k f). \end{aligned} \tag{56}$$

But $I(f^*) > 0$ and $f^*(g(x_0)) = f^*(g^{-1}(x_0))$ for each $g \in G$ so (55) soon unveils the relationship

$$f(g(x_0)) - K \leq \frac{h(g)}{I(f^*)} \leq \frac{\sum_{k\leq n} I(f_k \cdot f)[f^*(g_k^{-1}g(x_0)) + \epsilon]}{I(f^*)}.$$

Gx_0 is dense in X by G's weak transitivity, and so we deduce from this inequality

$$(57) \qquad f(x) - K \leq \frac{h(g)}{I(f^*)} \leq \sum_{k\leq n} \frac{I(f_k \cdot f)[f^*(g^{-1}(x)) + \epsilon]}{I(f^*)},$$

and this is for each $x \in X$ and $g \in G$.

Let

$$c_0 = \frac{I(f)}{I(f^*)} = \frac{I(\sum_{k\leq n} f_k \cdot f)}{I(f^*)} = \frac{\sum_{k\leq n} I(f_k \cdot f)}{I(f^*)}.$$

Then (57) is now

$$\begin{aligned} f(x) &\leq K + c_0[f^*(g^{-1})(x) + \epsilon] \\ &\leq K + c_0\epsilon + c_0 f^*(g^{-1}(x)), \end{aligned}$$

so

$$f(x) \leq (K + c_0\epsilon)\phi(x) + \frac{\sum_{k\leq n} I(f_k \cdot f)}{I(f^*)} f^*(g^{-1}(x)).$$

After all, we need only look at x's where $f(x) > 0$ and for such x's, $\phi(x) = 1$.

We now keep track of the relative size of nonzero test functions much like we did with the fractional covering functions in Chapter 8. Now if $f^\#$ is a test function, we define

$$N(f^\#, f^*) = \inf\left\{\sum_{k\leq n} c_k\right\},$$

where this infimum is taken over all $c_1, \dots, c_n$ such that for some $g_1^\#, \dots, g_n^\# \in G$,

$$f^*(x) \leq \sum_{k\leq n} c_k f^*(g_k^\# x).$$

Lemma 9.43 tells us that $N(f, f^*) < \infty$, and our work above soon translates into saying

$$N(f, f^*) \leq (K + \epsilon c_0) \cdot N(\phi, f^*) + c_0.$$

But $\epsilon > 0$ was arbitrary, so we can take it away at will—just let $\epsilon \searrow 0$, and we get

$$N(f, f^*) \leq K \cdot N(\phi, f^*) + c_0.$$

Now I is still G-invariant, and so any inequality such as,

$$f(x) \leq \sum_{k\leq n} c_k f^*(g_i^\# x),$$

soon becomes

$$I(f) \leq \sum_{k \leq n} c_k I(f^*).$$

It follows that

$$I(f) \leq N(f, f^*) I(f^*).$$

Here's what we have: Start with $x_0 \in X$ and $U \in \mathcal{U}$. Find $V \in \mathcal{U}$ so that

$$gV[x_0] \subseteq U[g(x_0)]$$

for each $g \in G$ with $\overline{V[x_0]}$ compact. Find a test function f^* that is positive at x_0, vanishes off $V[x_0]$, and is symmetric at x_0. Let I be any G-invariant integral. Let f be any test function with support contained in the compact set C. Let ϕ be a test function that is $[0,1]$-valued and is identically one on C. Then

$$N(f, f^*) - (\text{oscillation of } f \text{ on } \mathcal{U}) \cdot N(\phi, f^*) \leq \frac{I(f)}{I(f^*)} \leq N(f, f^*).$$

Now fix the test function f_0. Both f and f_0 are uniformly continuous, so for each n there is a $U_n \in \mathcal{U}$ such that

$$|f(x) - f(y)| < \frac{1}{n} \text{ and } |f_0(x) - f_0(y)| < \frac{1}{n},$$

so long as $(x, y) \in U_n$.

Applying the equicontinuity of G, find $V_n \in \mathcal{U}$ so that

$$gV_n[x_0] \subseteq U_n[gx_0]$$

for each $g \in G$, and $\overline{V[x_0]}$ is compact. Find a test function f_n^* that is positive at x_0, vanishes off $V_n[x_0]$ and is symmetric about x_0. Let C_0 and C be compact sets containing the support of f_0 and f_1, respectively, and let ϕ_0 and ϕ be $[0,1]$-valued test function with ϕ_0 identically one on C_0 and ϕ identically one on C. Then by our work summarized above we have

$$I(f_n^*)N(f, f_n^*) - \frac{1}{n}N(\phi, f_n^*)I(f_n^*) \leq I(f) \leq I(f_n^*)N(f, f_n^*) \tag{58}$$

and

$$I(f_n^*)N(f_0, f_n^*) - \frac{1}{n}N(\phi_0, f_n^*)I(f_n^*) \leq I(f_0) \leq I(f_n^*)N(f_0, f_n^*). \tag{59}$$

It's part of the nature of the counting function N that if f_1, f_2 and f_3 are test functions, then

$$N(f_1, f_3) \leq N(f_1, f_2)N(f_2, f_3).$$

Hence,

$$N(\phi_0, f_n^*) \leq N(\phi_0, f_0)N(f_0, f_n^*).$$

Now $N(f_0, f_n^*) > 0$ for each n since $I(f_0) > 0$.

If $N(\phi_0, f_n^*) = 0$, then

$$N(f_0, f_n^*) - \frac{1}{n} N(\phi_0, f_n^*) > 0.$$

If $N(\phi_0, f_n^*) > 0$ and $n > N(\phi_0, f_0)$, then

$$N(f_0, f_n^*) - \frac{1}{n} N(\phi_0, f_n^*) > N(f_0, f_n^*) - \frac{N(\phi_0, f_n^*)}{N(\phi_0, f_0)} \geq 0.$$

Hence whenever $n > N(\phi_0, f_0)$ we have (from (58) and (59))

$$\frac{I(f_n^*)[N(f, f_n^*) - \frac{1}{n} N(\phi, f_n^*)]}{I(f_n^*) N(f_0, f_n^*)} \leq \frac{I(f)}{I(f_0)} \leq \frac{I(f_n^*) N(f, f_n^*)}{I(f_n^*)[N(f_0, f_n^*) - \frac{1}{n} N(\phi_0, f_n^*)]},$$

or on cancelling $I(f_n^*)'s$ whenever possible,

$$\frac{N(f, f_n^*)}{N(f_0, f_n^*)} - \frac{1}{n} \frac{N(\phi, f_n^*)}{N(f_0, f_n^*)} \leq \frac{I(f)}{I(f_0)} \leq \frac{N(f, f_n^*)}{N(f_0, f_n^*)} \left[\frac{1}{1 - \frac{1}{n} \frac{N(\phi_0, f_n^*)}{N(f_0, f_n^*)}} \right].$$

Now regardless of n,

$$\frac{N(\phi_0, f_n^*)}{N(f_0, f_n^*)} \leq N(\phi_0, f_0) < \infty$$

and

$$\frac{N(\phi, f_n^*)}{N(f_0, f_n^*)} \leq N(\phi, f_0) < \infty.$$

Letting $n \to \infty$, we get

$$\limsup_n \frac{N(f, f_n^*)}{N(f_0, f_n^*)} \leq \frac{I(f)}{I(f_0)} \leq \liminf_n \frac{N(f, f_n^*)}{N(f_0, f_n^*)}.$$

Think about it! The f_n^*'s were chosen and the above inequality was shown—for any Haar integral. Thus if J is also a Haar integral,

$$\frac{J(f)}{J(f_0)} = \frac{I(f)}{I(f_0)}.$$

So for a fixed $f_0 \in \mathcal{K}^+(X)$, $f_0 \neq 0$,

$$J(f) = \frac{J(f_0)}{I(f_0)} I(f). \qquad \square$$

7. Notes and Remarks

The paper of R. Steinlage [**121**] included earlier work of S. Banach (see Chapter 4) and I. E. Segal [**118**]. The simple condition that the group G separate compact subsets of X was noted by E. Hewitt in his extremely positive review of Steinlage's paper.

D. Ross [**107**] spoke to invariance relative to a collection G of local homeomorphisms (closed under inverses) on a compact Hausdorff space X. Using a variation of the theme of separating compact sets, Ross showed the existence of a G-invariant Borel probability on X. His arguments are nonstandard.

P. Niemiec [**95**] studied G-invariant regular Borel probabilities on a compact Hausdorff space X, where G is an equicontinuous *semigroup* of maps from X to itself. Of particular interest is Niemiec's discovery that if the collection of G-invariant regular Borel probabilities is nonempty, then it is a Choquet simplex.

Chapter 10

Oxtoby's View of Haar Measure

1. Invariant Measures on Polish Groups

In this short chapter we broach the subject of invariant Borel measures in groups which are *not* locally compact. More precisely, we look at separable metrizable topological groups where the topology is Polish; i.e., the group topology is that of a complete metric space. In addition to the well-developed theory of Polish spaces, this class of topological groups seems regular enough to offer hope of a reasonable theory.

The three theorems presented here indicate that without local compactness, invariant measures can be quite pathological. These results are due to J. Oxtoby and are, we believe, beautiful and surprising.

In this discussion a Borel measure will mean a countably additive, $[0, \infty]$-valued set function defined on each Borel set, vanishing on singletons, and having at least one nontrivial value.

Our first result indicates that there are some surprises in store for us when discussing invariant measures in nonlocally compact topological groups.

Theorem 10.1 (Oxtoby and Ulam). *Let G be a complete, separable metrizable topological group that is* not *locally compact, and let m be any left-invariant Borel measure on G having at least one nontrivial value. Then every nonempty open set in G contains uncountably many disjoint congruent compact sets of equal finite positive m-measure.*

Proof. Suppose B is a Borel set for which $0 < m(B) < \infty$. Let U be a nonempty open subset of G.

Since G is separable and $\{xU : x \in G\}$ is an open cover of G, at least one of the sets xU intersects B in a set of positive m-measure, say $m(aU \cap B) > 0$. Then $a^{-1}B \cap U$ is a Borel subset of U having positive m-measure. It follows that $a^{-1}B \cap U$ contains a compact set C with $0 < m(C) < \infty$. In fact $\tilde{m}(E) = m(E \cap (a^{-1}B) \cap U)$ is a finite Borel measure on G so Ulam's theorem applies to ensure $\tilde{m}$ is regular.

Consider the subgroup $G_1 = \langle C \rangle$ generated by C. Keeping in mind the compactness of $C_1 \cdot C_2$ and C_1^{-1} given that of C_1 and C_2, we see that G_1 is the countable union of $\bigcup_n C_n$ of an ascending sequence of compact sets:

$$C_1 = C \cup C^{-1}, \quad C_2 = C_1 \cdot C_1, \quad C_3 = C_2 \cup C_2^{-1}, \quad C_4 = C_3 \cdot C_3, \ldots .$$

Now compact sets in nonlocally compact groups are nowhere dense, so G_1 is of the first category in G, and so too are all of G_1's left cosets. Let V be an open subset of G that contains the identity and satisfies $V \cdot C \subseteq U$. The left cosets of G_1 of the form vG_1, where $v \in V$, cover V. But V is of the second category, as is any open subset of a complete separable metric space, thanks to Baire's category theorem. Since each left coset of G_1 is of the first category, the cosets vG_1 $(v \in V)$ that are distinct (and hence disjoint) are uncountable in number.

Let's stop to admire what we've done: each of the left cosets vG_1 contains a set vC that's congruent to C and satisfies

$$vC \subseteq V \cdot C \subseteq U,$$

so each of the vC's is contained in U. The vC's are pairwise disjoint, congruent to C, uncountable in number, and have positive finite m-measure. □

Remarkable. And true.

Corollary 10.2. *If X is an infinite-dimensional separable Banach space, then there does NOT exist a translation invariant Borel measure m on X for which $m(B_X) < \infty$.*

The surprises found in the measure theory of invariant Borel measures on complete separable metric groups are not done yet. Here is a stunning refinement of the Oxtoby–Ulam Theorem.

Theorem 10.3 (Oxtoby). *Let G be a complete separable metric topological group that is not locally compact, and suppose m is a translation invariant Borel measure in G. Then every nonempty open set in G contains a compact* *set which is the union of uncountably many disjoint congruent compact sets of (equal) finite positive m-measure.*

Proof. Let U be a nonempty open subset of G. Apply the Oxtoby–Ulam result above to find a compact set $C \subseteq U$ with $0 < m(C) < \infty$, and again let $G_1 = \langle C \rangle$ be the subgroup of G generated by C. As before, pick V to be an open subset of G that contains the identity e of G and satisfies $V \cdot C \subseteq U$.

Then $C \cdot C^{-1}$ is compact, and so is nowhere dense in G.

It follows that for each $\epsilon > 0$, there is an open set $V_\epsilon \subseteq V$ of diameter less than ϵ, containing e with $V_\epsilon \not\subseteq C \cdot C^{-1}$; therefore, we can find a sequence (a_n) in G that converges to e with $a_n \notin C \cdot C^{-1}$ for each n.

Look at the collection of all points of the form

$$a_1^{j_1} a_2^{j_2} \cdots a_n^{j_n},$$

where $j_1, \ldots, j_n = 0$ or 1 and $n \in \mathbb{N}$. Let Γ be the closure of this countable collection. We plan to show how to force Γ to be a homeomorph of the Cantor set by judiciously controlling the rate at which (a_n) converges to e. Then, with a bit of added wisdom, we will show how to force Γ to be inside of V. Finally, playing Solomon to the hilt, we will arrange that

$$(\Gamma^{-1} \cdot \Gamma) \cap (C \cdot C^{-1}) = \{e\}.$$

Once we have done this, we will know that the different points of Γ generate disjoint translates of C: indeed, if $x, y \in \Gamma$ are different and $z \in (xC) \cap (yC)$, then there must be $c_x, c_y \in C$ so that

$$xc_i = z = yc_y,$$

which says that the member $y^{-1}x$ of $\Gamma^{-1} \cdot \Gamma$ is also

$$y^{-1}x = c_y c_x^{-1} \in C \cdot C^{-1},$$

and so $y^{-1}x = e$, and that is not as it ought to be. Of course, $\Gamma \cdot C$ is the desired compact subset of U. It is even the union of a *continuum* of disjoint translates of C.

We now have $e = \lim_n a_n$ where $a_n \notin C \cdot C^{-1}$. Let a_{i_1} be the first a_n so that the open ball

$$O\left(e; 2\rho\left(a_{i_1}, e\right)\right) \subseteq V,$$

centered at e of the radius twice the distance ρ from a_{i_1} to e is inside V.

Suppose $a_{i_1}, a_{i_2}, \ldots, a_{i_n}$ have been selected.

Let

$$A_n = \left\{ a_{i_1}^{j_1} \cdot a_{i_2}^{j_2} \cdots a_{i_n}^{j_n} : j_1, j_2, \ldots, j_n = 0 \text{ or } 1 \right\},$$

$$d_n = \min\left\{\rho(x,y)\ :\ x \neq y,\ x,\ y \in A_n\right\},$$

and

$$\delta_n = \frac{1}{4}\min\left\{\rho(a_{i_n}, C\cdot C^{-1}), \rho(a_{i_n}^{-1}, C\cdot C^{-1}), d_n\right\}.$$

Then $a_{i_{n+1}}$ is the first a_n with $i_{n+1} > i_n$ so that for any $a \in A_n^{-1}\cdot A_n$,

$$\rho\ (a,\ aa_{i_{n+1}}),\ \rho\ (a,\ a_{i_{n+1}}^{-1}a),\ \rho\ (a,\ a_{i_{n+1}}^{-1}aa_{i_{n+1}}) < \delta_n.$$

Remember $A_n^{-1}A_n$ is a finite set so we can arrange the choice of $a_{i_{n+1}}$ without much fuss.

Now Γ is the closure of the set

$$\left\{a_{i_1}^{j_1}\cdot a_{i_2}^{j_2}\cdots a_{i_n}^{j_n} : n \in \mathbb{N}, (j_k) \in \{0,1\}^n\right\}.$$

Enough already with the subscripts! We have done our pruning so let's just call $(a_{i_n}), (a_n)$, and proceed.

It is worth nothing that $e \in A_n^{-1}A_n$, so

$$\rho(e, a_{n+1}) = \rho(e, ea_{n+1}) < \delta_n,$$

which in turn implies that

$$\begin{aligned} d_{n+1} &= \min\ \{\rho(x,y) : x, y \in A_{n+1}, x \neq y\} \\ &\leq\ \rho\ (e, a_{n+1})\ \leq\ \delta_n \\ &\qquad\ \uparrow\ \uparrow \\ &\quad \text{each in } A_{n+1}. \end{aligned}$$

From this we see that

$$4\delta_{n+1} \leq d_{n+1} \leq \delta_n,$$

and the sequence (δ_n) tends to zero fast—geometrically, in fact.

Lets run with this for a bit. Take $(j_n)_n \in \{0,1\}^{\mathbb{N}}$. Look at the sequence

$$(x_n) = (a_1^{j_1}a_2^{j_2}\cdots a_n^{j_n}).$$

Notice that

$$\begin{aligned} \rho(x_n, x_{n+k}) &\leq \rho(x_n, x_{n+1}) + \rho(x_{n+1}, x_{n+2}) + \cdots + \rho(x_{n+k-1}, x_{n+k}) \\ &\leq \rho(a_1^{j_1}a_2^{j_2}\cdots a_n^{j_n},\ a_1^{j_1}a_2^{j_2}\cdots a_n^{j_n}a_{n+1}^{j_{n+1}}) \\ &+ \rho(a_1^{j_1}a_2^{j_2}\cdots a_n^{j_n}a_{n+1}^{j_{n+1}}, a_1^{j_1}a_2^{j_2}\cdots a_{n+1}^{j_{n+1}}a_{n+2}^{j_{n+2}}) \\ &+ \cdots + \rho(a_1^{j_1}\cdots a_{n+k-1}^{j_{n+k-1}}, a_1^{j_1}\cdots a_{n+k-1}^{j_{n+k-1}}a_{n+k}^{j_{n+k}}). \end{aligned}$$

A typical term herein is

$$\rho(a_1^{j_1}a_2^{j_2}\cdots a_j^{j_l},\ a_1^{j_1}a_2^{j_2}\cdots a_j^{j_l}a_{j+1}^{j_{l+1}}),$$

where $n \le l \le n+k-1$; this term is of the form

$$\rho(a, a\,a_{l+1}^{i_{l+1}})$$

for some $a \in A_n$ and $j_{l+1} = 0$ or 1. So it is either 0 or

$$\rho(a, a\,a_{l+1}).$$

Either way, it is less than or equal to δ_n. The result is

$$\rho(x_n, x_{n+k}) \le \delta_n + \delta_{n+1} + \cdots + \delta_{n+k-1} \le \delta_n + \frac{\delta_n}{4} + \frac{\delta_{n+1}}{4^2} + \cdots + \cdots \le \frac{4}{3}\delta_m.$$

It follows that the sequence $(a_1^{j_1} a_2^{j_2} \cdots a_n^{j_n})_n$ is a ρ-Cauchy sequence regardless of $(j_n)_n \in \{0,1\}^{\mathbb{N}}$. It follows that this sequence converges to some $x((j_n)) \in \Gamma$.

Now if $(j_n')_n$ is a different sequence in $\{0,1\}^{\mathbb{N}}$ and $y_n = a_1^{j_1'} \cdots a_n^{j_n'}$, then (y_n) converges to some $y((j_n')) \in \{0,1\}^{\mathbb{N}}$. If $j_k \ne j_k'$ while $j_1 = j_1', \ldots, j_{k-1} = j_{k-1}'$, then

$$4\delta_k \le d_k \le \rho(x_k, y_k) \le \rho(x_k, x) + \rho(x, y) + \rho(y, y_k) \le \frac{4}{3}\delta_k + \rho(x, y) + \frac{4}{3}\delta_{k_j}.$$

It follows that

$$\rho(x, y) \ge \frac{4}{3}\delta_k.$$

From this we see that Γ is homeomorphic to the Cantor set. After all, $\phi((j_n)_n) = (x(j_n))$ is a continuous, one-to-one surjection of $\{0,1\}^{\mathbb{N}}$ onto Γ. To see that

$$\Gamma \subseteq V,$$

we take any $a \in A_m$. Then for some $j_1, \ldots, j_n \in \{0,1\}$,

$$\begin{aligned}
\rho(e, a) &= \rho(e, a_1^{j_1} \cdots a_n^{j_n}) \\
&\le \rho(e, a_1^{j_1}) + \rho(a_1^{j_1}, a_1^{j_1} a_2^{j_2}) + \cdots + \rho(a_1^{j_1} \cdots a_{n-1}^{j_{n-1}}, a) \\
&\le \rho(e, a_1) + \delta_1 + \cdots + \delta_{n-1} \\
&\le \rho(e, a_1) + \frac{\rho(e_1 a_1)}{4} + \cdots + \frac{\rho(e_1 a_1)}{4^{n-1}},
\end{aligned}$$

by the definition of δ_k's and the presence of e in CC^{-1}. We have that

$$\rho(e, a) \le \rho(e_1 a_1)(1 + \frac{1}{4} + \cdots) = \frac{4}{3}\rho(e_1 a_1).$$

It follows that for any $\gamma \in \Gamma$

$$\rho(e, \gamma) \le \frac{4}{3}\rho(e, a_1) < 2\rho(e, a_1)$$

and $\gamma \in V$. Therefore $\Gamma \subseteq V$.

Now we can establish that

$$\Gamma^{-1}\Gamma \cap C\,C^{-1} = \{e\}.$$

Take $z \in \Gamma^{-1}\Gamma$. Then

$$z = \lim_n b_n,$$

where $b_n \in A_n^{-1}A_n$. If $z \neq e$, then sooner or later $b_k \neq e$ as well. Let's look at how b_k differs from e when $b_k \in A_k^{-1}A_k$:

$$b_k = (a_1^{i_1} \cdots a_k^{i_k})^{-1}(a_1^{j_1} \cdots a_k^{j_k}),$$

where $i_1, \ldots, i_k, j_1, \ldots, j_k \in \{0, 1\}$. So

$$b_k = a_k^{-i_k} \cdots a_1^{-i_1} a_1^{j_1} \cdots a_k^{j_k} j,$$

but if $i_1 = j_1, \ldots, i_k = j_k$, then $b_n = e$. So suppose b_k is the first b_j that is not e and so k is the first time $i_k \neq j_k$. At this step,

$$b_k = a_k^{-i_k} \cdots a_1^{-i_1} a_1^{j_1} \cdots a_k^{j_k} = \cdots = a_k^{-i_k} a_k^{j_k}.$$

But $i_k, j_k \in \{0, 1\}$ and $i_k \neq j_k$, so

$$b_k = a_k^{-i_k} a_k^{j_n} = a_k^{\pm 1}.$$

It follows that

$$\rho(b_k, z) \leq \delta_k + \delta_{k+1} + \cdots \leq \frac{4}{3}\delta_k \leq \frac{\rho(b_k, C \cdot C^{-1})}{3}.$$

Hence, if $z \in C \cdot C^{-1}$,

$$\rho(b_k, C \cdot C^{-1}) \leq \rho(b_k, z) \leq \frac{\rho(b_k, C \cdot C^{-1})}{3},$$

so if $z \in \Gamma^{-1} \cdot \Gamma$ and $z \neq e$, then $z \notin CC^{-1}$, and so

$$\Gamma^{-1} \cdot \Gamma \cap C \cdot C^{-1} = \{e\}.$$ □

Corollary 10.4. *If X is an infinite-dimensional separable Banach space, then there does not exist a translation invariant Borel measure on X that's finite on all compact subsets of X.*

On the positive side of the ledger, we have the following theorem:

Theorem 10.5 (Oxtoby). *In any complete separable metric group G which is dense in itself, there exists a left invariant Borel measure having a nontrivial value.*

Proof. We start by applying the Birkhoff–Kakutani metrization theorem to find a left invariant metric on the group. Now in so doing we may well lose the completeness, but we gain some flexibility. Let's agree that in this new left invariant metric. we will denote the distance between $x, y \in G$ by (x, y).

Now we will begin the construction in earnest. The knowledgeable and attentive reader will recognize similarities with the construction of Hausdorff measure.

Let (U_n) be a sequence of open sets whose diameters (with respect to $(\, , \,)$) tend to zero, and let (w_n) be a sequence of positive numbers which descends to zero. Let $\mathcal{U}$ denote the family of (open) subsets of G that are congruent to one of the U_n's.

Let $A \subseteq G$ and $r > 0$. We define

$$\lambda_r(A) = \inf\left\{\sum w_{n_i} : A \subseteq \bigcup b_i U_{n_i}, \operatorname{diam}(U_{n_i}) < r\right\}.$$

Notice that the smaller the r, the fewer the coverings of A via members of $\mathcal{U}$, and so the bigger is $\lambda_r(A)$. We also define

$$m^*(A) = \lim_{r \to 0} \lambda_r(A).$$

It may be that $m^*(A) = +\infty$, but we will just have to live with this.

It is reasonably straightforward to verify that m^* is a metric outer measure that vanishes at singletons and is left invariant.

Let m be the countably additive Borel measure on G generated by m^* à la Carathéodory. Now we are face-to-face with the crux of the matter: Does there exist an A such that $0 < m(A) < \infty$?

We will show how to find a Cantor set with measure 1. It will take some careful tinkering, but it is possible. So, we will construct a family

$$\left\{C(n; i_1, \ldots, i_n) : n \in \mathbb{N}, i_1, \ldots, i_n i\{0,1\}\right\}$$

of compact subsets of G so that

(i) for any m,
$C(n; i_1, \ldots, i_n) = C(n+1; i_1, \ldots, i_n, 0) \cup C(n+1; i_1, \ldots, i_k, 1)$;

(ii) all of the 2^n C's of rank n are congruent;

(iii) the diameters d_n of each of the rank n sets tend to zero;

(iv) any two of the sets of rank n are separated by $3d_n$ and so are, in particular, disjoint.

The resulting set C satisfies

$$C = C(1;0) \cup C(1;1) = \bigcup_{i_1, \ldots, i_n \in \{0,1\}} C(n; i_1, \ldots, i_n)$$

and must be homeomorphic to the Cantor set.

Since G is dense in itself, there exists a sequence (x_n) in G so that for each n

$$0 < (e, x_{n+1}) < \frac{(e, x_n)}{9}.$$

Let $i_1, i_2, \ldots, i_n \in \{0, 1\}$ and

$$a_{i_1,\ldots,i_n} = x_1^{i_1} x_2^{i_2} \cdots x_n^{i_n}.$$

We define

$$A(n; i_1, \ldots, i_n) = \{a_{i,\ldots,i_{n+p}} : p \geq 0\}.$$

So $A(n; i_1, \ldots, i_n)$ is the set of all $a_{i_1}, \ldots, a_{i_{n+p}}$ whose first indices are equal to $i_1, \ldots, i_n$.

If

$$a_{i_1,\ldots,i_{n+p}}, a_{j_1,\ldots,j_{n+q}} \in A(n; i_1, \ldots, i_n),$$

then

$$a_{i_1,\ldots,i_{n+p}} = x_1^{i_1} x_2^{i_2} \cdots x_n^{i_n} x_{n+1}^{i_{n+1}} \cdots x_{n+p}^{i_{n+p}}$$

and

$$a_{j,\ldots,j_{n+q}} = x_1^{j_1} x_2^{j_2} \cdots x_n^{j_n} x_{n+1}^{j_{n+1}} \cdots x_{n+q}^{j_{n+q}}$$

satisfy $i_1 = j_1, j_2 = j_2, \ldots, i_n = j_n$. We define $C(n; i_1, \ldots, i_n)$ by

$$C(n; i_1, \ldots, i_n) = \overline{A(n; i_1, \ldots, i_n)}.$$

Notice that

$$A(n; i_1, \ldots, i_n) = A(n+1; i_1, \ldots, i_n, 0) \cup A(n+1; i_1, \ldots, i_n, 1)$$

and

$$A(n; i_1, \ldots, i_n) = a_{i_1,i_2,\ldots,i_n} A(n; 0, 0, \ldots, 0).$$

Statements (i) and (ii) follow from their observations.

Next, let's play with estimating d_n. Using the left invariance of the metric $(\ ,\ )$, we see that

$$(e, y_1 y_2) \leq (e, y_1) + (y_1, y_1 y_2) = (e, y_1) + (e, y_2).$$

Further,

$$(e, y) = (y^{-1}, e) = (e, y^{-1});$$

so for $k = -1, 0, 1$ we have

$$(e, y^k) \leq (e, y).$$

Now a simple induction argument shows that for any $k_1, k_2, \ldots, k_n = -1, 0, 1$,

$$(e, y_1^{k_1} y_2^{k_2} \cdots y_n^{k_n}) \leq (e, y_1) + (e, y_2) + \cdots + (e, y_n).$$

Take two points of $A(n;0,0,\ldots,0)$, say $x_{n+1}^{i_{n+1}}\cdots x_{n+p}^{i_{n+p}}$ and $x_{n+1}^{j_{n+1}}\cdots x_{n+q}^{j_{n+q}}$, where the i's and j's are $-1,0$ or 1. Because the exponent 0 is allowed, we can assume $p=q$. How far apart are these points? Well

$$\begin{aligned}(x_{n+1}^{i_{n+1}}\cdots x_{n+p}^{i_{n+p}}, x_{n+1}^{j_1}\cdots x_{n+p}^{j_{n+p}}) &= (e, x_{n+1}^{-i_{n+1}}x_{n+2}^{-i_{n+2}}\cdots x_{n+p}^{-i_{n+p}}x_{n+1}^{j_1}\cdots x_{n+p}^{j_{n+p}})\\ &\le 2\Big[(e,x_{n+1})+(e,x_{n+2})+\cdots+(e,x_{n+p})\Big]\\ &\le 2\left[\frac{(e,x_n)}{9}+\frac{(e,x_n)}{9^2}+\cdots+\frac{(e,x_n)}{9^p}\right]\\ &= \left(\frac{2}{9}+\frac{2}{9^2}+\cdots+\frac{2}{9^p}\right)(e,x_n)\\ &< 2\sum_k 9^{-k}(e,x_n) < \frac{(e,x_n)}{4}.\end{aligned}$$

Therefore $d_n \le \frac{(e,x_n)}{4}$.

But both e and x_{n+1} are in $A(n;0,0,\ldots,0)$, so

$$(e,x_{n+1}) \le d_n$$

and

$$d_{n+1} \le \frac{(e,x_{n+1})}{4} \le \frac{d_n}{4},$$

and (iii) follows.

How far apart are distinct sets $A(n;i_1,\ldots,i_n)$ and $A(n;j_1,\ldots,j_n)$ of the same rank? Take the typical member of each of these sets, say $a_{i_1,\ldots,i_p}\in A(n;i_1,\ldots,i_n)$ and $a_{j_1,\ldots,j_q}\in A(n;j_1,\ldots,j_n)$, where once again, because all the i's and j's are $-1,0$, or, 1, we can assume $p=q$. Since the sets $A(n;i_1,\ldots,i_n)$ and $A(n;j_1,\ldots,j_n)$ are distinct, the indices of $a_{i_1,\ldots,i_p}$ and $a_{j_1,\ldots,j_p}$ differ at some index $k\le n$. We may as well assume that $i_k=0$ and $j_k=1$.

Look at

$$u = x_{k+1}^{i_{k+1}}\cdots x_p^{i_p}, \qquad v = x_k x_{k+1}^{j_{k+1}}\cdots x_p^{j_p}.$$

Of course

$$(u,v) = (a_{i_1,\ldots,i_p}, a_{j_1,\ldots,j_p})$$

and

$$\begin{aligned}
(e, x_k) &\le (e, u) + (u, v) + (v, x_k) \\
&= (e, x_{k+1}^{i_{k+1}} \cdots x_p^{i_p}) + (u, v) + (x_k x_{k+1}^{j_{k+1}} \cdots x_p^{j_p}, x_k) \\
&= (e, x_{k+1}^{i_{k+1}} \cdots x_p^{i_p}) + (u, v) + (x_{k+1}^{j_{k+1}} \cdots x_p^{j_p}, e) \\
&= (e, x_{k+1}^{i_{k+1}} \cdots x_p^{i_p}) + (u, v) + (e, x_{k+1}^{j_{k+1}} \cdots x_p^{j_p}) \\
&\le (u, v) + 2(e, x_k)\Big[(e, x_{k+1}) + \cdots + (e, x_p)\Big] \\
&\le (u, v) + 2(e, x_k)\left[\frac{1}{9} + \frac{1}{9^2} + \cdots + \frac{1}{9^{p-k}}\right] \\
&\le (u, v) + 2(e, x_k)\sum_n 9^{-n} \\
&\le (u, v) + \frac{(e, x_k)}{4}.
\end{aligned}$$

So $(u, v) \ge \frac{3}{4}(e, x_k) \ge \frac{3}{4}(e, x_n) \ge 3d_n$, and (iv) follows.

(Semi-)finally, each $C(n; i_1, \ldots, i_n)$ is the union of 2^{p-n} sets $C(p; i_1, \ldots, i_p)$, each having diameter d_p, where $d_p \to 0$. It follows that each $C(n; i_1, \ldots, i_n)$ is totally bounded.

Therefore $C(n; i_1, \ldots, i_n)$ is closed and totally bounded. If G were $(\ ,\ )$-complete, then we would be done. But we mentioned earlier that in this new metric $(\ ,\ )$, we had no assurances that G is complete. Did we throw out the baby with the bath water? Not if we follow Oxtoby's able leadership; we didn't.

Embed $(G, (\ ,\ ))$ into a complete metric space $\mathcal{G}$. Let $\mathcal{C}(n; i_1, \ldots, i_n)$ denote the closure in $\mathcal{G}$ of $C(n; i_1, \ldots, i_n)$. The point is just this: If we work a bit harder, we can rig things so that $C(n; i_1, \ldots, i_n)$ is already closed (hence, complete) in $\mathcal{G}$!

To be sure, denote by ρ the original complete metric that generates G's topology. With revisionist behavior so acceptable, take a careful look at our choice of (x_n). We used the dense-in-itself character of G to pick (x_n) so that for each n, $0 < (e, x_{n+1}) < \frac{(e, x_n)}{9}$. But the dense-in-itself character of G is topological, and so we could also have insisted, regardless of $i_1, \ldots, i_n = 0, 1$, that

$$\rho(a_{i_1,\ldots,i_n}, a_{i_1,\ldots,i_n} x_{n+1}) < 2^{-n} \tag{60}$$

also. But (60) has added benefits; indeed, regardless of the sequence (i_n) of indices $(i_n = 0, 1)$, the sequence $(a_{i_1,\ldots,i_n})$ is ρ-Cauchy! After all,

$$a_{i_1,\ldots,i_k} = x_1^{i_1} x_2^{i_2} \cdots x_n^{i_n},$$

so

$$\rho(a_{i_1,\ldots,i_n}, a_{i_1,\ldots,i_n,i_{n+1}}) = \rho\left(x_1^{i_1}x_2^{i_2}\cdots x_n^{i_n}, x_1^{i_1}x_2^{i_2}\cdots x_n^{i_n}x_{n+1}^{i_{n+1}}\right).$$

But

$$\rho\left(x_1^{i_1}x_2^{i_2}\cdots x_n^{i_n}, x_1^{i_1}x_2^{i_2}\cdots x_n^{i_n}x_{n+1}^{i_{n+1}}\right)$$
$$= \begin{cases} \rho(x_1^{i_1}x_2^{i_2}\cdots x_n^{i_n}, x_1^{i_1}x_2^{i_2}\cdots x_n^{i_n}), & \text{if } i_{n+1} = 0, \\ \text{or} & \\ \rho(x_1^{i_1}x_2^{i_2}\cdots x_n^{i_n}, x_1^{i_1}x_2^{i_2}\cdots x_n^{i_n}x_{n+1}), & \text{if } i_{n+1} = 1. \end{cases}$$

Whichever the case, the result is less than 2^{-n} by our judicious revisionist choices. It follows that if $m > n$, then

$$\rho(a_{i_1,\ldots,i_n}, a_{i_1,\ldots,i_m}) < 2^{n-m},$$

and $(a_{i_1,\ldots,i_n})_{n=1}^{\infty}$ is ρ-Cauchy regardless of the choice of indices (i_n).

Hence $(a_{i_1,\ldots,i_n})$ converges to a member of G modulo the adjustments made above in selecting (x_n). But any point of accumulation of C in G is the limit of just such a sequence and so must be in G. So, too, is all of C.

So we may suppose that C is complete (in G) and totally bounded, and C is, indeed, a Cantor set.

As yet we have not specified the weights (w_n); we will do so in such a way that $m(C) = 1$.

By (iv) we can find a sequence (ϵ_n) of positive numbers descending to zero so that for any two distinct components $C(n; i_1, \ldots, i_n)$ and $C(n; j_1, \ldots, j_n)$ of C of the same rank n we have

$$\operatorname{dist}_{(\ ,\ )}\Big(C(n, i_1, \ldots, i_m), C(n, j_1, \ldots, j_n)\Big) > d_n + 2\epsilon_n.$$

Let

$$U_n = \Big\{x \in G : d_{(\ ,\ )}(x, C(n; 0, 0, \ldots, 0)) < \epsilon_n\Big\},$$

and let $w_n = 2^{-n}$. We know the (,)-diameter of U_n is less than or equal to

$$\operatorname{diam}_{(\ ,\ )} C(n; 0, 0, \ldots, 0) + 2\epsilon_n \leq d_n + 2\epsilon_n \to 0.$$

It follows that any set congruent to U_n can overlap at most one set $C(n; i_1, \ldots, i_n)$ of rank n.

Let $r > 0$. Choose n so that diam $U_n < r$.

From (iii), we know that C can be covered by 2^n sets congruent to U_n; hence,

$$\lambda_r(C) \leq 2^n \cdot w_n = 1.$$

On the other hand, if we cover C by a sequence of sets each of the form $b_iU_{n_i}$, then we can always assume the covering is finite:

$$C \subseteq b_1U_{n_1} \cup \cdots \cup b_kU_{n_k}.$$

Set

$$p = \max\{n_1, \ldots, n_k\},$$

and let q_i be the number of components of $C(p; i_1, \ldots, i_p)$ that are overlapped by $b_iU_{n_i}$. Since

$$C \subseteq b_1U_{n_1} \cup \cdots \cup b_kU_{n_k},$$

$$q_1 + q_2 + \cdots + q_k \geq 2^p.$$

After all, to cover C, you have to cover each of the 2^p components of $C(p; i_1, \ldots, i_p)$. But $b_iU_{n_i}$ can overlap at most one component of rank n_i, and so at most 2^{p-n_i} components of rank p. It follows that

$$2^{p-n_i} \geq q_i.$$

So

$$w_{n_i} = 2^{-n_i} = \frac{2^{p-n_i}}{2^p} \geq \frac{q_i}{2^p}$$

and

$$\sum w_{n_i} \geq 2^{-p} \sum q_i \geq 2^{-p}2^p = 1.$$

Hence,

$$\lambda_r(C) \geq 1.$$

Therefore,

$$\lambda_r(C) = 1 \text{ for all } r > 0$$

and

$$m(C) = 1. \qquad \square$$

2. Notes and Remarks

We have presented only a few of the results found in Oxtoby's wonderful paper [**97**]. Here are a couple of others worthy of close study.

Theorem 10.6 (Oxtoby). *Let H be a Borel subset of the metric group G, and suppose m is a Borel measure on H such that*

$$m(xA) = m(A)$$

for any Borel set $A \subseteq H$ and any $x \in G$ with $xA \subseteq H$. Then there exist two left invariant Borel measures λ and ν on G, both of which are extensions of m such that if μ is any other such extension of m, then

$$\lambda(A) \leq \mu(A) \leq \nu(A)$$

for each Borel subset A of G.

Theorem 10.7. *If a metric group G contains a locally compact subgroup H that is dense in itself, then there exists a left invariant Borel measure on G which is an extension of the Haar measure on H.*

This classic paper of Oxtoby is chock-full of delicate examples, delicious observations, and curious results. We highly recommend it to youthful readers.

Corollary 10.4 might lead one to believe that the study of invariance of Borel sets in (separable) infinite-dimensional Banach spaces leads to a dead end. Such is not the case.

Motivated by classical results related to differentiability properties of convex functions and Lipschitz maps on $\mathbb{R}^n$, a number of mathematicians have isolated various notions of smallness that have been useful in the classification of Banach spaces—both linear and nonlinear.

Let B denote a separable infinite-dimensional Banach space. P. Mankiewicz [**74**] introduced cube measures. A *cube measure* on X is the distribution of B-valued random variables of the form

$$x + \sum_n X_n e_n,$$

where $x, e_1, \ldots, e_n \in B$,

$$\sum_n ||e_n|| < \infty,$$

and the sequence (X_n) is a sequence of uniformly distributed independent $[0,1]$-valued random variables (on some probability space). *Cube null sets* are those Borel subsets of B that are null sets for every nondegenerate cube measure on B.

Next, N. Aronszajn [**2**], in his study of Gateaux differentiability of Lipschitz mappings, considered the following: given $e_1, \ldots, e_n, \ldots \in B$, let $A(e_1, \ldots, e_n, \ldots)$ consist of those Borel sets E which are the union $\bigcup_n F_n$ of Borel sets E_n such that E_n is "null" on every line in the direction e_n. So for each $x \in B$, the Lebesgue measure of the set

$$\{t \in \mathbb{R} : a + te_n \in E_n\}$$

is zero. A Borel set is *Aronszajn null* if E belongs to $A(e_1, \ldots, e_n, \ldots)$ whenever the linear span of the e_n's is dense in B.

Again, R. R. Phelps [**99**] isolated the *Gaussian null sets* as those Borel sets in B that are null for every Gaussian measure.

Each notion of nullity comes with accompanying results about the differentiability of Lipschitz maps defined on B. There are similarities, to be sure.

In M. Csörnyei [**22**] these similarities were explained when she proved that in every B, the σ-ideals of Aronszajn null sets, Gaussian null sets, and cube null sets coincide.

Another notion of smallness has gained considerable interest—Haar null sets, a notion from the fertile mind of J. P. R. Christensen [**19**]. A Borel set A in an abelian Polish group G is said be *Haar null* if there is a Borel probability μ on G such that $\mu(x + A) = 0$ for each $x \in G$. Christensen establishes some basic (expected) properties of the Haar null sets, but each is shown only with considerable difficulty. The abelian nature of the group *simplifies* things.

Haar null sets have been studied in more general Polish groups and related to classical results. Almost always the arguments are nontrivial and intricate. We refer those interested to the papers of P. Dodos [**30**] and [**29**], É. Matheron and M. Zelený [**79**], and S. Solecki [**120**] for a good start in this fascinating subject.

Of related interest to the work of this chapter (indeed, to all we've been discussing in these deliberations) is work initiated by G. W. Mackey [**73**] in the late 1950s. Mackey was looking for a proper home for "duality", similar to what is encountered in the case of commutative locally compact groups with the Pontrjagin–van Kampen duality. He came to believe the proper setting for such a duality might be found in classical descriptive set theoretical ideas.

Mackey reformulates ideas from that theory to fit his approach to general representation theory.

To be precise, a *Borel space* is a set S along with a σ-field $\mathcal{B}$ of subsets of S. Members of $\mathcal{B}$ are called *Borel sets.* The Borel space is called *standard* if the σ-field $\mathcal{B}$ is isomorphic (as a σ-field) to the Borel σ-field of an uncountable Borel subset of a Polish space (a homeomorph of a separable complete metric space).

Along with Borel spaces comes Borel functions: if S and T are Borel spaces and f is a function from S to T such that the preimage of any Borel subset of T is a Borel subset of S, the f is called a *Borel function.*

Cementing ties with the classical theory, he establishes that a a Borel space has a countable family of sets that generates $\mathcal{B}$ precisely when $(S, \mathcal{B})$ is isomorphic to a subspace of the Cantor set with its natural collection of Borel sets.

Again, following the path of the classical theory, Mackey isolates the analytic Borel spaces as those countably generated Borel spaces that are the image of a standard Borel space under a Borel map. Another link to the classical theory: a necessary and sufficient condition that a Borel space be analytic is that it be Borel isomorphic to the Borel space generated by the topology of an analytic subset of the reals.

Groups? Suppose G is a Borel space which is also a group; it is called a *Borel group* if multiplication and taking inverses are Borel functions. Standard Borel groups and analytic Borel groups have the natural meanings.

Measures? If S is a Borel space, a *Borel measure* on S is a σ-finite, countably additive, nonnegative, extended real-valued map on the Borel σ-field of S. A Borel measure is *standard* if it is supported by a standard Borel set.

A *measure class* is the collection C of all Borel measures on a Borel space having the same null sets as a fixed, nontrivial Borel measure. If G is a Borel group and C is a measure class, C said to be *right (left) invariant* if whenever μ is in C so is each measure μ_g given by

$$\mu_g(E) = \mu(Eg)(g_\mu(E) = \mu(gE)).$$

To whet the appetite of those who've enjoyed our discussions of Haar measures, here's a gorgeous theorem of Mackey that generalizes the earlier theorem of Weil.

Theorem 10.8. *Suppose G is an analytic group, and assume G admits a left invariant measure class. Then G has a unique locally compact topology whose Borel sets are the given ones and under which G is a topological group.*

This was but the start of Mackey's work. He went on to apply these descriptive set theoretic ideas to the representation theory of locally compact groups satisfying the second axiom (thereby encompassing all Lie groups) and to develop a duality theory for such groups that was to bring to bear measure-theoretic ideas to supplant the topological basis of duality in the commutative case.

To give a synopsis of Mackey's work is beyond the capabilities of the present authors. A clear overview can be found in the special volume of *Contemporary Mathematics* [**32**], paying tribute to his work.

Appendix A

In 1933 Alfred Haar gave the first abstract construction of a countably additive translation invariant measure on a locally compact separable metrizable topological group. This appendix is devoted to an exposition of Haar's construction in the compact case. In this setting Haar's construction is particularly elegant and transparent.

Haar's construction starts with the introduction of his covering function, which gauges the relative size of a given subset of the compact group *vis-à-vis* a nonempty open subset of the group. Next, to put all sets on a level playing ground, he averages against shrinking sequences of open balls. This averaging procedure is shown to be especially effective when applied to open sets whose boundary is "null" with respect to the notion of content (a natural generalization of the Jordan content). These open sets are called *intervals*, and they are used much as intervals in Euclidean spaces are in the generation of an outer measure there.

Let G be a compact metrizable topological group. Fix a metric on G that generates its compact metrizable topology.

If A and B are subsets of G, we say that A and B are *congruent* if $B = Ax$ for some $x \in G$. Of course, if A and B are congruent, then they are open together or closed together being homeomorphic.

For $B \subseteq G$ and U an open set in G we define $h(\bar{B}, \bar{U})$ to be the smallest number of sets congruent to $\bar{U}$ needed to cover $\bar{B}$. Since B is totally bounded, $h(\bar{B}, \bar{U}) < \infty$.

Theorem A.1. *Some pertinent features of* $h(\cdot, \cdot)$ *follow:*

(i) $h(\bar{U}, \bar{U}) = 1$.

(ii) *If* $C \subseteq B$, *then* $h(\bar{C}, \bar{U}) \leq h(\bar{B}, \bar{U})$.

(iii) *If* $B_1, B_2 \subseteq C$, *then* $h(\bar{B}_1 \cup \bar{B}_2, \bar{U}) \leq h(\bar{B}_1, \bar{U}) + h(\bar{B}_2, \bar{U})$.

(iv) *If* B_1 *and* B_2 *are congruent, then* $h(\bar{B}_1, \bar{U}) = h(\bar{B}_2, \bar{U})$.

(v) *If* U *and* V *are open, then* $h(\bar{B}, \bar{U}) \leq h(\bar{B}, \bar{V})h(\bar{V}, \bar{U})$.

(vi) *If* (U_n) *is a descending sequence of open sets with exactly one point in common, then whenever* B *and* C *are open sets with* $\bar{B} \cap \bar{C} = \emptyset$, *there is an* n_0 *so that for* $n \geq n_0$,

$$h(\bar{B} \cup \bar{C}, \bar{U_n}) = h(\bar{B}, \bar{U_n}) + h(\bar{C}, \bar{U_n}).$$

The proofs are straightforward with the proof of the last statement following Banach's variation of Haar's initial approach, so we omit them.

Fix a sequence (U_n) of open balls with common center and with radius $\frac{1}{n}$. Let B be an open set in G. Define

$$l_n(\bar{B}) = \frac{h(\bar{B}, \bar{U_n})}{h(\bar{G}, \bar{U_n})}.$$

Notice that

$$\frac{1}{h(G, \bar{B})} \leq l(\bar{B}) \leq h(\bar{B}, G).$$

So if B is a nonempty open set in G, then $(l_n(\bar{B}))$ is a bounded sequence of real numbers, each greater than $\frac{1}{h(G,\bar{B})}$.

Thus there is an increasing sequence (n_k) of positive integers so that $(l_{n_k}(\bar{B}))$ converges (to a positive number between $\frac{1}{h(G,\bar{B})}$ and $h(\bar{B}, G)$). What's more, if C is congruent to B, then

$$\lim_k l_{n_k}(\bar{B}) = \lim_k l_{n_k}(\bar{C}).$$

Now we choose a countable collection $\{B_m\}$ of nonempty open sets as follows. Choose a countably dense sequence (a_n) in G, and look at any open ball centered at a_m with rational radius. The collection of all such sets is countable. If we look at the collection formed by taking all finite unions of this collection, then we arrive at our desired collection $\{B_m : m \in \mathbb{N}\}$.

A straightforward diagonal argument allows us to find a sequence (n_k) so that

$$\lim_k l_{n_k}(\bar{B}_n)$$

exists for each n. For simplicity, we assume $n_k = k$.

A subset N of G is called a *null set* if given $\epsilon > 0$ there is an open set $\mathcal{U}_\epsilon$ containing N so that $\overline{\lim}_k l_k(\bar{\mathcal{U}}_\epsilon) \leq \epsilon$.

Subsets of a null set are null sets, finite unions of null sets are null sets, and sets congruent to a null sets are null sets. We plan to show that

$$\lim_k l_k(\bar{C})$$

exists whenever the boundary $\partial C = \bar{C} \setminus C$ of the open set C is a null set.

First though we show that such open sets C with null set boundaries exist at all.

Take a typical a_k from our countable dense subset of G, and let $\mathcal{B}(r)$ be the open ball centered at a_k with radius r. If r is rational, then $\mathcal{B}(r)$ is among the B_m's for which $\lim_k l_k(\bar{B}_m)$ exists. In any case,

$$\lambda_n(r) = l_n(\overline{\mathcal{B}(r)})$$

is nondecreasing in r. As a consequence, the monotone functions $\lambda_n(r)$ converge at all but countably many r's to a function $\lambda(r)$. λ is also nondecreasing and so, except for countably many r's, is continuous. If r is such that $\lim_n \lambda_n(r) = \lambda(r)$ exists and r is a point of continuity of λ, then the open ball $\mathcal{B}(r)$ has a boundary that is a null set. In fact, by the continuity of λ at such r's, if $\epsilon > 0$, then there are r', r'', and r''' such that $r' < r'' < r < r'''$ and

$$0 \leq \lambda(r''') - \lambda(r') < \epsilon.$$

On the one hand, the closed sets $\overline{\mathcal{B}(r''')} \setminus \mathcal{B}(r'')$ and $\overline{\mathcal{B}(r')}$ are disjoint, so for n large enough

$$h([\overline{\mathcal{B}(r''')} \setminus \mathcal{B}(r'')] \cup \overline{\mathcal{B}(r')}, \bar{\mathcal{U}}_n) = h(\overline{\mathcal{B}(r''')} \setminus \mathcal{B}(r''), \bar{\mathcal{U}}_n) + h(\overline{\mathcal{B}(r')}, \bar{\mathcal{U}}_n).$$

But now

$$h([\overline{\mathcal{B}(r''')} \setminus \mathcal{B}(r'')] \cup \overline{\mathcal{B}(r')}, \bar{\mathcal{U}}_n) \leq h(\overline{\mathcal{B}(r''')}, \bar{\mathcal{U}}_n)$$

so that

$$h(\overline{\mathcal{B}(r''')} \setminus \mathcal{B}(r''), \bar{\mathcal{U}}_n) \leq h(\overline{\mathcal{B}(r''')}, \bar{\mathcal{U}}_n) - h(\overline{\mathcal{B}(r')}, \bar{\mathcal{U}}_n).$$

It follows that

$$\limsup_n \frac{h(\overline{\mathcal{B}(r''')} \setminus \mathcal{B}(r''), \bar{\mathcal{U}}_n)}{h(G, \bar{\mathcal{U}}_n)} \leq \lambda(r''') - \lambda(r') < \epsilon.$$

Since the compact set $\overline{\mathcal{B}(r''')} \setminus \mathcal{B}(r'')$ contains the boundary of $\mathcal{B}(r)$, this boundary is indeed a null set. A few moments of reflection reveals that we've shown not only that open sets whose boundaries are null sets exist, but each open ball $\mathcal{B}$ with center one of the points of the dense set $\{a_n : n \in \mathbb{N}\}$—with but countably many exceptions—has $\lim_n \lambda_n(\bar{\mathcal{B}})$ existing as well.

Let U be any open set, and let F be a closed subset of U. Then the boundary of U, $\partial U = \bar{U} \setminus U$, and F are disjoint, and the distance between ∂U and F is positive, say $\operatorname{dist}(\partial U, F) = \eta > 0$. Look at points of $\{a_n : n \in \mathbb{N}\}$. The open balls centered at points of $\{a_n : n \in \mathbb{N}\}$ that lie inside F and have radii between $\frac{\eta}{2}$ and $\frac{3\eta}{4}$ cover F. So then F is covered by a finite union of such open balls. We can even assume the balls have rational radii and have boundary that is a null set, and so we can assume that the finite union that covers F is one of the B_m's. What we have (because of the judicious choice of possible radii being between $\frac{\eta}{2}$ and $\frac{3\eta}{4}$) is the existence of m with

$$F \subseteq B_m \subseteq \bar{B}_m \subseteq U.$$

Along with this, we can assume

$$\lim_n l_n(\bar{B}_m)$$

exists.

Here's what we have at this stage: If U is open and F is a closed set, $F \subseteq U$, then there exists an open set C, $F \subseteq C \subseteq \bar{C} \subseteq U$ with ∂C a null set and there exists an open set B so that $F \subseteq B \subseteq \bar{B} \subseteq U$ where $\lim_n l_n(\bar{B})$ exists.

We plan to show that when the boundary of an open set U is a null set, then the sequence $(l_n(U))$ is convergent, and we will define $\lim_n l_n(\bar{U})$ to be the *content*, $I(U)$, of U.

Let C be an open set whose boundary is a null set. By definition for each $\epsilon > 0$ there is an open set U_ϵ containing ∂C such that

$$\limsup_n \frac{h(\bar{U}_\epsilon, \bar{U}_n)}{h(G, \bar{U}_n)} \leq \epsilon.$$

Let $C'' = C \cap U_\epsilon$ and $C' = C \setminus C''$. Then C'' is open and C' is closed. Since C'' is a subset of U_ϵ

$$\limsup_n \frac{h(\bar{C''}, \bar{U}_n)}{h(G, \bar{U}_n)} \leq \epsilon.$$

We know there is an open set B containing $\bar{C'}$ so $\bar{B} \subseteq C$ and

$$\lim_n l_n(\bar{B})$$

exists. Now

$$h(\bar{C}, \bar{U}_n) \geq h(\bar{B}, \bar{U}_n),$$

and so it follows that

$$\limsup_n \frac{h(\bar{C}, \bar{U}_n)}{h(G, \bar{U}_n)} \geq \lim_n l_n(\bar{B}).$$

Since $C \subseteq B \cup C''$, we have

$$h(\bar{C}, \bar{U}_n) \leq h(\bar{B}, \bar{U}_n) + h(\bar{C''}, \bar{U}_n),$$

and so

$$\frac{h(\bar{C},\bar{U}_n)}{h(G,\bar{U}_n)} \le \frac{h(\bar{B},\bar{U}_n)}{h(G,\bar{U}_n)} + \frac{h(\bar{C''},\bar{U}_n)}{h(G,\bar{U}_n)}.$$

Therefore

$$\begin{aligned}\limsup_n \frac{h(\bar{C},\bar{U}_n)}{h(G,\bar{U}_n)} &\le \limsup_n \frac{h(\bar{B},\bar{U}_n)}{h(G,\bar{U}_n)} + \limsup_n \frac{h(\bar{C''},\bar{U}_n)}{h(G,\bar{U}_n)}\\ &\le \lim_n l_n(\bar{B}) + \epsilon.\end{aligned}$$

On applying the theory of subtraction, we see

$$\limsup_n \frac{h(\bar{C},\bar{U}_n)}{h(G,\bar{U}_n)} - \liminf_n \frac{h(\bar{C},\bar{U}_n)}{h(G,\bar{U}_n)} \le \epsilon.$$

Therefore

$$\lim_n \frac{h(\bar{G},\bar{U}_n)}{h(G,\bar{U}_n)} = \lim_n l_n(\bar{C})$$

exists and is $I(C)$.

The content $I(C)$ exists for any nonempty open set C whose boundary is a null set. The following are more or less clear features of the content.

Fact A.2. (i) $I(C) > 0$.

(ii) If C_1 and C_2 are congruent sets with one of them open and having boundary that is a null set, then so too is the other, and $I(C_1) = I(C_2)$.

(iii) If $C_1, C_2, \ldots, C_n$ are open sets with boundaries that are null sets, then so too are $C_1 \cap C_2 \cap \cdots \cap C_n$ and $C_1 \cup C_2 \cup \cdots \cup C_n$.

A bit more complicated:

Fact A.3. If C_1 and C_2 are disjoint open sets with boundaries that are null sets, then $C_1 \cap C_2$ is an open set with boundary that is a null set and

$$I(C_1 \cup C_2) = I(C_1) + I(C_2).$$

More generally, if C_1 and C_2 are simply open sets with null set boundaries, then

$$I(C_1 \cup C_2) + I(C_1 \cap C_2) = I(C_1) + I(C_2).$$

Proposition A.4. *If C_1 and C_2 are disjoint open sets, each with a null set for boundary, then $C_1 \cup C_2$ is likewise an open set with null set for boundary and*

$$I(C_1 \cup C_2) = I(C_1) + I(C_2).$$

Proof. Let $\epsilon > 0$ be given. We've seen that there exists an open set U with $\bar{U} \subseteq C$ for which $\lim_n l(\bar{U}_n)$ exists and for which

$$l(C_1) \leq \lim_n l_n(\bar{U}) + \epsilon.$$

Now $\bar{U}$ and $\bar{C}_2$ are disjoint closed sets, and so for n large enough,

$$(61) \qquad h(\bar{U} \cup \bar{C}_2, \bar{U}_n) = h(\bar{U}, \bar{U}_n) + h(\bar{C}_2, \bar{U}_n).$$

Since $\bar{U} \cup \bar{C}_2 \subseteq \bar{C}_1 \cup \bar{C}_2$, we have

$$h(\bar{U} \cup \bar{C}_2, \bar{U}_n) \leq h(\bar{C}_1 \cup \bar{C}_2, \bar{U}_n) \leq h(\bar{C}_1 \bar{U}_n) + h(\bar{C}_2 \bar{U}_n),$$

and for n big enough for (61) to hold,

$$h(\bar{U}, \bar{U}_n) + h(\bar{C}_2, \bar{U}_n) \leq h(\bar{C}_1, \bar{U}_n) + h(\bar{C}_2, \bar{U}_n).$$

From this and our placement of U within C_1, we have

$$I(C_1) + I(C_2) \geq I(C_1 \cup C_2) \geq I(C_1) - \epsilon + I(C_2).$$

The usual epsilonics finishes the proof. □

Fact A.5. Suppose (C_n) is a descending sequence of open sets, each of which has a null set for a boundary. If $\bigcap_n C_n = \{c\}$, then

$$\lim_n I(C_n) = 0.$$

Let H be an open set with boundary a null set. Take distinct points $h_1, \ldots, h_k \in H$. Look at the maps

$$x \mapsto xc^{-1}h_p$$

for $p = 1, 2, \ldots, k$ from G onto itself. Suppose $C_n^{(1)}, \ldots, C_n^{(k)}$ are the images of C_n under these maps so $C_n^{(1)}, \ldots, C_n^{(k)}$ are mutually congruent. For n sufficiently large, $C_n^{(1)}, \ldots, C_n^{(k)}$ are pairwise disjoint, and so (since $I(C_n) = I(C_n^{(1)}) = \cdots = I(C_n^{(k)})$)

$$kI(C_n) = \sum_{p=1}^{k} I(C_n^{(p)}) = I(C_n^{(1)} \cup \cdots \cup C_n^{(p)}) \leq I(H).$$

It follows that for n large enough,

$$I(C_n) \leq \frac{I(H)}{k}$$

for each k.

We now introduce an analogue of Lebesgue measure in G in which congruent sets have the same measure. To emphasize the analogy, we will call any nonempty open set whose boundary is a null set an *interval.*

Let $M \subseteq G$. Define the *outer measure* $m^*(M)$ by

$$m^*(M) = \inf\left\{ \sum_n I(C_n) \right\},$$

where the infimum is taken over all coverings $\{C_1, C_2, \ldots, C_n, \ldots\}$ of M by intervals. Since M is totally bounded, it can be covered by finite collections of intervals, so it is plain that

$$0 \leq m^*(M) < \infty.$$

Further, if $M \subseteq M' \subseteq G$, then

$$m^*(M) \leq m^*(M').$$

Also, m^* is countably subadditive (something that's easy to see from its definition). Since sets congruent to intervals are intervals, and a set congruent to M can be covered by intervals that are congruent to intervals that cover M, it is plain that $m^*(M) = m^*(M')$ whenever M and M' are congruent.

Fact A.6. The outer measure $m^*(C)$ of an interval C is its content $I(C)$.

Indeed, let $\epsilon > 0$ be given. Given the interval C, there is an interval C' such that $\bar{C}' \subseteq C$ and $I(C') > I(C) - \epsilon$. Cover C, and so $\bar{C}'$, with intervals $C_1, C_2, \ldots, C_n$, so that

$$I(C) - \epsilon < I(C') \leq I(C_1) + I(C_2) + \cdots + I(C_n),$$

and conclude that $I(C) - \epsilon \leq m^*(C)$. Clearly, $I(C) \leq m^*(C)$ follows. Since $m^*(C) \leq I(C)$ is plain, our claim is established.

If C is an interval, then $m^*(\bar{C}) = I(C)$. In fact, if $\epsilon > 0$ is given, then there is an interval C'' containing $\bar{C}$ so that $I(C'') < I(C) + \epsilon$. Now

$$m^*(\bar{C}) \leq m^*(C'') = I(C'') < I(C) + \epsilon.$$

Epsilonics über alles!

Next, we note that

Fact A.7.

$$m^*(M) = \inf\left\{ \sum_n I(C'_n) \right\},$$

where $C'_1, \ldots, C'_n, \ldots$ are pairwise disjoint intervals with $M \subseteq \bigcup_n \bar{C}'_n$.

Suppose $M \subseteq C_1 \cup C_2 \cup \cdots \cup C_n \cup \cdots$, with all the C_k's intervals. We "disjointify": look at $C_2 \setminus C_1$. Take a countable dense family $\{x_n : n \in \mathbb{N}\}$ of points of $C_2 \setminus C_1$. Around x_1 there is an interval $\mathcal{D}$ that is contained in $C_2 \setminus C_1$. Look to the next x_k that is in $C_2 \setminus (C_1 \cup \mathcal{D})$. Around that x_k there is an interval that is contained in $C_2 \setminus (C_1 \cup \mathcal{D})$. Continue. We exhaust

$C_2 \setminus C_1$ with disjoint intervals. We now turn our attention to $C_3 \setminus (C_1 \cup C_2)$ and continue, ending up with a sequence (C'_n) of disjoint intervals such that $M \subseteq \bigcup_n C'_n$.

Our next task is to isolate what "measurability" means. For any subset M of G, define the *inner measure* of M to be

$$m_*(M) = I(G) - m^*(G \setminus M).$$

A set $M \subseteq G$ is *measurable* if and only if $m_*(M) > m^*(M)$. It is classical (and an excellent exercise) to show that the collection of measurable subsets of G is a σ-field and that on this σ-field m^* (or m_*) is a countably additive measure that assigns to any set M' congruent to M the same value of m^*.

Appendix B

In this final appendix we broach the possibility of extending Haar measure to a larger domain while preserving its countable additivity and its translation invariance. We restrict our attention to the situation where G is an infinite compact metrizable topological group. In the main, we follow the path blazed by S. Kakutani and J. Oxtoby for the reals and generalized by E. Hewitt and K. Ross to the current setting. In truth, out exposition is but a slight embellishment of that found in the text *Abstract Harmonic Analysis, volume I* [**55**] of Hewitt and Ross.

Note B.1. If G is a nondiscrete locally compact topological group, then G's cardinality is at least $\mathbf{c}$, the cardinality of the continuum.

Proof. To see this, we follow a Cantor-inspired path.

Let U be a nonempty open set in G with $\overline{U}$ compact. Then U must be infinite since G's topology is nondiscrete. Let $x_0, x_1 \in U, x_0 \neq x_1$. Envelop x_0 and x_1 in open sets U_0 and U_1, where $\overline{U_0} \cap \overline{U_1} = \emptyset$ and $\overline{U_0} \cup \overline{U_1} \subseteq U$. Each of U_0, U_1 is infinite. Let x_{00}, x_{01} be distinct points of U_0, and let x_{10}, x_{11} be distinct points of U_1.

Envelop x_{00} and x_{01} in open sets U_{00} and U_{01}, where $\overline{U}_{00} \cap \overline{U}_{01} = \emptyset$ and $\overline{U}_{00} \cup \overline{U}_{01} \subseteq \overline{U_0}$. Similarly envelop x_{10} and x_{11} in open sets U_{10} and U_{11}, where $\overline{U}_{10} \cap \overline{U}_{11} = \emptyset$ and $\overline{U}_{10} \cup \overline{U}_{11} \subseteq \overline{U}_1$.

Continue in this Cantor-like fashion.

For $(\epsilon_n) \in \{0,1\}^{\mathbb{N}}$, let

$$\Delta_m = \bigcup \{\overline{U}_{\epsilon_1,\ldots,\epsilon_m} : \epsilon_1, \ldots, \epsilon_m \in \{0,1\}\}.$$

Then each Δ_m is compact and Δ_m contains Δ_{m+1}. Let

$$\Delta = \bigcap_m \Delta_m.$$

Since the Δ_m's are descending nonempty compact sets, $\Delta \neq \emptyset$.

Moreover, if $(\epsilon_n) \in \{0,1\}^{\mathbb{N}}$, then

$$\bigcap_n \overline{U}_{\epsilon_1,\ldots,\epsilon_n} \neq \emptyset.$$

Further if $(\epsilon_n) \neq (\epsilon_n')$, then

$$\left(\bigcap_m \overline{U}_{\epsilon_1,\ldots,\epsilon_m}\right) \cap \left(\bigcap_m \overline{U}_{\epsilon_1',\ldots,\epsilon_m'}\right) = \emptyset,$$

and so Δ contains a set of cardinality at least $\mathbf{c}$. □

Note B.2. Now Suppose G is an infinite compact metrizable topological group. We know that G has at least $\mathbf{c}$ many members. In fact, the cardinality of G is exactly $\mathbf{c}$. After all, G is separable, and so G contains a countably infinite dense subset D. Each member of G is the limit of a sequence of points from D, and so G has no more than $\mathbf{c}$ many elements.

Not only does G have exactly $\mathbf{c}$ many elements, *G has exactly $\mathbf{c}$ many distinct compact sets having cardinality $\mathbf{c}$*. Indeed, any open subset of G is the union of (countably many) sets in a countable open base for its topology, so the totality of compact subsets of G (which is the same as the totality of closed subsets of G) is at most $\mathbf{c}$. That there are exactly $\mathbf{c}$ distinct compact subsets of G having cardinality of the continuum follows from the nature of the Cantor set. As noted in the Notes and Remarks section of Chapter 2, a classical theorem of Alexandrov and Hausdorff assures us that G contains a homeomorphic copy of the Cantor set $\Delta = \{0,1\}^{\mathbb{N}}$ which is itself homeomorphic to

$$\Delta \times \Delta = \{0,1\}^{\mathbb{N}} \times \{0,1\}^{\mathbb{N}}.$$

It is this last version of Δ that tells us the truth if we but wish to know it: if $\epsilon \in \Delta = \{0,1\}^{\mathbb{N}}$, then $\{\epsilon\} \times \{0,1\}^{\mathbb{N}}$ is homeomorphic to Δ, and so

$$\{\epsilon\} \times \{0,1\}^{\mathbb{N}} \subseteq \Delta = \{0,1\}^{\mathbb{N}} \times \{0,1\}^{\mathbb{N}}.$$

We can find $\mathbf{c}$ many distinct compact sets $\{\{\epsilon\} \times \{0,1\}^{\mathbb{N}} : \epsilon \in \Delta\}$, each of cardinality $\mathbf{c}$.

Next, recall the Fréchet–Nikodym metric space: Ω is a set, Σ a σ-field of subsets of Ω, and μ is a probability measure on Σ. For $E, F \in \Sigma$ we define the pseudo-metric d_μ as follows

$$d_\mu(E,F) = \mu(E\Delta F),$$

where $E\Delta F = (E\backslash F) \cup (F\backslash E)$ is the symmetric difference of E and F. If we identify sets E, F with $d_\mu(E,F) = 0$, then we get a complete metric space $(\Sigma/\mu^{\leftarrow}(\{0\}), d_\mu)$. The density character of this metric space is called the *character* of (Ω, Σ, μ); it is the smallest cardinal $\mathbf{m}$ such that there is a subset $\mathcal{S}$ of Σ with the cardinality of $\mathcal{S}$ equal to $\mathbf{m}$ so that given $\epsilon > 0$ and $E \in \Sigma$ there is $S \in \mathcal{S}$ with

$$d_\mu(S,E) = \mu(S\Delta E) < \epsilon.$$

Let G be a fixed infinite compact metrizable topological group, and let Σ be the σ-field of all μ-measurable subsets of G, where μ is the normalized Haar measure on G. It is plain that the character of (G, Σ, μ) is $\aleph_0$.

We now state the stunning theorem of Kakutani and Oxtoby á la Hewitt and Ross.

Theorem B.3 (Kakutani and Oxtoby). *There is an extension (G, Σ^*, μ^*) of (G, Σ, μ) such that the character of (G, Σ^*, μ^*) is $\mathbf{2^c}$ and μ^* is a left-, right-, and inverse-invariant countably additive probability measure on Σ^* which extends μ.*

Proof. The proof will entail careful bookkeeping and devilishly clever-yet-simple constructs.

We will denote by $\mathcal{F}$ the collection of all functions $f : G \to G$ of the form

$$f(x) = ax, \quad f(x) = xa, \quad \text{or} \quad f(x) = x^{-1}.$$

Step 1. We extend (G, Σ, μ) to (G, Σ', μ'), where if $A \subseteq G$ has fewer than $\mathbf{c}$ many members, then $A \in \Sigma'$ and $\mu'(A') = 0$. This extension will be a suitably invariant and complete probability on Σ'.

How? Let $\mathcal{P}$ denote the family of all $P \subseteq G$ such that P has cardinality less than $\mathbf{c}$.

If $P \in \mathcal{P} \cap \Sigma$, then $\mu(P) = 0$. After all, if $\mu(P) > 0$, then P contains a Borel set B with $\mu(B) > 0$. In turn then B contains a compact set K with $\mu(K) > 0$. Such a compact set K with $\mu(K) > 0$ must be uncountable, so it contains a homeomorphic copy of the Cantor set, by a classical result of Alexandrov and Hausdorff.

Next note that if $P_1, P_2, \ldots, P_n \cdots \in \mathcal{P}$, then

$$\bigcup_n P_n \in \mathcal{P}.$$

This is a consequence of König's theorem in the theory of sets.

Let Σ' consist of all those sets $M' \subseteq G$ such that $M'\Delta M \in \mathcal{P}$ for some $M \in \Sigma$. So $M' \in \Sigma'$ precisely when $M' = (M \cap A^c) \cup B$ for some $M \in \Sigma$ and $A, B \in \mathcal{P}$.

Note that Σ' is a σ-field:

If $M \in \Sigma$, then for any $M' \in \Sigma'$ we have

$$(M')^c \Delta M^c = M'\Delta M,$$

and so $M' \in \Sigma'$ is equivalent to $(M')^c \in \Sigma'$.

Suppose (M'_n) is a sequence of members of Σ'. Let (M_n) be a corresponding sequence in Σ so that

$$M'_n \Delta M_n \in \mathcal{P}.$$

Then

$$\left(\bigcup_n M'_n\right) \Delta \left(\bigcup_n M_n\right) \subseteq \bigcup_n (M'_n \Delta M_n) \in \mathcal{P},$$

so $\bigcup_n M'_n \in \Sigma'$.

Let $M' \in \Sigma'$. Then $M'\Delta M \in \mathcal{P}$ for some $M \in \Sigma$. Naturally, we would like to define $\mu'(M')$ by

$$\mu'(M') = \mu(M), \text{ if } M'\Delta M \in \mathcal{P}.$$

Is this legal? Well suppose $M_1, M_2 \in \Sigma$, and suppose

$$M'\Delta M_1, M'\Delta M_2 \in \mathcal{P}.$$

Then

$$M_1 \Delta M_2 \subseteq (M_1 \Delta M') \cup (M_2 \Delta M') \in \mathcal{P}.$$

So $M_1 \Delta M_2 \in \mathcal{P} \cap \Sigma$, and, as noted earlier, $\mu(M_1 \Delta M_2) = 0$. In particular,

$$\mu(M_1) = \mu(M_2),$$

and so

$$\mu'(M') = \mu(M)$$

for $M'\Delta M \in \mathcal{P}$ is not only legal but ethical as well.

Now we have M defined on Σ', let's show μ' is also countably additive thereupon. Suppose (M'_n) is a sequence of pairwise disjoint members of Σ'. Let (M_n) be a corresponding sequence of members of Σ with $M'_n \Delta M_n \in \mathcal{P}$ for each n. We disjointify the M_n's: Let

$$Q_1 = M_1, Q_2 = M_2 \setminus M_1, \ldots, Q_n = M_n \setminus (M_1 \cup \cdots \cup M_{n-1}).$$

Then (Q_n) is a sequence of pairwise disjoint members of Σ, and since

$$M'_n \Delta Q_n \subseteq \bigcup_{k=1}^{n} (M'_k \Delta M_k) \in \mathcal{P},$$

$M'_n \Delta Q_n \in \mathcal{P}$. What's more,

$$\left(\bigcup_n M'_n\right) \Delta \left(\bigcup_n Q_n\right) \in \mathcal{P}.$$

Hence

$$\mu'\left(\bigcup_n M'_n\right) = \mu\left(\bigcup_n Q_n\right) = \sum_n \mu(Q_n) = \sum_n \mu'(M'_n),$$

so μ''s countable additivity on Σ' is secure.

To see that (G, Σ', μ') is a complete probability space, suppose $\mu'(N) = 0$ and $N_1 \subseteq N$. Since $N \in \Sigma'$, there is an $M \in \Sigma$ such that $N \Delta M \in \mathcal{P}$. That $\mu'(N) = 0$ says we have $\mu(M) = 0$. Now $N_1 \cap M \subseteq M$, and (G, Σ, μ) is a complete probability space, so $\mu(N_1 \cap M) = 0$ with $N_1 \cap M \in \Sigma$. Since

$$N_1 \Delta (N_1 \Delta M) \subseteq N \Delta M,$$

and $N \Delta M \in \mathcal{P}$, we have

$$N_1 \Delta (N_1 \Delta M) \in \mathcal{P}$$

also. But this ensures us that $N_1 \in \Sigma'$ with $\mu'(N_1) = \mu(N_1 \cap M) = 0$. The completeness of (G, Σ', μ') is established.

Finally, μ' is invariant on Σ', left-, right-, and inversion-invariant. After all, if $P \in \mathcal{P}$ and $f \in \mathcal{F}$, then $f(P) \in \mathcal{P}$. If $M' \in \Sigma'$ and $M \in \Sigma$ is chosen so that $M' \Delta M \in \mathcal{P}$, then, thanks to the bijective nature of members of $\mathcal{F}$,

$$f(M') \Delta f(M) = f(M' \Delta M) \in \mathcal{P}.$$

It follows that $f(M') \in \Sigma'$ (each $f(M) \in \Sigma$) and

$$\mu'(f(M')) = \mu(f(M)) = \mu(M) = \mu'(M'),$$

thanks to the invariance properties of μ.

Our first step has been successfully taken.

An observation regarding members of Σ' of positive μ' measure: *if $M' \in \Sigma'$ and $\mu'(M') > 0$, then M' contains* **c** *many distinct compact sets each having cardinality* **c**. Indeed, if $M' \in \Sigma$ and $\mu'(M') > 0$, then there is $M \in \Sigma$ so that $M' \Delta M \in \mathcal{P}$ and $\mu(M) = \mu'(M') > 0$. Now $M' \Delta M \in \mathcal{P}$ just says $M' \Delta M$ has fewer than **c** members. Since $M \in \Sigma$, M contains a Borel set of positive μ-measure, which is necessarily uncountable. Hence, again by the Alexandrov–Hausdorff result, this set contains a homeomorphic copy of the Cantor set. But M must itself contain a continuum of pairwise disjoint compact sets each of cardinality of the continuum (look at our preliminary remarks). Now each of these lies in M, and so they are each disjoint from $M' \Delta M$; that is, each is in $M' \cap M$. How about that?

Step 2. Let $\boldsymbol{\omega_c}$ *denote the first ordinal having cardinality* $\mathbf{c}$*, and let* $(F_{\boldsymbol{\alpha}} : 1 \leq \boldsymbol{\alpha} \leq \boldsymbol{\omega_c})$ *be a corresponding well-ordering of order type* $\boldsymbol{\omega_c}$ *of all the distinct compact subsets of* G *having the cardinality of the continuum. Then for any* $M' \in \Sigma'$ *with* $\mu'(M') > 0$ *and any ordinal* $\boldsymbol{\alpha} < \boldsymbol{\omega_c}$*, there is an ordinal* $\boldsymbol{\beta}$ *with* $\boldsymbol{\alpha} < \boldsymbol{\beta} < \boldsymbol{\omega_c}$ *so that* $F_{\boldsymbol{\beta}} \subseteq M'$*.*

After all, we just saw that each $M' \in \Sigma'$ with $\mu'(M') > 0$ contains $\mathbf{c}$ many (distinct) compact subsets of G each having cardinality of the continuum. Since $\boldsymbol{\omega_c}$ is just the *first* ordinal having cardinality $\mathbf{c}$, for any $\boldsymbol{\alpha} < \boldsymbol{\omega_c}$, there must be an $F_{\boldsymbol{\beta}}$ following $F_{\boldsymbol{\alpha}}$ with $F_{\boldsymbol{\beta}} \subseteq M'$.

Let $A \subseteq G$ and define $\overline{\mu}(A)$ by

$$\overline{\mu}(A) = \inf\{\mu'(M') : A \subseteq M' \in \Sigma'\}.$$

We say that A is $\overline{\mu}$–$\mathcal{F}$ *invariant* if for any $f \in \mathcal{F}$ we have

$$f(A)\Delta A \in \Sigma' \text{ and } \mu'(f(A)\Delta A) = 0.$$

We are obviously going to try to mimic our first step to continue our construction.

Step 3. There exists a family $\{X_\nu : \nu \in \mathbb{R}\}$ *of pairwise disjoint subsets of* G *with*

$$\overline{\mu}(X_\nu) = 1$$

for each $\nu \in \mathbb{R}$*, and so that for any* $\mathbb{R}_0 \subseteq \mathbb{R}$*,*

$$\bigcup_{\nu \in \mathbb{R}_0} X_\nu$$

is $\overline{\mu}$–$\mathcal{F}$ *invariant.*

Let $(F_{\boldsymbol{\alpha}} : 1 \leq \boldsymbol{\alpha} \leq \boldsymbol{\omega_c})$ be as in Step 2: a well-ordered set of all the $\mathbf{c}$ many compact subsets of G each having cardinality of the continuum, so that if $M' \in \Sigma'$ has positive μ' measure and $\boldsymbol{\alpha}$ is an ordinal less than $\boldsymbol{\omega_c}$, there is a $\boldsymbol{\beta}$, $\boldsymbol{\alpha} \leq \boldsymbol{\beta} < \boldsymbol{\omega_c}$ so that $F_{\boldsymbol{\beta}} \subseteq M'$.

Well-order $\mathcal{F}$ with order type $\boldsymbol{\omega_c}$. For each $x \in G$ and ordinal $\boldsymbol{\alpha} < \boldsymbol{\omega_c}$, let $C_{\boldsymbol{\alpha}}(x)$ be the collection of all members of G of the form

$$f_{\boldsymbol{\beta_1}}^{\sigma_1} \circ f_{\boldsymbol{\beta_2}}^{\sigma_2} \circ \cdots \circ f_{\boldsymbol{\beta_n}}^{\sigma_n}(x),$$

where $\mathbf{1} \leq \boldsymbol{\beta_1}, \ldots, \boldsymbol{\beta_n} \leq \boldsymbol{\alpha}$ (the $\boldsymbol{\beta}$'s need not be distinct) and $\sigma_1, \ldots, \sigma_n \in \{\pm 1\}, f_{\boldsymbol{\beta_1}}, \ldots, f_{\boldsymbol{\beta_n}} \in \mathcal{F}, n \in \mathbb{N}$.

It is plain to see that $x \in C_{\boldsymbol{\alpha}}(x)$, and that if $\mathbf{1} \leq \boldsymbol{\beta} \leq \boldsymbol{\alpha} < \boldsymbol{\omega_c}$, then $f_{\boldsymbol{\beta}}(C_{\boldsymbol{\alpha}}(x)) = C_{\boldsymbol{\alpha}}(x)$. Moreover, the cardinality of $C_{\boldsymbol{\alpha}}(x)$ is no more than the maximum of the cardinality of $\boldsymbol{\alpha}$ and $\aleph_0$, and so $C_{\boldsymbol{\alpha}}(x)$ has fewer than $\mathbf{c}$ members, and $C_{\boldsymbol{\alpha}}(x) \in \Sigma'$.

We will now construct a transfinite doubly infinite sequence $(x^{\boldsymbol{\alpha}}_{\boldsymbol{\beta}} : \mathbf{1} \le \boldsymbol{\beta} \le \boldsymbol{\alpha} < \boldsymbol{\omega_c})$ satisfying

(i) $x^{\boldsymbol{\alpha}}_{\boldsymbol{\beta}} \in F_{\boldsymbol{\alpha}}$ and

(ii) $\{C_{\boldsymbol{\alpha}}(x^{\boldsymbol{\alpha}}_{\boldsymbol{\beta}}) : \mathbf{1} \le \boldsymbol{\beta} \le \boldsymbol{\alpha} < \boldsymbol{\omega_c}\}$ is a disjoint collection of subsets of G.

We will call on the lexicographic ordering $\prec$ of pairs $(\boldsymbol{\alpha}, \boldsymbol{\beta})$ of ordinals where $\mathbf{1} \le \boldsymbol{\beta} \le \boldsymbol{\alpha} < \boldsymbol{\omega_c}$ and define $(x^{\boldsymbol{\alpha}}_{\boldsymbol{\beta}})$ by transfinite induction.

Pick $x^{\mathbf{1}}_{\mathbf{1}} \in F_{\mathbf{1}}$.

Suppose $\mathbf{1} \le \boldsymbol{\beta} \le \boldsymbol{\alpha} < \boldsymbol{\omega_c}$, and suppose that $x^{\boldsymbol{\gamma}}_{\boldsymbol{\delta}}$ has been suitably chosen for all $(\boldsymbol{\gamma}, \boldsymbol{\delta}) \prec (\boldsymbol{\alpha}, \boldsymbol{\beta})$ with (i) and (ii) in mind.

Consider the set

$$D(\boldsymbol{\alpha}, \boldsymbol{\beta}) = \bigcup_{(\boldsymbol{\gamma}, \boldsymbol{\delta}) \prec (\boldsymbol{\alpha}, \boldsymbol{\beta})} C_{\boldsymbol{\alpha}}(x^{\boldsymbol{\gamma}}_{\boldsymbol{\delta}}).$$

The cardinality of $D(\boldsymbol{\alpha}, \boldsymbol{\beta})$ is no more than

$$(\#(\boldsymbol{\alpha}, \boldsymbol{\beta})) \cdot (\# C_{\boldsymbol{\alpha}}(x^{\boldsymbol{\gamma}}_{\boldsymbol{\delta}})),$$

which is strictly less than or equal to

$$(\#\boldsymbol{\alpha})^2 \cdot \max\{\#\boldsymbol{\alpha}, \aleph_0\} < \mathbf{c}.$$

Since $F_{\boldsymbol{\alpha}}$ has a continuum of members it is *not* a subset of $D(\boldsymbol{\alpha}, \boldsymbol{\beta})$. There must be a point $x^{\boldsymbol{\alpha}}_{\boldsymbol{\beta}} \in F_{\boldsymbol{\alpha}} \backslash D(\boldsymbol{\alpha}, \boldsymbol{\beta})$.

Suppose $(\boldsymbol{\gamma}, \boldsymbol{\delta}) \prec (\boldsymbol{\alpha}, \boldsymbol{\beta})$. Then

$$C_{\boldsymbol{\alpha}}(x^{\boldsymbol{\alpha}}_{\boldsymbol{\beta}}) \cup C_{\boldsymbol{\gamma}}(x^{\boldsymbol{\gamma}}_{\boldsymbol{\delta}}) = \emptyset.$$

In fact, if this intersection is nonempty it is because there are

$$\boldsymbol{\beta_1}, \ldots, \boldsymbol{\beta_n} \le \boldsymbol{\alpha}, \boldsymbol{\delta_1}, \ldots, \boldsymbol{\delta_m} \le \boldsymbol{\gamma}, \sigma_1, \ldots, \sigma_n, \sigma'_1, \ldots, \sigma'_m \in \{\pm 1\}$$

and

$$f_{\boldsymbol{\beta_1}}, \ldots, f_{\boldsymbol{\beta_n}}, f_{\boldsymbol{\delta_1}}, \ldots, f_{\boldsymbol{\delta_m}} \in \mathcal{F},$$

so that

$$f^{\sigma_1}_{\boldsymbol{\beta_1}} \circ \cdots \circ f^{\sigma_n}_{\boldsymbol{\beta_n}}(x^{\boldsymbol{\alpha}}_{\boldsymbol{\beta}}) = f^{\sigma'_1}_{\boldsymbol{\delta_1}} \circ \cdots \circ f^{\sigma'_m}_{\boldsymbol{\delta_m}}(x^{\boldsymbol{\gamma}}_{\boldsymbol{\delta}}).$$

It follows that

$$x^{\boldsymbol{\alpha}}_{\boldsymbol{\beta}} = f^{-\sigma_n}_{\boldsymbol{\beta_n}} \circ \cdots \circ f^{-\sigma_1}_{\boldsymbol{\beta_1}} \circ f^{\sigma'_1}_{\boldsymbol{\delta_1}} \circ \cdots \circ f^{\sigma'_m}_{\boldsymbol{\delta_m}}(x^{\boldsymbol{\gamma}}_{\boldsymbol{\delta}}) \in C_{\boldsymbol{\alpha}}(x^{\boldsymbol{\gamma}}_{\boldsymbol{\delta}}) \subseteq D(\boldsymbol{\alpha}, \boldsymbol{\beta}).$$

But $x^{\boldsymbol{\alpha}}_{\boldsymbol{\beta}}$ was chosen from $F_{\boldsymbol{\alpha}} \backslash D(\boldsymbol{\alpha}, \boldsymbol{\beta})$. It follows that

$$C_{\boldsymbol{\alpha}}(x^{\boldsymbol{\alpha}}_{\boldsymbol{\beta}}) \cup C_{\boldsymbol{\gamma}}(x^{\boldsymbol{\gamma}}_{\boldsymbol{\delta}}) = \emptyset,$$

whenever $(\boldsymbol{\gamma}, \boldsymbol{\delta}) \prec (\boldsymbol{\alpha}, \boldsymbol{\beta})$.

Now the collection of ordinals $\boldsymbol{\nu} < \boldsymbol{\omega_c}$ is equinumerous with $\mathbb{R}$, and so we will use this (well-ordered) collection to do some indexing. Let $\mathbf{1} \leq \boldsymbol{\nu} < \boldsymbol{\omega_c}$, and define $X_{\boldsymbol{\nu}}$ as

$$X_{\boldsymbol{\nu}} = \bigcup \{C_{\boldsymbol{\alpha}}(x_{\boldsymbol{\nu}}^{\boldsymbol{\alpha}}) : \boldsymbol{\nu} \leq \boldsymbol{\alpha} < \boldsymbol{\omega_c}\}.$$

By our earlier observation about the $C_{\boldsymbol{\alpha}}(x_{\boldsymbol{\beta}}^{\boldsymbol{\alpha}})$'s, the $X_{\boldsymbol{\nu}}$'s are pairwise disjoint.

Further, $\bar{\mu} X_{\boldsymbol{\nu}} = 1$ for each $\boldsymbol{\nu}$. In fact suppose $\bar{\mu} X_{\boldsymbol{\nu}} < 1$ for some $\boldsymbol{\nu}$. Then by definition of $\bar{\mu}$, there must be a $M' \in \Sigma$ so that $(X_{\boldsymbol{\nu}}) \subseteq M'$ and $\mu'(M') < 1$. This means, in particular, that $\mu'(M'^c) > 0$. So there is an ordinal $\boldsymbol{\alpha}, \boldsymbol{\nu} < \boldsymbol{\alpha} < \boldsymbol{\omega_c}$ with $F_{\boldsymbol{\alpha}} \subseteq (M')^c$. But $x_{\boldsymbol{\nu}}^{\boldsymbol{\alpha}} \in F_{\boldsymbol{\alpha}}$ so

$$x_{\boldsymbol{\nu}}^{\boldsymbol{\alpha}} \in C_\alpha(x_{\boldsymbol{\nu}}^{\boldsymbol{\alpha}}) \subseteq X_{\boldsymbol{\nu}} \subseteq M',$$

an impossible situation for a member of $F_{\boldsymbol{\alpha}}$ which is a disjoint from M'.

Finally, if $\mathbb{R}_0 \subseteq \{\boldsymbol{\nu} : \mathbf{1} \leq \boldsymbol{\nu} < \omega_c\}$, then

$$\bigcup_{\boldsymbol{\nu} \in \mathbb{R}_0} X_{\boldsymbol{\nu}}$$

is $\bar{\mu}$–$\mathcal{F}$ invariant. Indeed, pick any $f_\gamma \in \mathcal{F}$. Then

$$\bigcup_{\boldsymbol{\nu} \in \mathbb{R}_0} X_{\boldsymbol{\nu}} = \bigcup \{C_\alpha(x_{\boldsymbol{\nu}}^{\boldsymbol{\alpha}}); \boldsymbol{\nu} \in \mathbb{R}_0, \boldsymbol{\nu} \leq \boldsymbol{\alpha} < \boldsymbol{w_c}\}$$

is a disjoint union of its component sets $C_{\boldsymbol{\alpha}}(x_{\boldsymbol{\nu}}^{\boldsymbol{\alpha}})$, each of which is fixed by the action of f_γ whenever $\boldsymbol{\gamma} \leq \boldsymbol{\alpha}$.

Hence

$$\left(f_\gamma\left(\bigcup_{\boldsymbol{\nu} \in \mathbb{R}_0} X_{\boldsymbol{\nu}}\right)\right) \Delta \left(\bigcup_{\boldsymbol{\nu} \in \mathbb{R}_0} X_{\boldsymbol{\nu}}\right) \subseteq \bigcup_{\boldsymbol{\nu} \in \mathbb{R}_0, \boldsymbol{\nu} \leq \boldsymbol{\alpha} < \gamma} \left(C_{\boldsymbol{\alpha}}(x_{\boldsymbol{\nu}}^{\boldsymbol{\alpha}}) \cup f_\gamma(C_{\boldsymbol{\alpha}}(x_{\boldsymbol{\nu}}^{\boldsymbol{\alpha}})\right).$$

But the cardinality of this latter set is less than or equal to

$$\#\{(\boldsymbol{\alpha}, \boldsymbol{\nu}) : \boldsymbol{\nu} < \boldsymbol{\alpha} < \gamma\} \cdot \max\{\#\boldsymbol{\alpha}, \aleph_0\} \#(f_\gamma(C_{\boldsymbol{\alpha}}(x_{\boldsymbol{\nu}}^{\boldsymbol{\alpha}}))),$$

which is less than

$$(\#\boldsymbol{\alpha})^2 \cdot \boldsymbol{c} \cdot \boldsymbol{c} = \boldsymbol{c}.$$

So the set

$$\left(f_\gamma\left(\bigcup_{\boldsymbol{\nu} \in \mathbb{R}_0} X_{\boldsymbol{\nu}}\right)\right) \Delta \bigcup_{\boldsymbol{\nu} \in \mathbb{R}_0} X_{\boldsymbol{\nu}}$$

also has cardinality less than $\boldsymbol{c}$, and thus this set belongs to $\mathcal{P}$. The $\bar{\mu}$–$\mathcal{F}$ invariance of $\bigcup_{\boldsymbol{\nu} \in \mathbb{R}_0} X_{\boldsymbol{\nu}}$ is established, and we have taken Step 3 without serious mishap.

Step 4. We will prove a striking lemma of Tarski, tailored to our current purposes.

Lemma B.4. *There exists a family $\{N_\theta : \theta \in \Theta\}$ of distinct subsets of G each having cardinality* $\mathbf{c}$ *such that*

(i) $\#\Theta = 2^{\mathbf{c}}$ *and*

(ii) *given any sequence (θ_n) of distinct members of the family Θ and any sequence (σ_n) of $+1$'s or -1's we have*

$$\bigcap_n N_{\theta_n}^{\sigma_n} \neq \emptyset$$

where $A^{+1} = A$ and $A^{-1} = G \backslash A = A^c$ for $A \subseteq G$.

Proof. Some basic notational conveniences will be observed. For an underlying set we will choose $[0, 1)$, the half-closed, half-open interval from 0 to 1; Θ will be the collection of all subsets of $[0, 1)$ so (i) is thereby secure.

For $\theta \in \Theta$, let

$$B(\theta) = \theta \cup \left([1, 2) \cap \{x + 1 : x \in \theta\}^c \right),$$

so $B(\theta) \subseteq [0, 2]$ is just θ plus the translates of θ^c in $[1, 2)$. Notice that if θ_1, θ_2 are distinct members of Θ, then

$$B(\theta_1) \cap B(\theta_2)^c \neq \emptyset.$$

After all, if $x \in \theta_1 \cap \theta_2^c$, then $x \in B(\theta_1)$, since $\theta_1 \subseteq B(\theta_1)$ and $x \notin \theta_2$ so $x + 1$ is a member of $[1, 2)$ that is a translation of a member of θ_2^c. The case $x \in \theta_1^c \cap \theta_2$ is dealt with similarly.

Now let $\mathcal{C}$ be the family of all countable subsets of $[0, 2)$. Then the cardinality of $\mathcal{C}$ is $\mathbf{c}$. Let $\mathcal{N}$ be the family of all countable subsets of $\mathcal{C}$ so the cardinality of $\mathcal{N}$ is $\mathbf{c}$ also.

For any $\theta \in \Theta$ denote by the $\mathcal{C}_\theta$ the family of all countable subsets of $B(\theta)$, and denote by $\mathcal{N}_\theta$ the family of all countable subsets of $\mathcal{C} \cap \mathcal{C}_\theta^c$. Plainly, for any $\theta \in \Theta$

$$\mathcal{C}_\theta \subseteq \mathcal{C}, \text{ and } \mathcal{N}_\theta \subseteq \mathcal{N}.$$

Let (θ_n)be a sequence of distinct members of Θ, and let (σ_n) be a sequence of ± 1's.

Choose $\theta \in \Theta$ so that θ is not one of the θ_n's, and let $\sigma_0 \neq \sigma_1$. Consider $(\theta_i^{\sigma_i} : i \geq 0)$, and let

$$I = \{i \geq 0 : \sigma_i = +1\} \text{ and } J = \{i \geq 0 : \sigma_i = -1\}.$$

Since θ was chosen not to be in the list $\theta_1, \theta_2, \ldots$, each of I and J are nonempty. Moreover

$$I \cap J = \emptyset \text{ and } I \cup J = \{0, 1, 2, \ldots\}.$$

For each pair $(i,j) \in I \times J$, pick a point $x_{(i,j)} \in B(\theta_j)\backslash B(\theta_i)$. Let

$$C_j = \{x_{(i,j)} : i \in I\},$$

and let

$$\nu = \{C_j : j \in J\}.$$

It is plain that each $C_j \in \mathcal{C}$ and $\nu \in \mathcal{N}$. Moreover if $i \in I, j \in J$, then

$$x_{(i,j)} \in B(\theta_j),$$

and so

$$C_j \subseteq B(\theta_j).$$

Therefore

$$C_j \in \mathcal{C}_{\theta_j} \quad \text{and} \quad \nu \notin \mathcal{N}_{\theta_j}.$$

It follows that

$$\nu \in \bigcap_{j \in J} \mathcal{N}_{\theta_j}^{\sigma_i} = \bigcap_{j \in J} \mathcal{N}_{\theta_j}^{c}. \tag{62}$$

Also if $i \in I$ and $j \in J$, then

$$x_{(i,j)} \notin B(\theta_i).$$

So

$$C_j \nsubseteq B(\theta_i)$$

and

$$C_j \notin \mathcal{C}_{\theta_i} \quad \text{but} \quad \nu \in \mathcal{C}\backslash\mathcal{C}_{\theta_i}.$$

It follows that

$$\nu \in \mathcal{N}_{\theta_i},$$

and so

$$\nu \in \bigcap_{i \in I} \mathcal{N}_{\theta_i}. \tag{63}$$

Taking (62) and (63) into account along with the relationship of I and J, we see that

$$\nu \in \bigcap_{n} \mathcal{N}_{\theta_n}^{\sigma_n}.$$

We have successfully navigated Step 4. □

Our critical Step 5 will entail an upgrading of Tarski's lemma to incorporate $\bar{\mu}$–$\mathcal{F}$ invariance. Here is Step 5.

Step 5. There exists a family $\{E_\theta; \theta \in \Theta\}$ of distinct subsets of G such that

(i) $\#\Theta = \mathbf{2^c}$,

(ii) $\bigcap_{n=1}^{\infty} E_{\theta_n}^{\sigma_n}$ *is $\bar{\mu}$–$\mathcal{F}$ invariant for any sequence (θ_n) from Θ and every sequence (σ_n) of ± 1's, and*

(iii) $\bar{\mu}(\bigcap_{n=1}^{\infty} E_{\theta_n}^{\sigma_n}) = 1$ *for each sequence* (θ_n) *of distinct members of* Θ *and every sequence* (σ_n) *of* ± 1*'s.*

Let $\{X_\nu : \nu \in \mathbb{R}\}$ be as in Step 3, so $\{X_\nu : \nu \in \mathbb{R}\}$ is a family of pairwise disjoint subsets of G with

$$\bar{\mu}(X_v) = 1$$

for each ν, and regardless of $\mathbb{R}_0 \subseteq \mathbb{R}$,

$$\bigcup_{\nu \in \mathbb{R}_0} X_\nu$$

is $\bar{\mu}$–$\mathcal{F}$ invariant. In tandem with this we call in Step 4 to find the family $\{\mathcal{N}_\theta : \theta \in \Theta\}$. For any $\theta \in \Theta$, we let

$$E_\theta = \bigcup_{\nu \in \mathcal{N}_\theta} X_v.$$

The sets E_θ are all distinct since the X_ν's are pairwise disjoint.

Item (i) follows from the first claim in Tarski's lemma. By construction (Step 3) each E_θ is $\bar{\mu}$–$\mathcal{F}$ invariant, and so (ii) follow from the fact that the collection of $\bar{\mu}$–$\mathcal{F}$ invariant sets is a σ-field.

To see (iii), let $\nu_0 \in \bigcap_{n=1}^{\infty} N_{\theta_n}^{\sigma_n}$, a nonempty set thanks to Tarski's smarts. By definition

$$E_\theta = \bigcup \{X_\nu : \nu \in \mathcal{N}_\theta\},$$

and the X_ν's are pairwise disjoint, so

$$E_\theta^c = \left(\bigcup \{X_\nu : \nu \in \mathcal{N}_\theta\} \right)^c \supseteq \bigcup \{X_\nu : \nu \in \mathcal{N}_\theta^c\}.$$

Regardless of $\sigma = \pm 1$,

$$E_\theta^\sigma \supseteq \bigcup \{X_\nu : \nu \in \mathcal{N}_\theta^\sigma\}.$$

Hence

$$\begin{aligned} \bigcap_{n=1}^{\infty} E_{\theta_n}^{\sigma_n} &\supseteq \bigcap_{n=1}^{\infty} \left(\bigcup \left\{ X_\nu : \nu \in \mathcal{N}_{\theta_n}^{\sigma_n} \right\} \right) \\ &\supseteq \bigcup \left\{ X_\nu : \nu \in \bigcap_{n=1}^{\infty} \mathcal{N}_{\theta_n}^{\sigma_n} \right\} \\ &\supseteq X_{\nu_0}. \end{aligned}$$

Step 3 assures us that

$$1 = \bar{\mu}(X_{\nu_0}) \leq \bar{\mu} \left(\bigcap_{n=1}^{\infty} E_{\theta_n}^{\sigma_n} \right) \leq 1.$$

An aside: The crucial Step 5 hints at the enormity of a set having a continuum of elements. It gives proper respect to this event. We are now headed for the final turn.

Let $\mathcal{E}$ denote the family of all subsets E of G of the form

$$E = \bigcup_{\{\sigma_1,\ldots,\sigma_n\}} \left(\left(\bigcap_{k=1}^{n} E_{\theta_k}^{\sigma_k} \right) \cap M'_{\sigma_1,\ldots,\sigma_n} \right),$$

where $\theta_1, \ldots, \theta_n \in \Theta$ are distinct, $\sigma_1, \ldots, \sigma_n$ are ± 1's, and $M'_{\sigma_1,\ldots,\sigma_n} \in \Sigma'$. For each such E define

$$\lambda(E) = 2^{-n} \sum_{\sigma_1,\ldots,\sigma_n \in \{+1,-1\}} \mu'(M'_{\sigma_1,\ldots,\sigma_n}).$$

Then $\mathcal{E}$ is a field of sets that contains Σ', λ is well defined on $\mathcal{E}$ and finitely additive thereupon with $\lambda|_{\Sigma'} = \mu'$, and for any $E \in \mathcal{E}$ we have $f(E) \in \mathcal{E}$ with $\lambda(f(E)) = \lambda(E)$ wherever $f \in \mathcal{F}$.

This in hand we stand ready to finish the proof of the Kakutani–Oxtoby theorem.

Let Σ^* be the σ-field of subsets of the G generated by $\mathcal{E}$. To show λ can be extended to a countably additive measure μ^* on Σ^*, it is enough to show that if (E_p) is a sequence from $\mathcal{E}$ that is descending and $\lambda(E_p) \geq \alpha > 0$ for all p, then

$$\bigcap_p E_p \neq \emptyset.$$

We can (and do) suppose (θ_n) is a sequence of distinct members of Θ, (n_p) is a strictly increasing sequence in $\mathbb{N}$, and for each choice $\sigma_1, \ldots, \sigma_{n_p}$ of ± 1's,

$$E_p = \bigcup_{\sigma_1,\ldots,\sigma_{n_p} \in \{+1,-1\}} \left(\left(\bigcap_{k=1}^{n_p} E_{\theta_k}^{\sigma_k} \right) \cap M'_{\sigma_1,\ldots,\sigma_{n_p}} \right).$$

Replacing $M'_{\sigma_1,\ldots,\sigma_{n_p}}$ by

$$\bigcap_{q=1}^{p} M'_{\sigma_1,\ldots,\sigma_{n_q}}$$

does not affect the above equality since the E_p's are descending. So we suppose that

$$M'_{\sigma_1,\ldots,\sigma_{n_p},\sigma_{n_{p+1}}} \subseteq M'_{\sigma_1,\ldots,\sigma_{n_p}}$$

regardless of $\sigma_1, \ldots, \sigma_{n_p}, \sigma_{n_{p+1}}$ and p.

By definition

$$\lambda(E_p) = \frac{1}{2^{n_p}} \sum_{\sigma_1,\ldots,\sigma_{n_p} \in \{+1,-1\}} \mu'(M_{\sigma_1,\ldots,\sigma_{n_p}}) \geq \alpha,$$

so for some fixed $\sigma_1^{(p)}, \ldots, \sigma_{n_p}^{(p)} \in \{+1, -1\}$ we have

$$\mu'(M'_{\sigma_1^{(p)},\ldots,\sigma_{n_p}^{(p)}}) \geq \alpha,$$

and this we can do for each p.

It is time to apply the pigeon-hole principle.

There is a choice $\delta_1, \ldots, \delta_{n_1}$ of ± 1's so that

$$\mathbb{Z}_1 = \{q : \sigma_1^{(q)} = \delta_1, \ldots, \sigma_{n_1}^{(q)} = \delta_{n_1}\}$$

is infinite. After all, regardless of n_1's size, there are but 2^{n_1} possible choices of the first n_1 places in the sequence $\sigma_1^{(p)}, \ldots, \sigma_{n_1}^{(p)}, \sigma_{n_1+1}^{(p)}, \ldots$ for each p; at least one of those choices must be made infinitely often.

By the same token, we can choose $\delta_{n_1+1}, \ldots, \delta_{n_2}$ (each $+1$ or -1) so

$$\mathbb{Z}_2 = \{q \in \mathbb{Z}_1 : q \geq 2, \sigma_{n_1+1}^{(q)} = \delta_{n_1+1}, \ldots, \sigma_{n_2}^{(q)} = \delta_{n_2}\}$$

is infinite.

Etc., etc., etc.

So regardless of p, there is a $q \geq p$ so that

$$\begin{aligned} \mu'(M'_{\delta_1,\ldots,\delta_{n_p}}) &= \mu'(M'_{\sigma_1^{(q)},\ldots,\sigma_{n_p}^{(q)}}) \\ &\geq \lambda'(M'_{\sigma_1^{(q)},\ldots,\sigma_{n_q}^{(q)}}) \\ &\geq \alpha > 0. \end{aligned}$$

But $(M'_{\delta_1,\ldots,\delta_{n_p}})$ is a descending sequence in Σ' so

$$\mu'\left(\bigcap_p M'_{\delta_1,\ldots,\delta_p}\right) \geq \alpha > 0.$$

Since

$$\bar{\mu}\left(\bigcap_k E_{\theta_k}^{\delta_k}\right) = 1,$$

$$\left(\bigcap_p M'_{\delta_1,\ldots,\delta_p}\right) \cap \left(\bigcap_k E_{\theta_k}^{\delta_k}\right) \neq \emptyset.$$

It follows that

$$\bigcap_p E_p,$$

which contains this set, is also nonvoid.

Conclusion. λ can be extended to a countably additive measure μ^* on Σ^*.

Finally we show that (G, Σ^*, μ^*) has character $2^{\mathfrak{c}}$.

We know that if θ_1, θ_2 are distinct members of Θ, then

$$\lambda(E_{\theta_1} \cap E_{\theta_2}^c) = \frac{1}{4} = \mu^*(E_{\theta_1} \cap E_{\theta_2}^c).$$

It follows from this that if θ_1, θ_2 are distinct members of Θ, then

$$\mu^*(E_{\theta_1} \Delta E_{\theta_2}) = \frac{1}{2}.$$

Now should $\mathcal{S} \in \Sigma^*$ be a basis for the Fréchet–Nikodym metric space (Σ^*, d_{μ^*}), then for any $\theta \in \Theta$, there is an $S_\theta \in \mathcal{S}$ so that

$$\mu^*(E_\theta \Delta S_\theta) < \frac{1}{4}.$$

If $\theta_1 \neq \theta_2$ and $S_{\theta_1} = S_{\theta_2}$, then

$$\begin{aligned}
\frac{1}{2} &= \mu^*(E_{\theta_1} \triangle E_{\theta_2}) \\
&= d_{\mu^*}(E_{\theta_1}, E_{\theta_2}) \\
&\leq d_{\mu^*}(E_{\theta_1}, S_{\theta_1}) + d_{\mu^*}(S_{\theta_1}, S_{\theta_2}) + d_{\mu^*}(S_{\theta_2}, E_{\theta_2}) \\
&= d_{\mu^*}(E_{\theta_1}, S_{\theta_1}) + 0 + d_{\mu^*}(S_{\theta_2}, E_{\theta_2}) \\
&< \frac{1}{4} + \frac{1}{4} \\
&= \frac{1}{2}.
\end{aligned}$$

Hence if $\theta_1 \neq \theta_2$, then $S_{\theta_1} \neq S_{\theta_2}$. Now $\{S_\theta : \theta \in \Theta\}$ has $2^{\mathfrak{c}}$ members, and the character of (G, Σ^*, μ^*) is at least $2^{\mathfrak{c}}$. It cannot be any bigger than $2^{\mathfrak{c}}$ so we can conclude the character is exactly $2^{\mathfrak{c}}$.

Alas, the $\bar{\mu}$–F invariance of members of Σ^* is straightforward. As a matter of fact, if we look at the collection $\mathcal{B}$ of all those members B of Σ^* such that $f(B) \in \Sigma^*$ for each $f \in \mathcal{F}$ with $\mu^*(B) = \mu^*(f(B))$, then $\mathcal{B}$ contains $\mathcal{E}$. Moreover $\mathcal{B}$ is a monotone class of sets: if (B_n) is an ascending sequence of members of $\mathcal{B}$ and $f \in \mathcal{F}$, then

$$\left(\bigcup_n B_n\right) = \bigcup_n f\left(B_n\right),$$

and

$$\begin{aligned}
\mu^*\left(f\left(\bigcup_n B_n\right)\right) &= \mu^*\left(\bigcup_n f(B_n)\right) \\
&= \lim_n \mu^*(f(B_n)) \\
&= \lim_n \mu^*(B_n) \\
&= \mu^*\left(\bigcup_n B_n\right).
\end{aligned}$$

This ensures $\mathcal{B} = \Sigma^*$. This proof is complete. □

Bibliography

[1] E. M. Alfsen, *A simplified constructive proof of the existence and uniqueness of Haar measure*, Math. Scand. **12** (1963), 106–116.

[2] N. Aronszajn, *On the differentiablity of Lipschitzian mappings between Banach spaces*, Studia Math. **57** (1976), 147–190.

[3] Keith Ball, *An elementary introduction to modern convex geometry*, Flavors of Geometry, Vol. 31, MSRI Publications, Berkeley, CA, 1997.

[4] S. Banach, *Sur le probleme de la mesure*, Fund. Math. **4** (1923), 7–33.

[5] ______, *Sur les fonctionnelle linéaires I*, Studia Math. **1** (1929), no. 1, 211–216.

[6] ______, *Sur les fonctionnelle linéaires II*, Studia Math **1** (1929), no. 1, 233–239.

[7] ______, *Théorie des opérations linéaires*, Monogratje Matematyczne, Warsaw, 1932.

[8] Christoph Bandt, *Metric invariance of Haar measure*, Proc. Amer. Math. Soc. **87** (1983), no. 1, 65–69.

[9] Christoph Bandt and Gebreselassie Baraki, *Metrically invariant measures on locally homogeneous spaces and hyperspaces*, Pacific J. Math. **121** (1986), no. 1, 13–28.

[10] F. Bernstein, *Zur theorie der trigonomischen Reihe*, Sitz. Sachs, Akad. Wiss. Leipzig **60** (1908), 325–238.

[11] G. Birkhoff, *A note on topological groups*, Compositio Math. **3** (1936), 427–430.

[12] J. Braconnier, *Sur les groupes topologiques localement compacts*, J. Math. Pures Appl. (N.S.) **27** (1948), 1–5.

[13] C Carathéodory, *Übcr das lineare mass von punktmengen eine verallgemeinerung des längenbegriffs*, Gött. Nachr. (1914), 404–420.

[14] ______, *Vorlesungen Über reelle Funktionen*, Teubner, Berlin and Leipzig, 1918.

[15] H. Cartan, *Sur la mesure de Haar*, C. R. Acad. Sci Paris **211** (1940), 759–762.

[16] H. Cartan and R. Godement, *Théorie de la dualité et analyse harmonique dans les groupes abéliens localement compacts*, Ann. Sci. École Norm. Sup. (3) **64** (1947), 79–99.

[17] C. Chevalley and O. Frink, *Bicompactness of cartesian products*, Bull. Amer. Math. Soc. **47** (1941), 612–614.

[18] Claude Chevalley, *Theory of Lie groups I*, (third printing, 1957 ed.), Princeton University Press, Princeton, N. J., 1946, 1957.

[19] J. P. R. Christensen, *On sets of Haar measure zero*, Israel J. Math. **13** (1972), 255–260.

[20] D. Cohn, *Measure theory*, Birkhäuser, Boston, 1980.

[21] Eusebio Corbacho, Vaja Tarieladze, and Ricardo Vidal, *Observations about equicontinuity and related concepts*, Topology Appl. **156** (2009), no. 18, 3062–3069.

[22] M. Csörnyei, *Aronszajn null and Gaussian null sets coincide*, Israel J. Math. **111** (1999), 191–201.

[23] Harry F. Davis, *A note on Haar measure*, Proc. Amer. Math. Soc. **6** (1955), 318–321.

[24] Andreas Defant and Klaus Floret, *Tensor norms and operator ideals*, North-Holland Mathematics Studies, no. 176, North-Holland Publishing Co, Amsterdam, 1993.

[25] J. Diestel, H. Jarchow, and A. Tonge, *Absolutely summing operators*, Cambridge Studies in Advanced Mathematics, vol. 43, Cambridge University Press, Cambridge, 1995.

[26] Joe Diestel, Jan H. Fourie, and Johan Swart, *The metric theory of tensor products, Grothendieck's Résumé revisited*, American Mathematical Society, Providence, RI, 2008.

[27] Joe Diestel, Hans Jarchow, and Albrecht Pietsch, *Operator ideals*, Handbook of the Geometry of Banach Spaces, vol. I, North-Holland, Amsterdam, 2001.

[28] J. Dieudonné, *Une généralisation des espaces compacts*, J. Math. Pures Appl. **9** (1944), no. 23, 65–76.

[29] P. Dodos, *The Steinhaus property and Haar null set*, Israel J. Math. **144** (2004), 15–28.

[30] ———, *Dichotomies of the set of test measures of a Haar null set*, Bull. London Math. Soc. **41** (2009), 377–384.

[31] J. L. Doob, *Measure theory*, Graduate Texts in Mathematics, vol. 143, Springer-Verlag, New York, 1994.

[32] Robert S. Doran, Calvin C. Moore, and Robert J. Zimmer (eds.), *Group representations, ergodic theory, and mathematical physics: A tribute to George W. Mackey*, Contemp. Math., vol. 449, 2007.

[33] V. Drinfeld, *Finitely-additive measures on S^2 and S^3, invariant with respect to rotations*, Funktsional. Anal. i Prilozhen. **18** (1984), no. 3, 7. (Russian)

[34] Nelson Dunford and Jacob T. Schwartz, *Linear operators I: general theory*, Pure and Applied Mathematics, vol. 7, Interscience Publishers, Inc., New York, 1958.

[35] A. Dvoretzky, *Some results on convex bodies and Banach spaces*, Proc. Symp. Linear Spaces (1961), 123–160.

[36] P. Erdős and D. R. Mauldin, *The nonexistence of certain invariant measures.*, Proc. Amer. Math. Soc. **59** (1976), no. 2, 321–322.

[37] Herbert Federer, *Dimension and measures*, Trans. Amer. Math. Soc. **62** (1947), 536–547.

[38] ———, *Geometric measure theory*, Classics in Mathematics, Springer-Verlag, New York, 1996.

[39] T. Figiel, J. Lindenstrauss, and V. D. Milman, *The dimension of almost spherical sections of convex bodies*, Acta Math. **139** (1977), 53–94.

[40] H. Freudenthal, *Sätze über topologische Gruppen*, Ann. of Math. **37** (1936), 46–56.

[41] R. J. Gardner, *The Brunn-Minkowski inequality*, Bull. Amer. Math. Soc. (N.S.) **39** (2002), no. 3, 355–405.

[42] I. Gelfand and D. Raikov, *Irreducible unitary representations of locally bicompact groups*, Rec. Math. [Mat. Sbornik] (N.S.) **55** (1943), 301–316.

[43] R. Godement, *Les fonctions de type positif et la théorie des groupes*, Trans. Amer. Math. Soc. **63** (1948), 1–84.

[44] ———, *Sur la théorie des représentations unitaires*, Ann. of Math. (2) **53** (1951), 68–124.

[45] A. Grothendieck, *Résumé de la théorie métrique des produits tensoriels topologiques. [Summary of the metric theory of topological tensor products]*, Reprint of Bol. Soc. Mat. Sao Paulo **8** (1953), 1–79. (French)

[46] A. Gurevič, *Unitary representations in Hilbert space of a compact topological group*, Mat. Sb. (N.S.) **13** (1943), 79–86.

[47] Alfred Haar, *Der massbegriff in der Theorie der kontinuierichen Gruppe*, Ann. of Math. (2) **34** (1933), 147–169.

[48] H. Hadwiger and D. Ohmann, *Brunn-Minkowskischer satz und isoperimetrie*, Math. Z. **66** (1956), 1–8.

[49] H. Hahn, *Über linearer gleichungssysteme in linearer Räumen*, J. Reine Angew. Math. **157** (1927), 214–229.

[50] P. Halmos, *Measure theory*, Graduate Texts in Mathematics, vol. 18, Springer-Verlag, New York, 1980.

[51] Paul R. Halmos and Herbert E. Vaughan, *The marriage problem*, Amer. J. Math. **72** (1950), 214–215.

[52] F. Hausdorff, *Bemerkung über dan Inhalt von Punktmengen*, Math. Ann. **75** (1914), 428–433.

[53] ———, *Dimension und ausseres Mass*, Math. Ann. **79** (1919), 157–179.

[54] ———, *Set theory*, Chelsea Publishing, New York, 1957 (translated from the third edition of his Mengenlehre).

[55] E. Hewitt and K. Ross, *Abstract harmonic analysis*, Die Grundlehren der Mathematischen Wissenschaften in Einzeldarstellungen, vol. 1, Springer-Verlag, Van Nostrand, Princeton, 1963.

[56] P. J. Higgins, *An introduction to topological groups*, Vol. 15, London Math Soc. Lecture Note Series, 1974.

[57] K. H. Hoffman and S. A. Morris, *The structure of compact groups*, 2nd ed., vol. 25, de Grupter Studies in Mathematics, 2006.

[58] W. Hurewicz and H. Wallman, *Dimension theory*, 4, Princeton Mathematical Series, 1941, revised edition (1948).

[59] Shizuo Kakutani, *On the uniqueness of Haar's measure*, Proc. Imp. Acad. **14** (1938), no. 2, 27–31.

[60] ———, *Über die Metrisation der topologischen Gruppen*, Proc. Imp. Acad. **12** (1938), 82–84.

[61] ———, *Concrete representation of abstract (M)-spaces. (a characterization of the space of continuous functions)*, Ann. of Math. **42** (1941), 999–1024.

[62] ———, *A proof of the uniqueness of Haar's measure*, Ann. of Math. (2) **49** (1948), 225–226.

[63] Shizuo Kakutani and Kunihiko Kodaira, *Uber das Haarsche mass in der lokal bikompakten gruppe*, Proc. Imp. Acad. Tokyo **20** (1944), 444–450.

[64] Shizuo Kakutani and John C. Oxtoby, *Construction of a non-separable invariant extension of the Lebesgue measure space*, Ann. of Math. (2) **52** (1950), 580–590.

[65] D. A. Každan, *On the connection of the dual space of a group with the structure of its closed subgroups*, Funkcional. Anal. i Priložen. **1** (1967), 71–74. (Russian)

[66] J. Kelley, *General topology*, Graduate Texts in Mathematics, Vol. 27, Springer-Verlag, New York, 1955.

[67] V. Klee, *Invariant extensions of linear functionals*, Pacific J. Math. **4** (1954), 37–46.

[68] Paul Koosis, *An irreducible unitary representation of a compact group is finite dimensional*, Proc. Amer. Math. Soc. **8** (1957), 712–715.

[69] F. Leja, *Sur la notion du groupe abstrait topologique*, Fund. Math. **9** (1927), 37–44.

[70] J. Lindenstrauss, *Almost spherical sections: their existence and their applications*, Jber. de Dt. Math. Verein (1992), 39–61.

[71] J. Lindenstrauss and A. Pelczyński, *Absolutely summing operators in L_p-spaces and their applications*, Studia Math. **29** (1968), 275–326.

[72] L. A. Lyusternik, *Die Brunn-Minkowskische Unglenichung für beliebige messbare Mengen*, C. R. Acad. Sci. USSR **8** (1935), 55–58.

[73] G. W. Mackey, *Borel structures in groups and their duals*, Trans. Amer. Math. Soc. **85** (1957), 134–165.

[74] P. Mankiewicz, *On the differentiablity of Lipschitz maps on Banach spaces*, Studia Math. **45** (1973), 15–29.

[75] P. Mankiewicz, *Compact groups on proportional quotients of l_1^n*, Israel J. Math. **109** (1999), 75–91.

[76] P. Mankiewicz and N. Tomczak-Jaegermann, *A solution of the finite-dimensional homogeneous Banach space problem*, Israel J. Math. **75** (1991), 129–159.

[77] G. A. Margulis, *Some remarks on invariant means*, Monatsh. Math. **90** (1980), no. 3, 233–235.

[78] ———, *Finitely-additive invariant measures on Euclidean spaces*, Ergodic Theory Dynam. Systems **2** (1982), no. 3-4, 383–396.

[79] É. Matheron and M. Zelený, *Descriptive set theory of families of small sets*, Bull. Symbolic Logic **13** (2007), 482–537.

[80] Pertti Mattila, *Geometry of sets and measures in Euclidean spaces: Fractals and rectifiability*, Cambridge Studies in Advanced Mathematics, Cambridge University Press, 1999.

[81] B. Maurey, *Théorèmes de factorisation pour les opérateurs linéaires à valeurs dans les espaces L^p*, Astérisque **11** (1974), 1–163.

[82] V. Milman, *Dvoretzky's theorem—thirty years later*, Geometric and Functional Analysis **4** (1992), 455–479.

[83] V. Milman and G. Schechtman, *Aymptotic theory of finite dimensional normed spaces*, Lecture Notes in Mathematics, Vol. 1200, Springer, 2001.

[84] D. Montogomery and L. Zippin, *Topological transformation groups*, no. 1, Interscience Tracts in Pure and Applied Mathematics, 1955.

[85] K. Morita, *Star-finite coverings and the star-finite property*, Math. Japonicæ. **1** (1948), 60–68.

[86] S. A. Morris, *Pontrjagin Duality and the Stucture of Locally Compact Abelian Groups*, Vol. 29, London Math Soc. Lecture Note Series, 1977.

[87] M. E. Munroe, *Introduction to measure and integration theory*, Addison-Wesley, Cambridge, MA, 1953.

[88] L. Nachbin, *On the finite dimensionality of every irreducible representation of a comact group*, Proc. Amer. Math. Soc. **12** (1961), 11–12.

[89] ______, *The Haar integral*, Van Nostrand, Princeton, 1965.

[90] M. G. Nadkarni and V. S. Sunder, *Hamel bases and measurability*, Mathematics Newsletter **14** (2004), no. 3, 1–3.

[91] M. A. Naimark and A. I. Štern, *Normed rings*, P. Noordhoff N.v.: Groningen, 1964.

[92] ______, *Theory of Group Representations*, A Series of Comprenhensive Studies in Mathematics, Springer-Verlag, Berlin, 1982.

[93] Masahiro Nakamura and Shizuo Kakutani, *Banach limits and the Čech compactification of a countable discrete set*, Proc. Imp. Acad. Tokyo **19** (1943), 224–229.

[94] I. P. Natanson, *Theory of Functions of a Real Variable*, Frederick Ungar Publishing Co., New York, 1955.

[95] Piotr Niemiec, *Invariant measures for equicontinuous semigroups of continuous transformations of a compact Hausdorff space*, Topology Appl. **153** (2006), no. 18, 3373–3382.

[96] J. Oxtoby, *Measure and category: A survey of the analogies between topological and measure spaces.*, 2nd ed., Graduate Texts in Mathematics, vol. 2, Springer-Verlag, New York, 1980.

[97] John C. Oxtoby, *Invariant measures in groups which are not locally compact*, Trans. Amer. Math. Soc. **60** (1946), 215–237.

[98] Andrzej Pelc, *Invariant measures on abelian metric groups*, Colloq. Math. **54** (1987), no. 1, 95–101.

[99] R. R. Phelps, *Gaussian null sets and differentiability of Lipschitz maps on Banach space*, Pacific J. Math. **77** (1978), 523–531.

[100] A. Pietsch, *Absolut p-summierende Abbildungen in normierten Räumen*, Studia Math **28** (1966/1967), 333–353. (German)

[101] Gilles Pisier, *Factorization of linear operators and geometry of Banach spaces*, CBMS Regional Conference Series in Mathematics, Published for the Conference Board of the Mathematical Sciences, Washington, DC; by the American Mathematical Society, Providence, RI, Providence, RI, 1986.

[102] ______, *Grothendieck's theorem, past and present.*, Bull. Amer. Math. Soc. (N.S.) **49** (2012), no. 2, 237–323.

[103] L. Pontrjagin, *The theory of topological commutative groups*, Ann. of Math. (2) **35** (1934), no. 2, 361–388.

[104] L. Pontrjagin, *Topological groups*, Princeton Mathematical Series, Vol. 2, Princeton University Press, Princeton, 1939 (translated from the Russian by Emma Lehmer).

[105] C. A. Rogers, *Hausdorff measures*, Cambridge University Press, Cambridge, 1988.

[106] Joseph Rosenblatt, *Uniqueness of invariant means for measure-preserving transformations*, Trans. Amer. Math. Soc. **265** (1981), no. 2, 623–636.

[107] David Ross, *Measures invariant under local homeomorphisms*, Proc. Amer. Math. Soc. **102** (1988), no. 4, 901–905.

[108] W. Rudin, *Fourier analysis on groups*, Wiley Classics Library, Wiley-Interscience, New York, 1990.

[109] C. Ryll-Nardzewski, *Generalized random ergodic theorems and weakly almost periodic functions*, Bull. Acad. Polon. Sci. Sér. Sci. Math. Astronom. Phys. **10** (1962), 271–275.

[110] ______, *On fixed points of semigroups of endomorphisms of linear spaces*, Proc. Fifth Berkeley Sympos. Math. Statist. and Probability, Vol. II (Berkeley, CA), Contributions to Probability Theory, Part I, Univ. Calif. Press, 1967, pp. 55–61.

[111] S. Saks, *Theory of the integral*, vol. VII, Monografie Matematyczne, Warazawa-łlwów, 1937.

[112] ______, *Integration in abstract metric spaces*, Duke Math. J. **4** (1938), 408–411.

[113] ______, *Theory of the integral*, second edition, Dover, 2005.

[114] I. Schur, *Neue Begründung der Theorie der Gruppencharaktere*, Sitz. Pr. Akad. Wiss Berlin (1905), 406–432.

[115] Issai Schur, *Neue anwendungen der intergralrechnung auf probleme der invarianten theorie, I*, Sitz. Preuss. Acad. Wiss. phys-math-kl (1924), 189–208.

[116] ______, *Neue anwendungen der intergralrechnung auf probleme der invarianten theorie, II*, Sitz. Preuss. Acad. Wiss. phys-math-kl (1924), 297–321.

[117] ______, *Neue anwendungen der intergralrechnung auf probleme der invarianten theorie, III*, Sitz. Preuss. Acad. Wiss. phys-math-kl (1924), 346–353.

[118] I. E. Segal, *Invariant measures on locally compact spaces*, J. Indian Math. Soc.(N.S.) **13** (1949), 105–130.

[119] K. Shiga, *Representations of a compact group on a Banach space*, J. Math. Soc. Japan **7** (1955), 224–248.

[120] S. Solecki, *Size of subsets of groups and Haar null sets*, Geom. Funct. Anal. **15** (2005), 246–275.

[121] R. C. Steinlage, *On Haar measure in locally compact T_2 spaces*, Amer. J. Math. **97** (1975), 291–307.

[122] A. H. Stone, *Paracompactness and product spaces*, Bull. Amer. Math. Soc. **54** (1948), 977–982.

[123] Karl Stromberg, *The Banach-Tarski paradox*, American Math Monthly **86** (1979), no. 3, 151–161.

[124] R. Struble, *Metrics in locally compact groups*, Compositio Math. **28** (1974), no. 3, 217–222.

[125] Dennis Sullivan, *For $n > 3$ there is only one finitely additive rotationally invariant measure on the n-sphere defined on all Lebesgue measurable subsets*, Bull. Amer. Math. Soc. (N.S.) **4** (1981), no. 1, 121–123.

[126] E. Szpilrajn, *La dimension et la mesure*, Fund. Math. **25** (1937), 81–89.

[127] N. Tomczak-Jaegermann, *Banach-Mazur distances and finite dimensional operator ideals*, Pitman Monographs and Surveys in Pure and Applied Mathematics, Harlow and Wiley, New York, 1989.

[128] S. Ulam, *Zur masstheorie in der allgemeinen Mengen lehre*, Fund. Math. **16** (1930), 141–150.

[129] D. van Dantzig, *Über topologisch homogene Kontinua*, Fund. Math. **15** (1930), no. 1, 102–105.

[130] ______, *Zur topologisch algebra i. komplettierungstheorie*, Ann. of Math. **107** (1932), 587–626.

[131] E. R. van Kampen, *Locally bicompact abelian groups and their character groups*, Ann. of Math. (2) **36** (1935), 448–463.

$$\lambda_n(E \backslash \bigcup_m E_m) = 0,$$

$$\limsup_n y_n \le z \le \liminf_n x_n.$$

$$\text{RAve}_A(\text{LAve}_B f - q) = \text{RAve}_A\text{LAve}_B f - q$$

$$\text{LAve}_B(\text{RAve}_A f - p) = \text{LAve}_B\text{RAve}_A f - p.$$

$$\begin{aligned}
|p-q| &= |p - \text{RAve}_A\text{LAve}_B f(x) + \text{RAve}_A\text{LAve}_B f(x) - q| \\
&\leq |p - \text{RAve}_A\text{LAve}_B f(x)| + |\text{RAve}_A\text{LAve}_B f(x) - q| \\
&= |p - \text{LAve}_B\text{RAve}_A f(x)| + |\text{RAve}_A\text{LAve}_B f(x) - q| \\
&= |\text{LAve}_B(p - \text{RAve}_A f(x))| + |\text{RAve}_A(\text{LAve}_B f(x) - q)| \\
&\leq |p - \text{RAve}_A f(x)| + |\text{LAve}_B f(x) - q| \\
&\quad (\text{since } |\text{LAve}_B f| \leq |f| \text{ and } \text{RAve}_B f| \leq |f|) \\
&< \frac{\epsilon}{2} + \frac{\epsilon}{2} = \epsilon,
\end{aligned}$$

$$\mu(G_n \setminus E_n) < \frac{\epsilon}{2^n}.$$

$$\bar{\int} \sup_n f_n = \sup_n \bar{\int} f_n.$$

$$(44) \qquad R = \frac{c_f}{c_\omega}.$$

$$(46) \qquad c_\omega \mu_\phi(h) \geq 1 - \delta.$$

$$\lambda(B) \le \nu^h(B). \qquad \square$$

[132] Guiseppe Vitali, *Sul probleme della misura dei gruppi di punti di una retta*, Gamberani e Parmeggiani, Bologna, 1905.

[133] J. von Neumann, *Die Einführung analytisches Parameter in topologischen Gruppen*, Ann. of Math. **34** (1933), 170–190.

[134] John von Neumann, *Invariant measures*, American Mathematical Society, Providence, RI, 1999.

[135] André Weil, *Sur les espaces à structure uniforme et sur la topologie générale*, Actual. Sci. Ind., no. 551, Hermann et Cie., Paris, 1938.

[136] ———, *L'intégration dans les groupes topologiques et ses applications*, Actual. Sci. Ind., no. 869, Hermann et Cie., Paris, 1940.

[137] H. Weyl, *The classical groups, their invariants and representations*, Princeton University Press, Princeton, 1939.

[138] H. Weyl and F. Peter, *Die Vollständigkeit der primitive Darstellungen einen geschlossenen kontinuierlichen Gruppe*, Math. Ann. **97** (1927), 737–755.

Author Index

Subject Index

Selected Published Titles in This Series

150 **Joe Diestel and Angela Spalsbury,** The Joys of Haar Measure, 2013
149 **Daniel W. Stroock,** Mathematics of Probability, 2013
148 **Luis Barreira and Yakov Pesin,** Introduction to Smooth Ergodic Theory, 2013
147 **Xingzhi Zhan,** Matrix Theory, 2013
146 **Aaron N. Siegel,** Combinatorial Game Theory, 2013
145 **Charles A. Weibel,** The K-book, 2013
144 **Shun-Jen Cheng and Weiqiang Wang,** Dualities and Representations of Lie Superalgebras, 2012
143 **Alberto Bressan,** Lecture Notes on Functional Analysis, 2013
142 **Terence Tao,** Higher Order Fourier Analysis, 2012
141 **John B. Conway,** A Course in Abstract Analysis, 2012
140 **Gerald Teschl,** Ordinary Differential Equations and Dynamical Systems, 2012
139 **John B. Walsh,** Knowing the Odds, 2012
138 **Maciej Zworski,** Semiclassical Analysis, 2012
137 **Luis Barreira and Claudia Valls,** Ordinary Differential Equations, 2012
136 **Arshak Petrosyan, Henrik Shahgholian, and Nina Uraltseva,** Regularity of Free Boundaries in Obstacle-Type Problems, 2012
135 **Pascal Cherrier and Albert Milani,** Linear and Quasi-linear Evolution Equations in Hilbert Spaces, 2012
134 **Jean-Marie De Koninck and Florian Luca,** Analytic Number Theory, 2012
133 **Jeffrey Rauch,** Hyperbolic Partial Differential Equations and Geometric Optics, 2012
132 **Terence Tao,** Topics in Random Matrix Theory, 2012
131 **Ian M. Musson,** Lie Superalgebras and Enveloping Algebras, 2012
130 **Viviana Ene and Jürgen Herzog,** Gröbner Bases in Commutative Algebra, 2011
129 **Stuart P. Hastings and J. Bryce McLeod,** Classical Methods in Ordinary Differential Equations, 2012
128 **J. M. Landsberg,** Tensors: Geometry and Applications, 2012
127 **Jeffrey Strom,** Modern Classical Homotopy Theory, 2011
126 **Terence Tao,** An Introduction to Measure Theory, 2011
125 **Dror Varolin,** Riemann Surfaces by Way of Complex Analytic Geometry, 2011
124 **David A. Cox, John B. Little, and Henry K. Schenck,** Toric Varieties, 2011
123 **Gregory Eskin,** Lectures on Linear Partial Differential Equations, 2011
122 **Teresa Crespo and Zbigniew Hajto,** Algebraic Groups and Differential Galois Theory, 2011
121 **Tobias Holck Colding and William P. Minicozzi II,** A Course in Minimal Surfaces, 2011
120 **Qing Han,** A Basic Course in Partial Differential Equations, 2011
119 **Alexander Korostelev and Olga Korosteleva,** Mathematical Statistics, 2011
118 **Hal L. Smith and Horst R. Thieme,** Dynamical Systems and Population Persistence, 2011
117 **Terence Tao,** An Epsilon of Room, I: Real Analysis, 2010
116 **Joan Cerdà,** Linear Functional Analysis, 2010
115 **Julio González-Díaz, Ignacio García-Jurado, and M. Gloria Fiestras-Janeiro,** An Introductory Course on Mathematical Game Theory, 2010
114 **Joseph J. Rotman,** Advanced Modern Algebra, Second Edition, 2010
113 **Thomas M. Liggett,** Continuous Time Markov Processes, 2010